JT

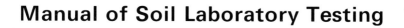

Manual of Soil Laboratory Testing

Manual of Soil Laboratory Testing

Volume 1: Soil Classification and Compaction Tests

Second Edition

K. H. Head, *MA (Cantab)*, *C.Eng, FICE, FGS*

Halsted Press: an Imprint of
JOHN WILEY & SONS, INC.
NEW YORK – TORONTO

Published in the United States and Canada
by Halsted Press, an Imprint of
John Wiley & Sons, Inc.

1st edition 1980
Reprinted 1984
2nd edition 1992

Library of Congress Cataloging in Publication Data available.

ISBN 0-470-21842-8

Printed in Great Britain

Preface to second edition

This revised edition takes into account the changes and additions to BS 1377 brought about by the publication of the 1990 revision of that Standard and some subsequent amendments. This volume relates to Parts 1, 2, 3 and 4 of BS 1377:1990, and covers most of the requirements and procedures given therein. Forthcoming revised editions of Volumes 2 and 3 will similarly relate to Parts 5, 6, 7 and 8.

Much new material has been added, including references to NAMAS requirements for calibration, and tests introduced into the new British Standard such as those related to the Moisture Condition Value. Some obsolete procedures have been removed. Known mistakes have been corrected, and in some areas I have tried to improve the presentation. For the benefit of users of ASTM Standards, the requirements of the latest available ASTM Standards have been covered more extensively than in the first edition. Otherwise the scope and objectives remain the same as before.

I am very grateful to Mr J. R. Masters for his useful suggestions and careful scrutiny of the revised manuscript. I wish to thank Messrs J. D. Harris, D. R. Norbury, A. W. Parsons and P. T. Sherwood for their valuable comments. My thanks are also due to those numerous readers who have made comments and suggestions from time to time since the first edition was published. I am grateful to Mrs R. M. Smith for her painstaking word-processor typing of the amendments and new material for this edition.

I hope that this updated volume will provide a useful and helpful complement to the new Standards.

K. H. Head

Acknowledgements

The author would like to express his appreciation to the following organisations and individuals for their assistance in the preparation of this book.

British Standards Institution, for permission to reproduce diagrams from BS 1377:Part 4:1990 for Figs 3.20, 6.23, 6.28 and 6.29.

The Geological Society for permission to reproduce Table 7.1 from Engineering Geology Special Publication No. 1, Site Investigation Practice.

Transport and Road Research Laboratory, for permission to reproduce Fig 6.34, and to quote from their Laboratory Reports.

Building Research Establishment, for permission to quote from their Current Papers CP3/68, CP23/77 and Digest No. 250, from which Table 5.6 has been prepared.

Shell Chemicals U.K. Ltd., for supplying data on industrial waxes.

Mr R. Glossop for his review and valuable criticism of the original text, and Mr R. C. Rooney for comments on Chapter 5.

Miss M. A. Sawyer for typing the calculation sheets, and Mr D. Roberts for preparing most of the drawings.

For photography: ELE International Limited, Hemel Hempstead; Soil Mechanics Ltd., Bracknell; Laing Technology Group, Mill Hill, London; Hatfield Polytechnic.

The author also wishes to thank his former colleagues at Soil Mechanics Ltd., and at ELE International Limited, for their continued assistance and co-operation.

Preface to first edition

This book is intended primarily as a working manual for laboratory technicians and others engaged on the testing of soils in a laboratory for building and civil engineering purposes. It is based on my own experience over many years both in managing large soil testing laboratories and in the instruction of technicians and engineers in test procedures. I have made a special effort to explain those points of detail which are often the cause of difficulty or misunderstanding. The step-by-step presentation of procedures, the use of flow diagrams, and the setting out of test data and calculations are provided for this purpose, especially for the newcomer to soil testing.

Volume 1 presents details of the methods and equipment used in soil classification and compaction tests, the former including relevant chemical tests. Most of these procedures are covered by British Standards, the most important being BS 1377:1975, 'Methods of test for Soils for civil engineering purposes', which this manual augments by its essentially practical approach. Reference is also made to certain US (ASTM) Standards.

A knowledge of mathematics, physics and chemistry up to GCE 'O' level standard is assumed, but some of the basic principles which are significant in soil testing are explained where appropriate. I hope that the sections giving background information, general applications and basic theory will enable technicians to obtain a better appreciation of the purpose and significance of the tests they perform. The inclusion of a chapter on soil description is intended as an introduction to that important aspect of soil mechanics, and may perhaps stimulate an interest in the broader topic of geology.

I should like to express my sincere thanks to Dr A. C. Meigh, who initiated this series of publications, and to Mr N. B. Hobbs for his most helpful and detailed comments on the original manuscript. In addition, I greatly appreciate the continued support and encouragement provided by Mr T. G. Clark, Mr I. K. Nixon and their colleagues at ELE International Ltd. (formerly Engineering Laboratory Equipment Ltd.) during the final stages of preparation of this work. Many others, too numerous to mention individually, have willingly given their support in various ways, and to them all I am very grateful.

I hope that this book will be well used in the laboratory, and I should welcome any comments and criticisms from those who use it.

K. H. Head

Contents

1	**EQUIPMENT, TECHNIQUES AND SAFETY**	1
1.1	Introduction	1
1.2	Laboratory equipment	5
1.3	Techniques	33
1.4	Care of samples	41
1.5	Preparation of disturbed samples for testing	44
1.6	Safety in the laboratory	48
1.7	Calibration	53
	Bibliography	58
2	**MOISTURE CONTENT AND INDEX TESTS**	59
2.1	Introduction	59
2.2	Definitions	60
2.3	Theory	60
2.4	Applications	68
2.5	Moisture content tests	71
2.6	Liquid and plastic limit tests	78
2.7	Shrinkage tests	99
2.8	Empirical index tests	109
	Bibliography	114
3	**DENSITY AND PARTICLE DENSITY**	116
3.1	Introduction	116
3.2	Definitions	117
3.3	Theory	118
3.4	Applications	128
3.5	Density tests	130
3.6	Particle density tests	140
3.7	Limiting density tests	149
	Bibliography	157
4	**PARTICLE SIZE**	159
4.1	Introduction	159
4.2	Definitions	160
4.3	Theory	160
4.4	Applications	165
4.5	Practical aspects	169
4.6	Sieving procedures	175
4.7	Sedimentation theory	201
4.8	Sedimentation procedures	205
	Bibliography	233
5	**CHEMICAL TESTS**	235
5.1	Introduction	235
5.2	Definitions and data	241
5.3	Theory	243
5.4	Applications	252
5.5	Tests for pH	254
5.6	Sulphate content tests	260
5.7	Organic content tests	272
5.8	Carbonate content tests	278
5.9	Chloride content tests	290
5.10	Miscellaneous tests	297
	Bibliography	301

6   COMPACTION TESTS   302
6.1   Introduction   302
6.2   Definitions   303
6.3   Theory   303
6.4   Applications   313
6.5   Compaction test procedures   315
6.6   Moisture condition tests   339
6.7   Chalk crushing value   352
6.8   Compactibility test for graded aggregates   354
      Bibliography   356

7   DESCRIPTION OF SOILS   357
7.1   Introduction   357
7.2   Definitions   360
7.3   Identification of soils   360
7.4   Description of coarse (granular) soils   361
7.5   Description of fine soils   366
7.6   Description of other soil types   370
      Bibliography   373

APPENDIX: UNITS AND SYMBOLS   375
A.1   Metric (SI) units   375
A.2   Symbols   381
A.3   Useful data   382
A.4   Comparison of BS and ASTM sieve aperture sizes   383

INDEX   384

Summary of test procedures described in Volume 1

Test designation	Section	Abbreviated reference (*see below*)
Chapter 2		
Moisture content:		
Oven drying	2.5.2	BS: Part 2: 3.2
Sand-bath	2.5.3	(BS 1377: 1975*)
Saturation moisture content of chalk	2.5.4	BS: Part 2: 3.3
Liquid limit:		
Cone penetrometer	2.6.4	BS: Part 2: 4.3
One-point penetrometer	2.6.5	,, 4.4
Casagrande	2.6.6	,, 4.5
One-point Casagrande	2.6.7	,, 4.6
Plastic limit	2.6.8	,, 5.3
Shrinkage limit:		
TRRL method	2.7.2	,, 6.3
ASTM method	2.7.3	,, 6.4
Linear shrinkage	2.7.4	,, 6.5
Puddle clay	2.8.2	Nixon (1956)
Free swell	2.8.3	Gibbs & Holtz (1956)
Sticky limit	2.8.4	Terzaghi & Peck (1948)
Chapter 3		
Density:		
Measurement	3.5.2	BS: Part 2: 7.2
Tube	3.5.3	Soil Mechanics Ltd.
Water displacement	3.5.4	BS: Part 2: 7.4
Immersion in water	3.5.5	,, 7.3
Particle density:		
Small pyknometer	3.6.2	,, 8.3
Gas jar	3.6.4	,, 8.2
Large pyknometer	3.6.5	,, 8.4
Maximum density:		
Sands	3.7.2	BS: Part 4: 4.2
Silty soils	3.7.3	Soil Mechanics Ltd.
Gravelly soils	3.7.4	BS: Part 4: 4.3
Minimum density:		
Sands	3.7.5	,, 4.4
Gravelly soils	3.7.6	,, 4.5
Chapter 4		
Sieving:		
Dry: simple	4.6.1	}
composite	4.6.2	}BS: Part 2: 9.3
very coarse soils	4.6.3	}
Wet: fine soils	4.6.4	}
gravelly soils	4.6.5	}BS: Part 2: 9.2
cohesive soils	4.6.6	}
'Boulder clay'	4.6.7	Soil Mechanics Ltd.
Sedimentation:		
Pipette	4.8.2	BS: Part 2: 9.4
Hydrometer	4.8.3	,, 9.5

Test designation	Section	Abbreviated reference (see below)
Chapter 5		
pH value:		
Indicator papers	5.5.1	Supplier
Electrometric	5.5.2	BS: Part 3: 19
Colorimetric	5.5.3	(BS 1377: 1975*)
Lovibond	5.5.4	Supplier
Sulphate content:		
Total sulphates—acid extraction	5.6.2	BS: Part 3: 5.2
Water-soluble sulphates—extraction	5.6.3	,, 5.3
Groundwater	5.6.4	,, 5.4
Gravimetric analysis	5.6.5	,, 5.5
Ion-exchange analysis	5.6.6 & 5.6.7	,, 5.6
Organic content:		
Dichromate oxidation	5.7.2	,, 3.4
Peroxide oxidation	5.7.3	(BS 1377: 1975*)
Carbonate content:		
Rapid titration	5.8.2	BS: Part 3: 6.3
Gravimetric	5.8.3	,, 6.4
Calcimeter: standard	5.8.4	Colling (1906)
simplified	5.8.5	Collins (1906)
Chloride content:		
Qualitative	5.9.2	BS: Part 3: 7.2.3.3
Water-soluble	5.9.3	,, 7.2
Mohr's method	5.9.4	Bowley (1979)
Acid soluble	5.9.5	BS: Part 3: 7.3
Total dissolved salts	5.10.2	,, 8.3
Loss on ignition	5.10.3	,, 4.3
Indicator papers	5.10.4	Supplier
Chapter 6		
'Light' compaction (1 litre mould)	6.5.3	BS: Part 4: 3.3
'Heavy' compaction (1 litre mould)	6.5.4	,, 3.4
Compaction in CBR mould	6.5.5	,, 3.5 & 3.6
ASTM compaction	6.5.7	ASTM D698 & D1557
Compaction by vibration	6.5.9	BS: Part 4: 3.7
Harvard miniature compaction	6.5.10	ASTM STP 479
Moisture Condition Value:		
MCV as received	6.6.3	BS: Part 4: 5.4
Moisture Condition Calibration (MCC)	6.6.4	,, 5.5
Rapid Assessment	6.6.5	,, 5.6
Chalk Crushing Value	6.7	,, 6.4
Compactability of aggregates	6.8	Pike (1972); Pike & Acott (1975)

* Superseded standard
NOTES
  BS: implies BS 1377: 1990 unless otherwise stated
  ASTM: American Society for Testing and Materials
  Supplier: Supplier's or manufacturer's instruction leaflets
  References are listed in full at the end of each chapter.

Chapter 1

Equipment, techniques and safety

1.1 INTRODUCTION

1.1.1 Soil Mechanics

Soil Mechanics is that branch of engineering science which applies the principles of mechanics, hydraulics and geology to the solution of engineering problems in soils. It is one aspect of the earth science known as geotechnics, or géotechnology (the French word is geotechnique), which also encompasses rock mechanics, geophysics, hydrology and engineering geology. The usual definition of soil, in the engineering sense, is given below (Section 1.1.6).

The study of soil mechanics covers the investigation, description, classification, testing and analysis of soils to determine their inter-reaction with structures built in or upon them, or built with them. Soil mechanics is the youngest discipline of civil engineering, although soil is the oldest construction material used by man, and is also the most plentiful

1.1.2 Soils as Engineering Materials

Soils, in the geotechnical sense, can be regarded as engineering materials. Their physical characteristics can be determined by experiment, and the application of methods of analysis enables these properties to be used to predict their likely behaviour under defined working conditions. But unlike other engineering materials such as metals and concrete, over which control can be exercised during manufacture, soils are naturally occurring materials, which more often than not have to be used in their natural condition. Even when some kind of processing is possible, either *in situ* or by using excavated material, the soil can be modified only to a limited extent by relatively simple procedures on site. Perhaps the most important exception is the use of certain types of clay for the manufacture of bricks, but this is outside the scope of geotechnology.

The variety of soils is very wide indeed, and no two sites have identical soil conditions. It is therefore necessary to evaluate the physical properties and engineering behaviour of the soils present at every site that is developed in any way. Many of the procedures used for determining soil characteristics consist of empirical methods derived from practical experience.

1.1.3 Purpose of Soil Testing

The physical properties of soils are usually determined by carrying out tests on samples of soil in a laboratory. These tests can be divided into two main categories:

(1) Classification tests, which indicate the general type of soil and the engineering category to which it belongs.
(2) Tests for the assessment of engineering properties, such as shear strength, compressibility and permeability.

The parameters determined from laboratory tests, taken together with descriptive data relating to the soil, are required by soil engineers for many purposes. The more usual applications are as follows:

(1)   Data acquired from classification tests are applied to the identification of soil strata when the subsurface conditions of a site are being investigated — the process known as site investigation.
(2)   Other test data enable the engineering properties of soils to be quantified in numerical terms, which can then be used as the basis of analysis on which the recommendations of a site investigation report are based.
(3)   Test data may be used for the confirmation of assumptions which have been based on previous experience and engineering judgement.
(4)   Criteria for the acceptance of a soil used in construction (possibly after a processing operation) can be drawn up in the light of available test results.
(5)   Laboratory tests are needed as part of the control measures which are applied during construction of earthworks or excavations, especially for ensuring that the design criteria are met.
(6)   The findings of a site investigation can be supplemented by further testing as construction proceeds, as, for instance, when new ground is being opened up.

The laboratory test procedures which have been evolved for the classification of soils, especially in Britain, are the subject of Volume 1. Test for the determination of engineering properties will be dealt with in Volumes 2 and 3.

1.1.4   Advantages of Laboratory Testing

In a site investigation for a construction project, the field operations, which include studies of the geology and history of the site, subsurface exploration and *in situ* testing, are of prime importance. The determination of the ground characteristics by *in situ* testing can take into account large-scale effects, such as soil fabric, structure and discontinuities of strata, which cannot be represented in small laboratory specimens. Nevertheless the measurement of soil properties by means of laboratory tests offers a number of advantages, as follows:

(1)   Full control of the test conditions, including boundary conditions, can be exercised.
(2)   Laboratory testing generally permits a greater degree of accuracy of measurements than does field testing.
(3)   Control can be exercised over the choice of material which is to be tested.
(4)   A test can be run under conditions which are similar to, or which differ from, those prevailing *in situ*, as may be appropriate.
(5)   Changes in conditions can be simulated, as can the conditions which are likely to occur during or after completion of construction.
(6)   Tests can be carried out on soils which have been broken down and reconstituted, or processed in other ways.

1.1.5   General Applications

During the last half-century the evaluation of soil properties from reliable test procedures has led to a closer understanding of the nature and probable behaviour of soils as engineering materials. Some of the resulting benefits in the realm of civil engineering construction have been:

(1)   Reduction of uncertainties in the analysis of foundations and earthworks.
(2)   Economies in design due to the use of lower factors of safety.
(3)   Exploitation of difficult sites.

(4) Erection of structures, and below-ground construction, which would not have been feasible without this knowledge.

(5) Increased economy in the use of soils as construction materials (for instance, in earth dams and embankments).

1.1.6 Definitions

The following definitions are based on those found in the *Shorter Oxford English Dictionary* and the *Concise Oxford Dictionary*, expanded to cover usage in this book.

TEST (from Latin *testum*, tile or earthen pot (cupel) originally used for treating or trying metals in, especially gold or silver alloys). 1. Critical examination or trial by which the quality of anything may be determined. 2. The action or process of examining a substance under known conditions in order to determine its identity or that of one of its constituents. 3. The action by which the physical properties of materials are tested in order to determine their ability to satisfy particular requirements.

Definition (3) is the one which is applicable to the testing of soils, but (1) is also relevant, and (2) applies to classification and chemical testing.

LABORATORY (from Latin *laborare*, labour). Room or building set apart for experiments in natural science (originally in chemistry).

SOIL (from Latin *solum*, ground). The earth or ground; the face or surface of the earth.

Soil can be defined in different ways for different purposes, and the above definition is too broad for engineering applications. An accepted definition of soil in the geotechnical sense is as as follows: Any naturally occurring deposit forming the outer part of the earth's crust, consisting of an assemblage of discrete particles (usually mineral, sometimes with organic matter) that can be separated by gentle mechanical means, together with variable amounts of water and gas (usually air).

This definition is discussed in greater detail in Chapter 7.

SAMPLE (From Middle English *essample*, example). 1. A small separated part which illustrates the properties of the mass from which it is taken. 2. A relatively small quantity of material from which the quality of the mass which it represents may be inferred.

Material taken from the ground as being representative of a particular deposit or stratum is referred to in this book as a 'sample'.

SPECIMEN (from Latin *specere*, to look at). 1. An example of something from which the character of the whole may be inferred. 2. A part of something taken as representative of the whole. 3. A part of portion of some substance serving as an example of the thing in question for purposes of investigation or scientific study.

A portion of the original sample which is actually used for testing purposes is usually referred to as a 'specimen' when the material remains virtually undisturbed. However, the words 'sample' and 'specimen' are often used synonymously.

1.1.7 Scope of Book

GENERAL

This Manual is concerned with the *testing* of *soils* in a *laboratory*, these words being used in the in the sense defined above. The laboratory may range from a large fully equipped establishment to a small rudimentary testing centre set up on the site of an investigation or a construction project. Tests which are carried out on soil *in situ* are not included in this book. Since it is intended as a working manual, it is addressed primarily to those who are responsible for carrying out the tests.

PROCEDURES COVERED

This volume deals with standard laboratory classification, chemical and compaction tests for soils. The tests are described in Chapters 2–6, and reference is made wherever possible to

British Standard 1377: Parts 1, 2, 3 and 4:1990 'Methods of test for soils for civil engineering purposes'. This important document is referred to in this volume as the British Standard, or BS. Other British Standards are referred to by their full title.

The test procedures described here are based on standard practice specified in the British Standard, where relevant. Where a test is not covered by the British Standard, reference is made to the appropriate source. Reference is also made to certain USA (ASTM) Standards, in which many test procedures are similar in principle to, but sometimes different in detail from, corresponding BS tests.

LIMITATIONS

The British Standard lays down standards of good practice which should be observed in soil testing. However, these procedures, and those described in this Manual, are based on normal British practice with sedimentary soils — that is, soils which were laid down under water and which form the majority of soils found in temperate zones. When other types of soil, such as residual soils as frequently encountered in tropical regions, are dealt with, special procedures may be necessary to obtain reliable and consistent test results. This applies particularly to the treatment of the soil before testing, and to the selection and preparation of a test specimen.

SUGGESTED APPROACH

It is essential that the laboratory technician be able to carry out tests with care and accuracy and to recognised standard procedures. This requires a knowledge of good testing techniques and an understanding of the correct procedures for the preparation of soil samples for testing. These topics are presented in Sections 1.3–1.5, and should be studied before the tests described in Chapters 2–6 are proceeded with. Important matters relating to safety in the laboratory are outlined in Section 1.6. Calibration of equipment is discussed in Section 1.7.

The summary of laboratory equipment given in Section 1.2 provides lists for reference purposes, together with comments on the characteristics and use of some of the items which are common to many different tests.

Chapter 7 provides an introduction to the description of soils in the laboratory. The engineering description of soils in an art which is acquired gradually over a period of years, but reasonable competence in giving reliable laboratory descriptions of soils can soon be gained by observing the correct basic principles and by applying common sense.

The Appendix provides a summary and a brief explanation of the metric (SI) units adopted by British Standards and used in this book. It also includes a summary of the symbols used, and a quick reference to other miscellaneous useful data. Units and symbols generally conform to those recommended by the International Society of Soil Mechanics and Foundation Engineering (ISSMFE).

GENERAL ARRANGEMENT

Each of the main chapters starts with a general introduction to the topic, followed by a list of definitions as applicable in this book. A section on theory presents sufficient theoretical background to enable the tests to be understood. This is followed by an outline, in general terms, of some of the more important applications of the results of the tests to engineering practice. The main emphasis of the book, however, is on the detailed procedures to be followed in preparing samples for and carrying out tests in the laboratory. Comments on general practical matters appertaining to the tests, details of the apparatus required, a list of the procedural stages, and step-by-step detailed procedures are included. Finally, the calculation and plotting of graphs and presentation of results are described, together with typical examples.

UNITS AND TERMINOLOGY

In this volume metric (SI) units of measurement are used throughout. In a few instances where obsolescent test equipment is referred to, the original Imperial measurements are also given.

In tests which are covered by a British Standard the terminology and symbols used are compatible with the BS. Otherwise the notation is as listed in the Appendix, except in those instances where it is separately defined.

REFERENCES

References are listed at the end of each chapter, under the names of authors arranged alphabetically. The British Standards listed at the end of Chapter 1 are quoted frequently throughout the book, but are not listed elsewhere.

1.2 LABORATORY EQUIPMENT

1.2.1 Scope

Equipment required for the laboratory tests covered by this volume is summarised in this section. Instruments used for making measurements of various kinds are listed first. Balances, ovens and ancillary items, which are required for almost all tests, are each discussed separately. Major items of mechanical and electrical equipment common to various tests are described, but reference is made to the appropriate chapter for apparatus needed only for a particular test.

Other items required in a soil testing laboratory are listed under glassware and ceramic ware; hardware (i.e. metalware, plastics ware, etc.); small tools; chemical reagents and indicators; miscellaneous materials; and cleaning materials.

Reference to the Sections and Tables listed below will provide a check list for all the equipment, consumables, reagents and other materials required for the tests described in this volume. Quantities are not suggested because they depend on individual circumstances.

Measurements:
length	Table 1.1
volume	Table 1.2; Sections 1.2.7 & 1.2.8
mass	Table 1.3; Section 1.2.3
time	Table 1.4
environment	Table 1.5
Balances	Section 1.2.3
Drying ovens etc.	Section 1.2.4
Other major items:	Section 1.2.5
Special apparatus for specific tests	Section 1.2.6
Glass and ceramic ware	Section 1.2.7
Hardware	Section 1.2.8
Small tools	Section 1.2.9
Chemical reagents and indicators	Section 1.2.10
Miscellaneous materials	Section 1.2.11
Cleaning materials	Section 1.2.11

1.2.2 Measuring Instruments

In soil testing, as in all laboratory work, it is necessary to take measurements of different kinds, and to record them. The devices used for making measurements in the performance of tests described in this volume are discussed in the following pages. However, measurements made with the use of instruments are not the only kind of observations necessary when testing soils, Visual observations requiring description in words can be equally important.

Table 1.1. INSTRUMENTS FOR MEASURING LENGTH

Instrument	Refer Fig. No.	Range	Resolution (mm)
Pocket tape	1.1(a)	2 m or 3 m	1
Metre stick	1.1(b)	1 m	1
Steel rule	1.1(c)	300 mm	0.5
Pocket steel rule	1.1(d)	150 mm	0.5
Slide (Vernier) calipers	1.1(e)	150 mm	0.1
Depth gauge		150 mm	0.1
External calipers	1.1(f)	Measured by steel rule	
Internal calipers	1.1(g)		
Micrometer	1.1(h)	25 mm	0.01
Micrometer		75–100 mm	0.01
Dial gauge	1.1(j)	15 mm	0.005
Dial gauge	1.1(k)	50 mm	0.01
Penetration gauge on cone LL apparatus	2.11	40 mm	0.1

Table 1.2. APPARATUS FOR MEASURING VOLUME AND FLUID DENSITY

Item	Refer to Fig. No.	Capacity	Resolution (mm)
Glass measuring cylinder	1.2(a)	2 litres	20
	1.2(b)	1 litre	10
	1.2(c)	500 ml	5
	1.2(d)	250	2
	1.2(e)	100	1
		50	0.5
	1.2(f)	25	0.5
Plastics measuring cylinder	1.2(g)	2 litres	
	1.2(h)	1 litre	
Glass beaker	1.2(j)	800 ml	limited
	1.2(k)	400	intermediate
	1.2(m)	200	markings
	1.2(n)	75	
Plastics beaker	1.2(p)	800 ml	
	1.2(q)	400	
Burette	1.2(r)	100 ml	0.2
Pipette	1.2(s)	50 ml	
	1.2(t)	25	no intermediate
	1.2(u)	10	markings
	1.2(v)	5	
		2	
Gas jar	1.2(w); 3.17	1 litre	
Density bottle	1.2(x); 3.15	50 ml	
Hydrometer (soil type)	4.36; 4.37	Range 0.995–1.030 g/ml	0.000 5 g/ml

Examples are the appearance and 'feel' of a soil (discussed in Chapter 7) and the behaviour of a sample during test. These aspects should not be overlooked, even though physical measurements are the main function of the tests.

The tests described in this book for the determination of soil properties involve measurements of the fundamental quantities of length, volume, mass, fluid density, time and temperature. The types of devices required for these measurements are summarised in Tables 1.1–1.5, which include the 'capacity' or 'range', and 'resolution', of most of them. Capacity

Table 1.3. BALANCES AND SCALES FOR MEASURING MASS (see Section 1.2.3)

Category	Type	Refer to Fig. No.	Capacity	Resolution	Accuracy (%)
Heavy	Platform scales	1.3	100 kg	50 g	0.05
Coarse	Semi-self-indicating scales	1.4	25 kg	1 g	0.004
Medium-coarse	Semi-self-indicating scales	1.5	7 kg	0.5 g	0.007
Automatic	Top pan balance	1.6	1200 g	0.01 g	0.0008
Fine	Analytical balance	1.7	160 g	0.1 mg	0.00006
Fine	Chain dial balance	1.8	250 g	1 mg	0.0004

Table 1.4. MEANS OF MEASURING TIME

Item	Scale range	Resolution
Stop-watch	30 min	0.2 s
Timer clock	1 h	1 s
Time switch (sieve shaker)	1 h	1 min
Wall clock	12 h	1 min
Calendar	1 year	1 day

Table 1.5. INSTRUMENTS FOR ENVIRONMENT MEASUREMENTS AND CONTROL

Item	Refer to Fig. No.	Range ($^\circ C$)	Resolution or accuracy ($^\circ C$)
Mercury thermometer	1.9(a)	0–250	1
	1.9(b)	0–110	1
	1.9(c)	0–50	0.5
Thermocouple		1000	20
Maximum/minimum thermometer	1.9(d)	−20–50	1
Wet and dry bulb thermometer	1.9(e)	−5–50	0.5
Water-bath control		15–50	0.1
Fortin barometer		670–820 mm Hg (approx.)	0.05 mm Hg
Vacuum gauge		zero to atmospheric	2 kPa or 10 mm Hg

or range indicates the maximum capacity for which the instrument is designed, or the extent of the available scale readings. Resolution is the size of the smallest interval marked on the scale of the instrument. Readings can often be estimated to within one-half or one-fifth of a marked division, but while this is sometimes necessary, the apparent gain in accuracy is not always justifiable if the instrument is not sensitive enough to respond to so small an increment.

Most of these instruments are shown in Figs. 1.1–1.9.

1.2.3 Balances

MEASUREMENT OF MASS

Mass can be measured to a greater degree of accuracy than can any other physical quantity in normal laboratory work. The accuracy of the balances referred to in Table 1.3, expressing

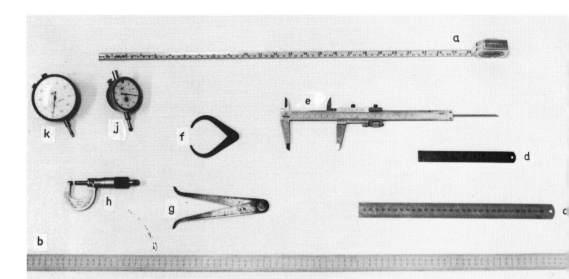

Fig. 1.1   *Instruments for measuring length* (*see Table* 1.1)

Fig. 1.2   *Apparatus for measuring volume*

Fig. 1.3 *Heavy balance: platform scales,* 100 *kg* × 50 *g*

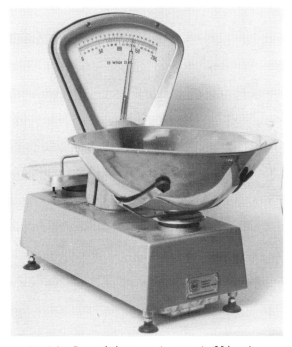

Fig. 1.4 *Coarse balance: semi-automatic,* 25 *kg* × 1 *g*

resolution as a percentage of the capacity, ranges from 0.05% (1 part in 2000) for platform scales to 0.000 06% (1 part in 1 600 000) for a sensitive analytical balance. Several different types of balance are necessary in a soil laboratory in order to cater for accurate weighing over the range from a fraction of a gram up to perhaps 100 kg.

Fig. 1.5   *Medium balance: semi-automatic,* 7 *g* × 0.5 *g*

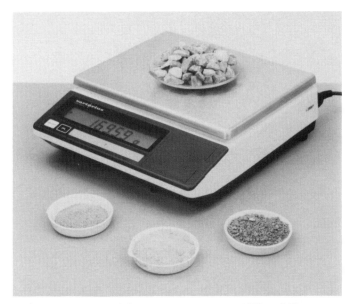

Fig. 1.6   *Fine balance: top pan automatic,* 1200 *g* × 0.01 *g*

SELECTION OF BALANCES

A selection of balances suitable for soil testing is represented by the first five items listed in Table 1.3, and illustrated in Figs. 1.3 to 1.7. Other types are available, the choice depending on circumstances. Many modern electronic balances with dual-range facilities have sufficient accuracy and resolution in each range to require only three, or perhaps two, types to cover the range of weighing needed for the majority of soil tests.

Electric balances and the fine analytical balance require connection to an electricity supply, either to the mains via a transformer or to a 12 V battery. A chain dial balance (Fig. 1.8)

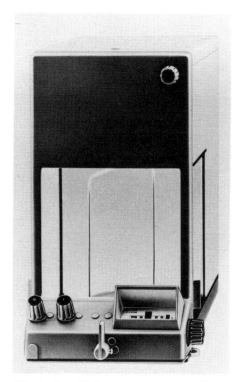

Fig. 1.7   Analytical balance: semi-automatic, 160 g × 0.1 mg

Fig. 1.8   Chain-dial balance, 250 g × 1 mg

is independent of electricity and is a relatively simple and inexpensive alternative to the analytical type of balance. While it is accurate enough for many soil testing purposes, and is suitable for use in a small site laboratory, it does not have the high sensitivity needed for chemical tests and for the pipette sedimentation test.

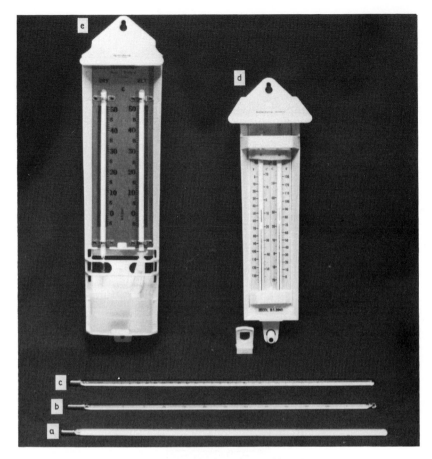

Fig 1.9   Thermometers (see Table 1.5)

1.2.4   Drying Ovens and Drying Equipment

CONVENTIONAL DRYING OVENS

Laboratory drying ovens provide a convenient means of drying material by heating to a predetermined and controlled temperature. Electrically heated ovens are normally used, and two sizes of oven are shown in Figs. 1.10 and 1.11. Ovens operating on town gas and bottled gas are also available, the latter being convenient for site use.

Before an oven is used, it is essential to calibrate the thermostat setting. The scale on the thermostat dial is not necessarily marked in °C but may be only a calibration number (often from 1 to 10, with intermediate divisions). The oven is calibrated by setting the thermostat to a certain mark and recording the temperature indicated by a thermometer when the temperature becomes steady. This is repeated for each major calibration mark on the scale, and a temperature–calibration mark curve (similar to Fig. 1.12) is plotted. For any required temperature the corresponding setting is read off the curve. The usual temperature setting for drying soil is 105–110°C, although for some soils a lower temperature (60–80°C) may be necessary (see Section 2.5.2). It is important to calibrate the oven in the surroundings in which it is to be used. If an oven is moved to a different part of the laboratory, a fresh calibration may be necessary. A full oven may give a different calibration from an empty oven, so the calibration should be done with 'average' contents in the oven.

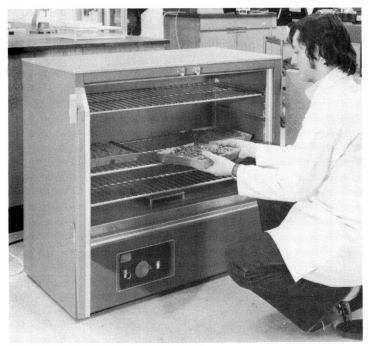

Fig. 1.10 *Large-capacity electric drying oven*

Fig. 1.11 *Bench oven with moisture containers and accessories*

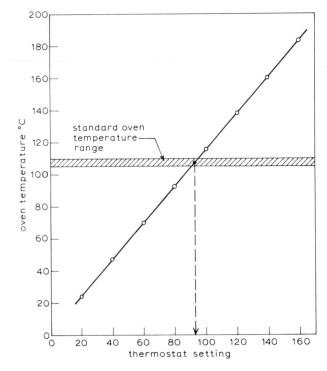

Fig. 1.12   *Calibration of oven*

A laboratory drying oven must be capable of maintaining a steady temperature which is substantially uniform over the whole of the drying space. Free circulation of air within the oven is essential. In a small oven this might be achieved by convection alone, but in a large oven convection is assisted by means of a fan. A good oven should show a temperature fluctuation at any one point over a period of time within $\pm 0.5°C$, and a maximum variation throughout the drying space to within 4°C. Temperatures should be checked under working conditions at enough locations to cover the whole of the working space. If at any location it is found that the temperature cannot be maintained within the specified limits, this area should not be used for tests requiring the specified temperature, and the restricted area or areas should be clearly designated. An electric thermocouple probe is required for inserting into all areas of an oven. Oven shelves should be a perforated plate or wire grid. The vent opening must be kept free both inside and outside, and the oven must never be so overloaded that the circulation of air is restricted.

Good insulation helps to maintain a uniform temperature and to economise on power consumption.

The thermostat unit must not be obstructed or knocked. Some ovens incorporate a secondary thermostat as a safety feature in case the main unit fails. Once the thermostat is set to the required temperature, it should be locked or held fast in that position, or at least marked clearly. Separate neon lamps indicate when the oven is switched on and when the heater is actually taking current.

Samples for drying should be placed in the oven in metal containers. Occasionally it is desirable to dry out a sample in a glass jar, in which case it should be handled with extreme care and placed on a high shelf away from the heating elements. Hot sample containers should always be handled with oven tongs or by wearing oven gloves and placed on a heat-insulated mat immediately on removal from the oven. Hot glassware should NEVER be placed on to a cold metal surface, as otherwise the glass is likely to crack. Containers should be transferred to a desiccator cabinet to cool with minimum delay.

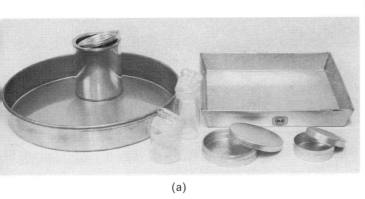

(a)   (b)

Fig. 1.13   Containers for drying soil. (a) Metal containers. (b) Glass weighing bottles

The time required to achieve complete drying at the specified temperature may be determined by removing the sample from the oven at intervals, cooling, weighing and returning to the oven. By plotting the mass of the sample against time it can be seen at what point the mass becomes constant. Alternatively, a difference between successive weighings of less than 0.1 % of the dry mass of soil used is an acceptable criterion for constant mass. Drying for 16–24 h (generally overnight) is usually sufficient, but a longer time may be necessary for a very large or very wet sample, or if the oven also contains a number of very wet samples.

CONTAINERS FOR DRYING SOIL

Some typical metal containers used for drying soil in an oven are shown in Fig. 1.13 (a), and glass weighing bottles in Fig. 1.13(b). Containers used specifically for the determination of moisture content are discussed in Section 2.5.2.

OTHER DRYING METHODS

Four other types of equipment are available for drying soils:

(1)   Infra-red drying cabinet (Fig. 1.14). This is useful for the rapid drying of large samples of granular soil in preparation for particle size or compaction tests. However, it is not suitable for determining moisture content, especially if organic matter is present, because there is no means of temperature control.
(2)   Microwave oven. Microwave heating with forced air convection gives a very rapid method of drying, but without temperature control. This method also is not suitable for the determination of moisture content.
(3)   Warm air blower. A hand-held electric blower (Fig. 1.15) with a built-in heater unit providing a stream of air at a nozzle temperature of about 100°C is useful for rapidly reducing the moisture content of small quantities of soil. It is not suitable for drying soil for the determination of moisture content.
(4)   Electric hot-plate (shown in Fig. 4.28). Electric hot-plates fitted with a heat control unit are available in several sizes. They are useful for the rapid drying of non-cohesive soils spread out on trays, but an uneven distribution of heat requires that the soil be constantly watched and agitated. Hot-plates may be used for the determination of moisture content of some soils by the sand-bath method. They also provide a convenient source of controlled heat for chemical tests, and for many purposes can be used in place of bunsen burners.

Fig. 1.14   Infra-red drying cabinet

Fig. 1.15   Warm air blower

1.2.5   Other Major Items of Equipment

(1) DESICCATORS

After a soil sample has been removed from a hot oven, it is important to allow it to cool in dry air. If exposed to a damp atmosphere, the soil can pick up moisture as it cools. A desiccator is a sealed enclosure in which the air is kept dry by means of a desiccant, and may consist of a conventional laboratory glass desiccator (vacuum or non-vacuum type; Fig. 1.16) or a small cabinet (Fig. 1.17).

The most convenient desiccant is silica gel, in the form of crystals 4–6 mm in size. When freshly dried and active, the crystals are bright blue, but they lose their colour as they absorb moisture, and turn pink when saturated. They can be completely reactivated by drying in the oven at 110°C for a few hours, but they must not be overheated.

Fig. 1.16   *Vacuum desiccator and protective cage*

Fig. 1.17   *Desiccator cabinet*

Fig. 1.18   *Constant-temperature bath with heater/thermostat/stirrer unit*

The crystals should be spread about 10 mm deep on a small tray under the perforated floor of the desiccator. A desiccator cabinet consisting of two or more separate enclosures must have desiccant in each enclosure.

Usual practice is to keep two trayfuls of silica gel for each desiccator enclosure, one in the desiccator and one drying in the oven at 105–110°C. They are interchanged daily so that the desiccant in use is always active. A convenient arrangement is to position the desiccator between the oven and the balances most often used for moisture content measurements. The flow sequence is balance–oven–desiccator–balance.

(2) CONSTANT-TEMPERATURE BATH

For tests in which it is necessary to maintain a uniform temperature for any length of time (such as a sedimentation test, or a particle density test using density bottles), a constant-temperature water-bath is essential. A bath for sedimentation tests needs to be deep enough to surround the suspension in the cylinder, and should have clear glass sides. A suitable size is about $600 \times 300 \times 380$ mm deep, as shown in Fig. 1.18. A smaller bath can be used for density bottles or, alternatively, a perforated tray supported from the sides of a deeper tank can be provided for supporting density bottles at the appropriate level.

An electrically operated unit consisting of a stirrer, heating coil and adjustable thermostat, together with indicator lamps, can be rested on or clamped to the sides of the bath. As with an oven, the thermostat setting should be calibrated against temperature in the environment in which the bath is to be used. The water temperature should be verified by inserting a calibrated thermometer reading to 0.2°C. The water level should be maintained at the correct level so that the heater and stirrer units remain fully submerged. Use of distilled or boiled water for topping up will minimise build-up of 'scale' on the sides of the tank. A cover placed over the tank keeps the water clean and reduces heat and evaporation losses. Small plastics balls which float on the surface are available as an alternative means of insulation.

A suitable constant temperature for most work is 25°C. If 20°C is chosen, cooling of the water will be necessary in hot weather unless the laboratory is air-conditioned to maintain this temperature.

To inhibit algae growth, a water stabiliser as used in aquarium tanks may be added to each fresh batch of water placed in the tank.

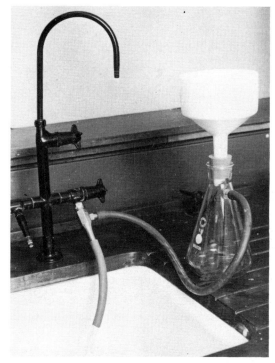

Fig. 1.19   *Filter pump and vacuum filtration flask*

(3) VACUUM

A simple way of providing a moderate vacuum is by using a filter pump attached to a running water tap (Fig. 1.19). With a good mains water pressure a vacuum of about 15 torr (15 mmHg) can be obtained. (Atmospheric pressure is about 760 mmHg, or 760 torr.) This arrangement is adequate for a simgle item requiring a moderate degree of vacuum, but is not suitable for multiple uses and is wasteful of water.

A vacuum of 1 torr or less can be produced by means of an electrically driven vacuum pump (Fig. 1.20). A pump of this kind, if of sufficient capacity, can be connected to a vacuum line fitted throughout the laboratory, from which connections are taken for a variety of purposes as required.

There are a few points to bear in mind if a system of this kind is to be used. A water-trap which can be drained off periodically should be installed near the pump to minimise the amount of condensation collecting in the pump. A similar trap is advisable at any outlet which is to be used extensively for vacuum filtration. Valves must be of a type suitable for use under vacuum, and should be included where they can isolate lengths of line when not in use. A vacuum gauge, reading in torr, should be fitted adjacent to the vacuum pump, and also at outlets where it is necessary to maintain a high vacuum such as one serving a vacuum desiccator used for de-airing density bottles (see Fig. 1.20).

Regular maintenance of a vacuum pump, in accordance with the manufacturer's instructions, is essential. The oil level should be topped up, and regular draining of water out of the reservoir chamber should not be overlooked. A properly maintained pump can be left running continuously for long periods. Before the pump is switched off, all connecting lines should first be isolated from it or opened to atmosphere.

Whenever a glass vessel such as a vacuum desiccator is to be connected to the vacuum line, it should first be covered by a metal cage designed for the purpose, as shown in Fig. 1.16, as a protective measure in case of implosion. A high vacuum should be applied gradually,

Fig. 1.20   *Electric vacuum pump with vacuum gauge and water trap connected to vacuum desiccator (courtesy Division of Civil Engineering, Hatfield Polytechnic)*

not suddenly by rapid opening of the connecting tap. Only vessels designed to withstand external atmospheric pressure should be put under vacuum.

The vacuum gauge should be calibrated at intervals not exceeding one year by a suitably accredited external organisation.

(4) WATER PURIFICATION

In many soil laboratory tests use of distilled or de-ionised water is called for. Ordinary tap-water contains dissolved solids and bacteria, which may lead to ion-exchange or other reactions taking place with the minerals in the soil and could affect the test results.

Distilled water is usually obtained from an electrically-heated water still. A modern type with a 2.75 kW heater and producing 3.5 litres per hour is shown in Fig. 1.21. Installation and operating instructions are provided by the manufacturer, but one vital precaution is to ensure that the water is always circulating when the power supply is switched on. The heating elements need to be descaled or replaced frequently in hard-water districts to maintain efficiency.

In Britain it is a legal requirement to notify the local Excise Officer of HM Customs and Excise when a still is installed. An Inspector may visit occasionally to confirm that only water is being distilled.

An alternative method for the purification of water is de-ionisation. The most convenient de-ioniser is the renewable cartridge type, such as the one shown in Fig. 1.22. The unit is connected to the water supply and enables de-ionised water to be drawn off at the turn of the tap. An electrical resistance meter, usually battery operated, provides a simple indication of the quality of the water being produced, and the cartridge is easily replaceable when

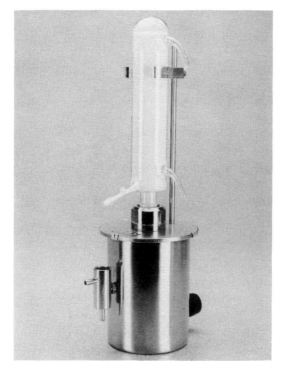

Fig. 1.21 *Electric water still*

necessary. De-ionised water is considered to be of a greater degree of purity than ordinary distilled water.

In this Manual, the term 'distilled water' also includes de-ionised water.

The requirements for distilled or de-ionised water given in BS1377: Part 1:1990 are as follows.

(a) Not more than 5 mg of non-volatile residue per litre, when tested in accordance with appendix A of BS3978:1987.
(b) pH value between 5.0 and 7.5.
(c) Specific conductance at 20°C not greater than 1 μS/mm. This requires a value not exceeding 0.2 μS/mm on leaving the apparatus.

(5) WAXPOT AND WAX

Coating with paraffin wax provides an effective and inexpensive means of sealing samples and of protecting them against moisture content changes when in storage. A low-melting-point wax (about 52–54°C) should be used, to avoid damage to samples while being coated. Waxes with a higher melting point tend to be more brittle. Microcrystalline wax is preferable to ordinary paraffin wax because its shrinkage is less.

Wax should be used at a temperature only just above its melting point, and should not be overheated, as otherwise its sealing properties will be impaired and it is likely to become brittle. A thermostatically controlled electrically heated waxpot or wax-bath (Fig. 1.23) will maintain the wax at the desired temperature, and is an essential requirement in a soil laboratory. Additional items needed are:

Ladle.

Fig. 1.22   *Water de-ioniser*

Paint brushes (say 10 mm and 50 mm).
Cutting rings, the same diameter as a U100 tube, for making wax discs.

(6) SOIL MIXER

An electrically operated mixer such as a Hobart mixer (Fig. 1.24), working on the same principle as a food mixer, facilitates thorough mixing of water with soil, It is particularly useful with non-cohesive soils. Clays may require a 'dough-hook' paddle, but machine mixing may not be practicable for clays of a firm to stiff consistency.

(7) PRESSURE VESSEL FOR DE-AIRING WATER

Tap-water or distilled water may be made free of air by applying a vacuum to the water in a suitable vessel, or by boiling followed by cooling in a sealed container. With either method, the container must be strong enough to resist the full external atmospheric pressure.

(8) RIFFLE BOX

A representative subdivision of a large sample of cohesionless soil may be obtained conveniently and rapidly by using a sample divider, or riffle box. Riffle boxes of several different sizes are available, ranging in capacity from 0.3 litre to 18 litres, the largest accepting gravel size particles up to 50 mm. A complete range of sizes is shown in Fig. 1.25, of which the three suggested as being the most useful for soil testing are listed in Table 1.6.

Fig. 1.23   *Electrically heated wax melting pot*

Fig. 1.24   *Soil sample mixer*

Fig. 1.25   *Riffle boxes*

Table 1.6.   RIFFLE BOXES

Maximum particle size (mm)	Slot width (mm)	No. of slots	Capacity (approx.) litres	Capacity (approx.) cm³
50	64	8	18	18 000
20	30	10	4.4	4 400
5	7	12	0.3	300

A riffle box consists of two separate containers beneath a row of slots; half the slots feed soil into one container, half into the other, arranged alternately. This ensures that each container receives an identical half of the original sample. Use of a third container quickens the process of successive subdivisions.

(9) SIEVE SHAKERS

A mechanical sieve shaker relieves the operator of physical effort, and ensures a uniform sieving procedure if properly used. There are two main types of shaker, each available in various sizes and with a built-in time switch. The automatic timer should be calibrated at regular intervals against a calibrated laboratory timer.

The 'Inclyno' type (Fig. 1.26) imparts a jarring action at a frequency of about 5 Hz to the nest of sieves, which is clamped to the shaking table a few degrees out of vertical. The table slowing rotates so that the soil is spread evenly over each sieve mesh. This type is useful for gravelly soils, and can be fitted with a means of washing during shaking. It is rather noisy in operation, and the cam mechanism is subject to appreciable wear. Three sizes are available, to accommodate sieves of 200, 300 and 450 mm diameters.

Fig. 1.26   Sieve shaker, Inclyno type

The 'Endecott' type (Fig. 1.27) operates at a higher frequency, and imparts both a vertical and a rotary motion to the sieves. It is more suitable for dry sieving of sand and fine gravel sizes after washing out the fines, and is relatively quiet in operation. Two sizes are available, to accommodate 200 mm or 300 mm diameter sieves.

(10) BOTTLE SHAKERS

During the pretreatment of fine-grained soils prior to a sedimentation test it is necessary to shake a suspension of soil in water for several hours. Continuous shaking is necessary also in some chemical tests. Vibratory shakers for use with conical flasks are available for this purpose from suppliers of chemical test apparatus. In BS1377, end-over-end shaking is suggested, using a device which rotates at 30 to 60 revolutions per minute. A purpose-made shaker of this kind is shown in Fig. 1.28. Up to eight tightly-stoppered bottles can be attached to a motor-driven rotatable drum using suitable clips. If there is an odd number of test specimens, balance is maintained by placing water in unused bottles.

Fig. 1.27   *Sieve shaker, Endecott type*

An alternative device is the gas jar shaker shown in Fig. 3.17, which would require adaptation to accept a number of bottles or flasks.

The rate of rotation should be checked periodically, with the shaker loaded, by using a calibrated laboratory timer.

(11) CENTRIFUGE

The centrifuge specified in BS1377: Part 2:1990 for sedimentation tests is one capable of holding bottles of 250 ml capacity. This is a large and expensive machine, and is not likely to be available in a soil testing laboratory. However, by splitting the test sample into smaller portions a machine of more reasonable size and cost, accepting 50 ml centrifuge tubes, may be used instead. A typical centrifuge of this type is shown in Fig. 1.29.

The rate of rotation should be calibrated at regular intervals using a suitable stroboscopic instrument.

A centrifuge is not an essential item, because filtration under vacuum, although slower, may be carried out instead.

(12) MUFFLE FURNACE

A muffle furnace is required in some chemical tests for ignition of material at a specified temperature. Temperatures up to 800°C are required, and the equipment should be capable of maintaining the temperature automatically to within the limits specified for the test (e.g.

Fig. 1.28   *End-over-end bottle shaker* (*Courtesy LTG Laboratories, London*)

±25°C). A muffle furnace should comply with BS4450, and should incorporate safety features such as automatic switch-off when the door is opened, and protection of the heating elements in the event of breakdown of the temperature sensor.

A calibrated thermocouple should be available so that the temperature reached by the furnace can be related to the indicated or pre-set value.

The muffle furnace temperature control should be calibrated at regular intervals by inserting a suitable calibrated thermocouple probe. Calibration should be continued long enough to verify that a specified temperature can be maintained for the period required by the test method. The services of a suitably accredited external organisation may be required.

1.2.6 Special Apparatus

Some items of equipment are required specifically for particular tests. These special items are listed below against the number of the chapter in which they are described more fully, so that a complete record of soil testing equipment may be presented within these pages. These lists include certain specialised items of glassware which are not repeated under General Glassware in Section 1.2.7.

Chapter 2:   Liquid limit cone penetrometer apparatus
              Cone sharpness gauge
              Metal cups for penetrometer test

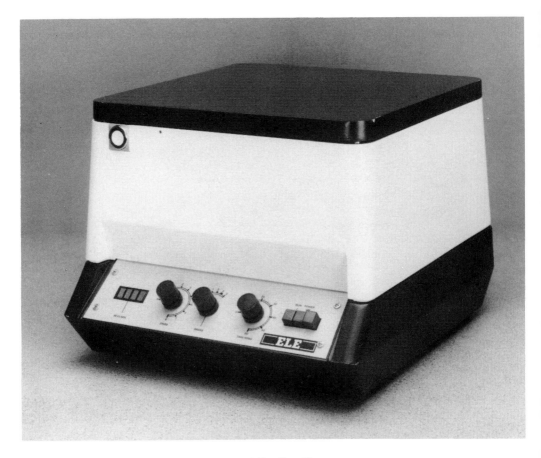

Fig. 1.29 *Centrifuge*

	Liquid limit apparatus, Casagrande type
	Casagrande grooving tool and gauge
	Gauge rod 3 mm diameter
	Shrinkage limit cell (TRRL apparatus)
	Shrinkage limit containers (ASTM apparatus)
	Shrinkage limit mould
	Shrinkage limit prong plate
	Mercury
	Linear shrinkage moulds
Chapter 3:	Sample extruder
	Sample tube adapters
	Thin-walled sampling tubes
	End-over-end shaker
	Magnetic stirrer
	Water displacement apparatus
	Balance attachments for weighing in water
	Vibrating hammer and tamping attachment
	Support for vibrating hammer
Chapter 4:	Sieves 450 mm diameter ⎫
	Sieves 300 mm diameter ⎬ all with lids and receivers
	Sieves 200 mm diameter ⎭

	Sieve brush
	High-speed stirrer
	Vibro-stirrer
	Sampling pipette and support frame
	Soil hydrometer
Chaoter 5:	BDH soil test kit
	Lovibond comparator
	Electric pH meter
	Ion-exchange column and constant-head device
	Collins calcimeter apparatus
	Carbon dioxide absorption apparatus
Chapter 6:	One-litre compaction mould, collar, baseplate
	Compaction rammer: 'light' (2.5 kg)
	'heavy' (4.5 kg)
	CBR mould, collar baseplate and tools
	Vibrating hammer
	Tamping attachments: 102 mm diameter
	145 mm diameter
	Automatic compaction machine
	Harvard compaction tamper
	Moisture Condition Value apparatus
	Spare fibre discs
	Mould extrusion jack with attachment for one-litre, CBR and MCV moulds

1.2.7 Glass and Ceramic Ware

GENERAL GLASSWARE

Beakers, Pyrex, with watch glass covers	100 ml
	250 ml
	400 ml
	600 ml
tall form	250 ml
Measuring cylinders graduated	10 ml
	25 ml
	50 ml
	100 ml
	250 ml
	500 ml
	1000 ml
	2000 ml
one mark, no spout	1000 ml
Volumetric flasks	250 ml
	500 ml
	1000 ml
Conical flasks	250 ml
Conical flasks, wide mouth	500 ml
	650 ml
	1000 ml
Pipettes, bulb	10 ml
	25 ml
	50 ml
	100 ml

Dropping bottle with pipette	
Burettes	25 ml
	50 ml
	100 ml
Extraction bottle	100 ml
Weighing bottles	25 mm dia × 50 mm
	50 mm dia × 30 mm
	40 mm dia × 80 mm
Desiccators: vacuum	200 mm diameter
	250 mm diameter
cabinet	300 × 300 × 300 mm
Funnels	50 mm diameter
	110 mm diameter
Filter flasks	500 ml
	1000 ml
Stirring rods	7 mm dia × 200 mm
	3 mm dia × 100 mm
Watch-glasses	50 mm & 75 mm dia
Gas jar with cover	1 litre
Density bottle with capillary vent stopper	50 ml
Pyknometer jar with brass cone	1 kg
Suction syringe	
Sedimentation tube	500 ml
Tubing	
Glass plate	500 × 500 × 10 mm
	300 × 300 × 10 mm
Bottle with watertight stopper	500 ml
Bottle, amber-coloured, with stopper	1000 ml

CERAMIC WARE

Evaporating dishes	100 mm diameter
	150 mm diameter
Buchner funnels	110 mm diameter
	150 mm diameter
Crucibles: silica	35 mm diameter
porcelain, with lid	25 ml
Mortar & pestle	200 mm diameter
Rubber-covered pestle	

1.2.8 Hardware

METAL WARE

Moisture content containers	75 g
	150 g
	500 g
	4 kg
Sample trays	250 × 250 × 40 mm
	300 × 300 × 40 mm
	600 × 600 × 60 mm
	760 × 760 × 60 mm

Quartering trays	$900 \times 900 \times 75$ mm
	$1000 \times 1000 \times 75$ mm
	$1200 \times 1200 \times 50$ mm
Spatula, steel blade	100×20 mm
	150×25 mm
	200×30 mm
Spatula, square ended, rubber or plastics	100×20 mm
Chattaway spatula	150×3 mm
	130×10 mm
Tongs: oven	400 mm
crucible	200 mm
bow	200 mm
rubber-coated	150 mm

Electric hotplate with heat control
Bunsen burner
Tripod stand
Wire gauze
Iron wire triangle
Heat-insulated mat
Bottled gas burner or paraffin stove
Spare gas cylinders or paraffin
Stands: burette
        retort
        funnel
Bosshead and clamps
Cheese grater
Trays
Trolley
Diamond-tipped pencil

PLASTICS WARE, ETC.

Beakers	250 ml
	600 ml
	1000 ml
Measuring cylinders	250 ml
	500 ml
	1000 ml
	2000 ml
Centrifuge bottles, polypropylene	250 ml
Funnels	115 mm diameter
	200 mm diameter
Wash bottle	500 ml
Polythene bottle with watertight screwtop	500 ml
	2000 ml
Aspirator bottle	4.5 litres

Rubber tubing
Rubber vacuum tubing
Plastics tubing
Rubber bungs; teat; 'policeman'
Filter pump
Water spray rose

Bucket	9 litres
Dustbin	50 litres

1.2.9   Small Tools

Scoop		Test-tube brushes	
Hand shovel, flat blade		Glass cutter	
Gardening trowel		Cork boring tool set	
Pointing trowel		Magnet	
Float, steel		Steel rules	150 mm
Shovel			300 mm
Trimming knife		Steel try-square	
Craft tool with replacement blades		Reference straight-edge	
Saw:		Mitre box	
medium to coarse teeth		Calipers:	
piano wire		gauging	
spiral wire		external	
Ball-pein hammers	120 g	internal	
	250 g	Vernier	
	500 g	Depth gauge	
Club hammer		Pliers:	
Geological hammers	500 g	flat:	
	1 kg	gas	
Hide mallet		electrical	
Straight-edge scraper	300 mm	Pincers	
Wire brush		Screwdrivers	
Soft hair brush		Files: hand	Smooth, and
Sieve brush		round	second cut,
Wax ladle		triangular	150 mm or 250 mm
Wax brush		Engineers' vice	

1.2.10   Chemical Reagents and Indicators

REAGENTS

Absorbent for carbon dioxide	Potassium chromate
Ammonia (0.880 g/cm^3)	Potassium dichromate
Ammonium ferric sulphate	Potassium hydrogen phthalate
Barium chloride	Potassium or ammonium thiocyanate
Barium sulphate	Pumice coated with anhydrous copper
Ferrous sulphate	sulphate
Hydrochloric acid (1.18 g/cm^3)	Silver nitrate
Hydrogen peroxide (20 vols.)	Sodium carbonate
Lead acetate	Sodium diphenylamine sulphonate
Magnesium perchlorate	Sodium hydroxide
Nitric acid	Sodium tetraborate
Orthophosphoric acid, 85% (1.70–1.75 g/cm^3)	Sulphuric acid (1.84 g/cm^3)
Potassium chloride	3,5,5-Trimethylhexan-1-ol

INDICATORS

Litmus paper:	Bomocresol green
red	Bromothymol blue
blue	Methyl red
pH indicator papers:	Thymol blue
universal	Soil indicator
narrow ranges (see Table 5.5)	Sodium diphenylaminesulphonate

Methyl orange (screened)   Special indicator strips
Buffer solution:
   pH 4.0 (potassium hydrogen phthalate)
   pH 9.2 (sodium tetraborate)

OTHER MATERIALS

Cationic exchange resin (Zeo-Karb 225 or   Filter papers:
   Amberlite IT-120)   Whatman No. 40
Mercury, redistilled   Whatman No. 42
Silica gel granules   Whatman No. 44
Sodium hexametaphosphate (Calgon)   Whatman No. 50
   Whatman No. 541

1.2.11 Miscellaneous Materials

GENERAL

Kerosene (paraffin)   Felt-tip markers
Leighton Buzzard silica sand, 63–600 μm   spirit base
Plasticine or putty   water base
Paraffin wax (melting point 52–54°C)   Clip-boards
Microcrystalline wax (melting point 60–63°C)   Glass sample jars
Silicone grease or petroleum jelly   Plastics adhesive tape
Vacuum grease   Cling-film wrapping
Labels: Aluminium foil
   tie on   Muslin
   adhesive   Polythene bags (various sizes)
   Protective clothing (see Section 1.6.11)

CLEANING IMPLEMENTS   CLEANING MATERIALS

Wire brush   Soap
Bottle brushes   Teepol
Test-tube brushes   Scouring powder
Soft hair brush   Woodwork polish
Washing-up brush   Metal polish
Viscose sponges   Cellulose cloths
Nylon pan scourer   'Brillo' pads
Tea towels   Steel wool
Hand towels   White spirit
Dusters   Acetone
Polishing cloths   Alcohol
Waste bin   Ether
   Slaked lime
   Flowers of sulphur

1.3 TECHNIQUES

1.3.1 General

Good laboratory practice depends first and foremost upon the development of correct techniques in performing tests, observing data and recording observations. Careful plotting of graphs where necessary, accuracy in calculations and correct reporting of results are other skills which a technician must acquire.

Procedures for carrying out the tests are detailed in the respective chapters. The techniques given here are those which are relevant to laboratory testing generally and which are required for most tests on soil, such as the correct use of balances. Some of these procedures may appear to be elementary, but it is essential to acquire the right techniques at the outset.

During the course of laboratory testing, the need for certain small items recurs continually. It is, therefore, convenient to carry items such as the following in the pockets, usually of the laboratory coat or overall:

Pencils (well sharpened)—HB grade for writing; H grade for plotting graphs.
Ball-point pen.
Rubber eraser.
Notebook.
Felt-tip pen (waterproof).
Small dusting brush.
Steel rule (150 mm).
Small spatula.
Pocket knife.
Hand lens.
Clean cloth.

A small pocket calculator is useful for on-the-spot calculations. Laboratory test data sheets, and graph and calculation sheets are conveniently carried on a clip-board.

1.3.2   Use of Calipers

Gauging calipers consist of a pair of hinged steel jaws which are used for measuring dimensions of solid objects where a scale rule cannot be applied directly. There are two types (illustrated at f and g in Fig. 1.1), for external and internal measurements, respectively.

When gauging calipers or vernier calipers are being used to measure the external diameter of a circular or cylindrical object, it is essential that a true diameter be measured, and that the measurement be made squarely and not on the skew (Fig. 1.30a). The true diameter is the *smallest* measurement attainable between the jaws. The jaws are closed until they just touch the object at the desired points of measurement, and should be brought up just tight, but not over-tightened. The distance between the jaws is then measured on a steel rule, to an accuracy of 0.5 mm.

The same principle applies when the internal diameter (of a sample tube, for instance) is being measured, as shown in Fig. 1.30(b). But here the true diameter is the *largest* measurement between the jaw extensions, provided that the measurement is made normal to the axis of the tube—that is, at a square end.

1.3.3   Using a Vernier Scale

A vernier scale is fitted to many types of measuring and surveying instruments. It is commonly used in soil testing as the basis of slide calipers, or vernier calipers (Fig. 1.1e). The vernier, invented by Pierre Vernier in the seventeenth century, enables measurements to be made by direct reading to 0.1 mm without having to estimate fractions of a division. The scales can be manufactured quite simply without requiring high-precision machine tools.

Vernier calipers consist of a steel scale marked in millimetres with a fixed jaw at one end, and a sliding jaw carrying a scale which is 9 mm long and divided into 10 equal parts (Fig. 1.31). A vernier division is therefore 0.9 mm long—that is, 0.1 mm shorter than a 1 mm scale division. When the sliding jaw is brought up to the object being measured, the nearest scale division to left of the zero on the vernier indicates the number of whole millimetres being measured, and the vernier mark which exactly coincides with a scale mark gives the number

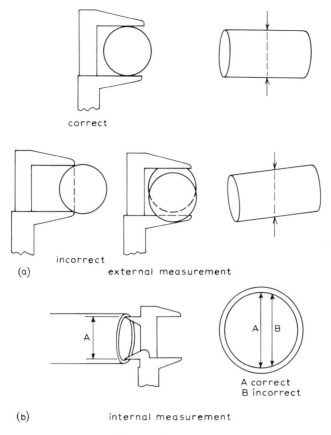

Fig. 1.30 *Use of measuring calipers*

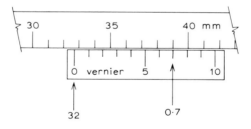

Fig. 1.31 *Reading a vernier scale*

of tenths of a millimetre to be added. In the example shown in Fig. 1.31 the vernier zero is between 32 and 33 mm, so the measurement is 32 mm plus a fraction. The seventh mark on the vernier lies directly opposite a scale mark (which is not read), so the fraction is 0.7 mm. The measurement is therefore 32.7 mm.

Slide calipers have jaws for both internal and external measurements. They should be kept clean and the sliding parts should be kept free from dust and soil particles.

Vernier calipers can be calibrated by placing calibrated gauge blocks between the jaws and comparing the scale reading with the block thickness. Gauging bars or 'long slips' are required for calibrating jaw openings exceeding 100 mm.

1.3.4 Use of Balances

The following notes provide a general guide to the proper use of balances and procedures for weighing. Details will vary with each individual type of balance. Six main types are listed in Table 1.3.

CHOICE OF BALANCE

To achieve sufficient accuracy, it is necessary before weighing to select the balance which is appropriate to the mass to be measured. This will usually be the balance with the highest sensitivity which has the required capacity. Under no circumstances should a balance be loaded beyond its stated capacity, otherwise damage may be caused to the mechanism. If in doubt, rough-weigh first on a coarse balance.

SETTING UP A BALANCE

The principles to be followed in setting up a balance are outlined below. Details will vary with each type of balance, and the manufacturer's instructions or recommendations should be followed.

(1) Balances should be sited in areas away from walkways, doors and other sources of vibration such as machinery and lifts; and preferably not adjacent to an external wall.
(2) Balances should not be placed next to a heater, radiator or oven, and must be protected from direct sunlight and from draughts. Balances of high accuracy are particularly sensitive to temperature changes.
(3) A sensitive balance should have an independent support. A good arrangement consists of brick or concrete piers mounted on a solid floor carrying a concrete or slate slab 50 mm thick — a paving slab is ideal (see Fig. 1.8). A layer of felt between the piers and slab will help to insulate the balance from vibration.
(4) Sufficient space should be allowed for an operator to sit (on a stool of the correct height) or stand at the balance, with enough bench space for recording data as well as for weights and samples.
(5) Wherever a balance is situated, it should stand on a firm and solid level surface.
(6) The balance should be levelled with the adjusting feet (if fitted), and it should be ensured that it stands firm.
(7) The balance should be protected from dust when not in use.

WEIGHING

(1) Check that the pans are clean, and that the reading is zero when unloaded (after switching on the indicator light if fitted).
(2) Set the tare adjustment to zero with the empty container on the pan, if this type of balance is used.
(3) Allow the sample to cool in a desiccator if weighing dry from the oven.
(4) Powdered substances (including soils) should never be placed directly on the pan of a sensitive balance, but on a watch-glass or in a weighing-bottle or moisture container. (This does not apply when scales fitted with a scoop-type pan designed for holding soil are used.)
(5) Place sample and container on the pan gently and add loose weights (if appropriate) until balanced (see (6) below). If a two-pan balance is being used, the object being weighed is placed in the left-hand pan and loose weights are added to the right-hand pan as in Fig. 1.8; but when a given mass of soil or other substance is to be weighed out, weights making up the given mass are first placed in the left-hand pan so that the material can be added to the right-hand pan.
(6) When a balance which incorporates a 'rest' position to take the weight of the pans off the

knife-edges is being used, the pans should be in this position whenever adjustments are made to the contents of the pans. Raising on to the knife-edges should be done steadily.

(7)   When the balance has steadied, read the indicator, and add to this the total value of any loose weights.
(8)   Write down this weight immediately in the appropriate space on the test sheet, while in front of the balance.
(9)   Check back to the indicated weight and loose weights, and confirm that the correct value has been written down.
(10)   Remove any loose weights and in doing so recheck the total. Replace them in the box.
(11)   Remove sample from pan, and clean the pan.
(12)   Switch off the indicator lamp.

GENERAL CARE

(1)   Before use check that the balance is level and firm.
(2)   Always keep a balance clean and dust-free. If it has a cover, replace it after use.
(3)   Use the tweezers for handling small loose weights.
(4)   Loose weights must always be replaced in the weights box.
(5)   Never use a highly sensitive balance for weighing out corrosive liquids.
(6)   A regular maintenance contract with one of the specialist firms who provide this service will ensure that balances are kept in good working order. Separate contracts may be necessary with two different firms to cover analytical and fine balances, and the heavier balances and scales.

CALIBRATION OF BALANCES

Balances should be calibrated at regular intervals using suitable certificated reference weights. Calibration may be carried out either by the contractor (preferably NAMAS accredited) responsible for maintenance of the balance immediately after servicing, as an extension to the maintenance contract; or in-house by senior laboratory staff. The latter should be carried out immediately after servicing.

There are numerous different types of balance, and the detailed calibration procedure will vary according to type. The following outline indicates the general principles. For a balance having dual-range scales, the process should be repeated within each range. Reference weights having valid and traceable calibrations should be used.

Reference weights approximating to 25%, 50%, 75% and 100% of the full scale reading of the balance should be added in turn, and the scale readings observed. At least two cycles of loading and unloading should be applied to determine repeatability as well as accuracy. In addition, weights corresponding to about 50% of full scale reading should be added and removed five times as a further repeatability check, observing both the loaded and the unloaded readings each time. All readings should be recorded. The observed accuracy, linearity and repeatability can then be assessed to determine whether or not the performance falls within the specification for the balance.

Calibration intervals should not be less than six months. In between calibrations, checks should be made regularly (e.g. once a week) by using working weights that have been calibrated against the reference weights. If any discrepancy is found the full calibration procedures should be followed.

1.3.5   Recording Test Data

TEST FORMS

The first essential is to record a full identification of the sample to be tested on the appropriate test form, together with an engineering description. Any special testing instructions should

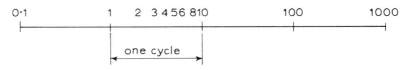

Fig. 1.32   Logarithmic graph paper

also be recorded on the form. All entries should be plainly and clearly written. Readings should be accurately recorded at the time they are taken, directly into the appropriate space on the form. Errors should be crossed out (not altered) and fresh figures entered above, so that there can be no possibility of misunderstanding what has been written.

The use of odd pieces of paper for the recording of data should be strictly forbidden. Mistakes are easily made when transferring data from one piece of paper to another, and odd sheets often get lost. Incorrect calculations are another source of error. Figures and calculations should be double-checked, preferably by different methods or by someone else.

On completion of the test, the operator should sign the record sheet and enter the date of the test. An example of a laboratory test form is given in Fig. 2.15.

METHODS OF PRESENTING DATA

All laboratory tests involve the recording of numbers, which are observed in various ways during the test. Sometimes these observations form part of the data required from the test; and sometimes the observed data must be processed by calculation or other means to arrive at the required results. In either case the observed or calculated figures must be presented in a form in which they can be clearly understood. This can be done in two ways: either by drawing up a table or by plotting a graph.

Tabulation of results is fairly straightforward and will be discussed under each test procedure. Graphs require the application of a few simple rules if the maximum benefit is to be derived from them.

1.3.6   Graphs

A graph is a means of representing a series of observed or calculated numerical data in diagrammatic form. The advantage of a graph over a table is that it links the data together and gives a general picture at a glance. It also enables values in between actual observed values to be estimated easily and reliably.

Most graphs are drawn on squared paper. One set of figures is represented along the horizontal axis (the abscissa) to a certain scale, and a second set along the vertical axis (the ordinate) to the same or a different scale. Each corresponding pair of observations or figures is plotted as a point, and a smooth curve or straight line is usually drawn through the points to represent the relationship between the two quantities. An example is shown in Fig. 2.15, in which penetration of the cone (mm) is plotted against moisture content (%).

TYPES OF GRAPH PAPER

Arithmetical graph paper is ruled up as a square grid. The sheets generally used are in 1 mm, 2 mm or 5 mm squares, sometimes with heavier rulings at 10 mm or 50 mm intervals.

Semi-logarithmic scales are used for some applications where a wide range of values is required. The ordinate is divided on an arithmetical scale, but the abscissa is divided into a number of cycles on a logarithmic scale. One cycle represents multiplication by a factor of 10. Thus the distances between 1 and 10, 10 and 100, 100 and 1000 units, etc., are all equal. Intermediate markings follow a fixed pattern, as shown in Fig. 1.32. A logarithmic scale enables very large numbers to be accommodated, while at the same time very small values can be plotted to the same degree of accuracy. Its main application is in particle size tests, for which are

Table 1.7.

Examples of recommended scales	Examples of inconvenient scales
1 cm = 1, 2, 5 units 10, 20, 50 units 0.1, 0.2, 0.5 units 1 mm = 0.001, 0.002, 0.005 units	1 cm = 4 units 25 units 3 units 1 mm = 0.004 units 0.025 units

plotted sizes ranging from 0.002 mm to 200 mm or more (a range of five cycles — i.e. a ratio of 100 000).

Semi-logarithmic graph paper is supplied in various forms with up to seven logarithmic cycles across the width of the sheet.

CHOICE OF SCALES

The choice of suitable scales for the axes of a graph on an arithmetical grid is important, and must ensure: ease of plotting; accuracy of plotting; ease of reading off intermediate values.

(1)   Select a *convenient* scale in accordance with the recommendations given below.
(2)   Use the available area of the paper to best advantage.

It is of no advantage to cover the whole sheet if an inconvenient or non-standard scale has to be used to achieve this. The scale used should be one in which each major division represents a single unit of the quantity being plotted, or is related to it by a factor of 2, 5, 10; 20, 50, 100; etc. Do not use a multiple of 4, and *never* a multiple of 3. The rule is to use factors of 1, 2, 5 or their multiples of 10 (or of 1/10). Some examples of suitable and inconvenient scales using graph paper divided in mm and cm squares are given in Table 1.7.

The selection of a scale for the arithmetical ordinate of a semi-logarithmic plot should follow the same rules as given above.

Whenever possible, use the same scale when comparing results of the same test on different samples. Otherwise, indicate very clearly that the scales are different.

1.3.7   Calculating and Checking

Calculations based on laboratory test data are normally carried out by using a pocket calculator, a desk-top computer or a computer terminal. The last two, and sometimes the first, can be programmed to perform the calculations automatically and to display or print out the final results. However, in order to understand the principles correctly it is necessary for the operator to begin by following through all stages of the calculations using ordinary arithmetic. To illustrate the principles involved, all calculations given in this volume are set out in full, in the same way as would be done in practice using conventional work-sheets without the aid of programmed calculators.

All calculations, whether computer-aided or not, should be checked independently by an experienced person. The results should be examined critically, and the question asked — do the results look sensible for this type of test on that type of soil? If there is any doubt after re-checking the calculations and any plotted data a repeat test should be carried out. Proper checking of data and assessment of the validity of results requires judgement based on experience.

1.3.8   Reporting

Select the correct form on which to present the results. Test results should always be written legibly and must be reported in a clearly understandable and unambiguous manner.

Table 1.8. RECOMMENDED ACCURACY FOR REPORTING LABORATORY TEST RESULTS

Item	Accuracy	Unit
Density	0.01	Mg/m^3
Moisture content	less than 10%: 0.1 more than 10%: 1	%
Liquid limit Plastic limit Plasticity index	1	%
Particle density	0.01	Mg/m^3
Clay, silt, sand, gravel fraction	generally: 1 less than 5%: 0.1	%
Voids ratio	0.01	—

Where there is no standard form, the information required for a complete record of the results consists of:

(1) Name and address of laboratory organisation
(2) Nature of test.
(3) Sample identification data (see Section 1.4.2).
(4) Name of job and location number.
(5) Name of client (if known).
(6) Description of test sample.
(7) Test specification and any special instructions.
(8) Test results as required by the specification and special instructions.
(9) Particular reference to anything unusual that occurred during the test or in the results.
(10) Signature and date of test.

Local laboratory rules may require that every test sheet be signed by the person who carried out the test, who should be responsible for ensuring that all information, including calculations and transfer of data, has been checked. The signed and dated forms with the completed results can then be passed to the supervisor.

The degree of accuracy to which results should be reported is indicated under 'Report results' at the end of each test description. These recommendations are summarised for reference in Table 1.8.

1.3.9 Laboratory Climate

Regular measurement of the climate in the laboratory is desirable and can be quite important for some tests, although those covered in this volume are not particularly sensitive to changes experienced under normal laboratory conditions. The items usually measured are:

Laboratory temperature.
Atmospheric pressure.
Relative humidity.

TEMPERATURE

To record the overall temperature changes in the laboratory, a maximum/minimum thermometer is required. The usual type is shown in Fig. 1.9(d). The pointers indicating the maximum and minimum readings can be easily reset to the mercury surface by use of the small magnet which is provided. The bottom of each pointer indicates the maximum and minimum readings, and the mercury level on either side indicates the prevailing temperature.

Typically, five readings are taken and recorded each day, as follows.

At 9 a.m. (or start of the day) read overnight minimum temperature, the previous day's maximum and the prevailing temperature at the time. Set the pointers to the mercury surface.

At mid-day, and at 5 p.m. (or end of the day), read the temperature.

If a self-recording apparatus is available, all that is necessary is to renew the record chart each week, or at the appropriate intervals.

The thermometer or recording apparatus must be sited away from sources of heat, and shielded at all times from direct sunlight.

ATMOSPHERIC PRESSURE

The most satisfactory apparatus for atmospheric pressure is a Fortin barometer, which reads to 0.05 mmHg (0.05 torr) by means of a vernier on the scale. The instrument must be carefully installed. Alternatively, an aneroid barometer may be used.

Readings should be taken twice daily — morning and evening. Self-recording instruments are also obtainable.

RELATIVE HUMIDITY

A wet-and-dry bulb thermometer (Fig. 1.9e) is used to determine relative humidity. It is essential to keep the small reservoir topped up with *distilled* water, and to ensure that the wick is always moist. It should be replaced if it goes hard or non-absorbent.

From the room temperature (indicated by the dry bulb thermometer) and the difference between the readings of the two thermometers, the relative humidity (%) can be read off from tables supplied with the instrument.

1.3.10 Cleaning

It is important to keep glassware and ceramic ware absolutely clean and free from grease, especially when used for chemical tests. One of the simplest and most effective means of cleaning is to wash with warm soapy water, or water containing a little synthetic detergent (such as Teepol), and wiping with blotting paper or a soft cloth, or with a long-handled brush for burettes and the like. It is essential to rinse thoroughly afterwards to ensure complete removal of the cleaning agent. The items are then allowed to drain on a drying rack and wiped dry with a clean cloth.

1.4 CARE OF SAMPLES

1.4.1 Observations

Soil samples provide some of the most important evidence from a subsurface investigation. They are costly to obtain, and should be treated and handled with proper care.

Inspection and description of samples are usually carried out by a soil engineer or geologist, but observations made when samples are prepared and tested in the laboratory can be equally important. The operator should record as much information as possible about the samples, by means of descriptions of what is seen, felt and smelled. There will be closer contact with samples in the laboratory than anywhere else. For instance, an engineer may initially inspect an undisturbed tube sample only at each end, but it is not until the sample is extruded for test that the whole length of the sample can be observed. If there are significant variations from what has been seen at the ends, these should be brought to the notice of the engineer before any tests are started.

A systematic procedure for description of soils is given in Chapter 7. The following sections are concerned with the general handling and storage of samples.

1.4.2 Identification

Every sample should be clearly identified by means of a unique number, normally comprising three parts:

Job reference number/Borehole number/Depth (m) below surface

In some identification systems a consecutive number, starting from 1 for each borehole, may be used in addition to, or in place of, the sample depth. The essential requirement is to provide every sample with a unique, unambiguous and understandable reference number. Where appropriate, instead of a borehole number the identification number of a pit, trench, adit or other sampling location may be used.

The above minimum information should be written with a waterproof marker on a suitable label attached to the sample container. Reliance should not be placed solely on a label or markings on a lid or end cap. These details should also be recorded clearly at the outset on the sample description sheet. If there is any doubt about the sample numbering, this should be queried immediately.

The full identification of a sample will include the following details:

Name of location.
Location reference number.
Borehole or pit number, or similar reference.
Sample number.
Depth below surface (top and bottom).
Type of sample (disturbed or undisturbed).
Container type and number.
Date sampled.
Visual description of sample.
By whom described.
Date of description.
Signature.

The laboratory visal description may be subsequently modified by the Engineer in the light of additional data provided by the laboratory test results.

1.4.3 Opening a Sample Container

Before opening up a sample for inspection or for testing, examine the container to see whether it was properly sealed. Record any observations regarding inadequacy of sealing or packing, especially of undisturbed samples. Also note any damage to or deterioration of the sample container.

Sample containers must be opened carefully to avoid disturbance and loss of material or fluid. The protective wax coating applied to undisturbed samples should be gently prised off. Undisturbed samples must at all times be handled with care, and not dropped or bumped on the bench.

Samples should always be labelled on the container itself, not on the lid or end caps. Nevertheless the lid or caps should remain close to the sample container when removed, so that they do not get mixed up.

1.4.4 Resealing Samples

When sufficient material from an undisturbed sample has been taken for testing, the portion which remains should be resealed and returned to the sample stores. Samples should be resealed as soon as possible after description or removal of material for test, to prevent drying out. The sealing procedure is as follows.

An unprotected sample should not be immersed in molten wax. A thin coating of just molten wax should first be applied by brush, making sure that there are no cavities. If the sample is porous, it is first covered with a layer of waxed paper. When the wax has hardened, the sample may be dipped in the wax-bath two or three times, allowing the wax to set between each dipping.

For protecting block samples of cohesive material, muslin or cheesecloth is first wrapped around the sample to reinforce and retain the wax on the sides and at corners.

When a sample is sealed in a U100 tube, a wax disc is first inserted adjacent to the end of the sample. Then two or three coats of wax are brushed on to seal the gap between the disc and the tube. Finally a thin layer of molten wax is poured on top of the disc, followed by a second layer of about 20 mm thickness when the first has hardened. To seal a smaller tube, a disc of paper is inserted and brushed with wax to seal it in, and then two or three layers of molten wax are poured on.

Jar and bag samples should be resealed to prevent loss of moisture and material. After screwing on the lid of a glass jar, it may be sealed with adhesive plastics tape or by brushing on a coating of wax.

Polythene bags are best sealed by applying plastics adhesive tape over the opening, after removing as much air as possible. Twisting the neck of the bag and tying with wire or tape does not seal in the moisture, but this method may be adequate for coarse-grained soils in which the moisture content is not important. Never tie the neck of the polythene bag in a knot; it is very difficult to untie.

All samples must carry clear identification labels before being returned to stores. Remove or delete any old labels.

Material which has been used for a test should not be put back with the original material, but repacked separately in a suitable container with a label identifying the sample and showing clearly the type of test for which it has been used.

1.4.5 Storage of Samples

Soil samples should be stored under cover in a cool room which is protected from extremes of heat and cold. Ideally, the store should be maintained at a high humidity, but this may not be practicable except for a small room for special samples.

Disturbed samples in glass jars (honey type) can be conveniently stored in milk bottle crates. Large disturbed samples in polythene bags ('bulk bag' samples) should not be piled one on top of another, but placed individually on shelves or racks. Pallet racking with the aid of a fork-lift truck is perhaps the most convenient method of storage, and enables the full height of a store shed to be used to maximum advantage. The layout of racks must allow sufficient manoeuvering space.

Undisturbed tube samples may be laid on racks designed for the purpose. However, tubes containing wet sandy or silty soil should be stored upright (suitably protected against being knocked over), to prevent possible slumping and segregation of water. The end caps of tube samples which are to be stored for any length of time should be sealed with tape or wax, in addition to the wax seal next to the sample itself. Any space between the wax and the end cap should be filled with packing material, such as sawdust. Sand or any other type of soil, and grass or straw, should not be used. These samples should be stored away from any possible source of heat, and not high up in the building, where warm air tends to collect.

Racks and shelves and all storage areas should be clearly numbered. It is essential to keep adequate records and to maintain an efficient filing system, so that any given sample can be readily located.

1.5   PREPARATION OF DISTURBED SAMPLES FOR TESTING

1.5.1   Soil Categories

This section deals with the general principles involved in the preparation of disturbed samples for the classification and compaction tests described in Volume 1. It is based on clause 7 of BS1377: Part 1:1990. The actual amount of material required for test is given with each test procedure, and depends upon the type of soil.

In BS1377:1990, soils are classified as either cohesive or cohesionless (non-cohesive), and each type is then divided into three categories, designated fine-grained, medium-grained and coarse-grained. These categories depend on the largest size of individual particles present in substantial proportion, i.e. more than 10% by mass. If the category is not obvious it should be assessed by sieving. The minimum mass required for assessment sieving is obtained by inspection and by reference to Table 1.9.

The above soil types and categories are defined as follows. All percentages are by mass.

COHESIVE SOIL   Soil which because of its fine-grained content will form a coherent mass at suitable moisture contents (e.g. clay).
COHESIONLESS SOIL   Granular soil consisting of particles which can be identified individually by the naked eye or with the aid of a hand lens (e.g. gravel, sand).
FINE-GRAINED SOILS contain not more than 10% of particles retained on a 2 mm test sieve.
MEDIUM-GRAINED SOILS contain more than 10% of particles retained on a 2 mm sieve but not more than 10% retained on a 20 mm sieve.
COARSE-GRAINED SOILS contain more than 10% of particles retained on a 20 mm sieve but not more than 10% retained on a 37.5 mm sieve.

Materials with more than 10% retained on a 37.5 mm sieve are not usually recognised as soils; some would be classed as aggregates, to which BS812 relates. The only useful soil tests which can be performed on these materials are particle size analysis (sieving), and moisture content and plasticity tests on the fine fraction (if present).

Since the percentages quoted above are used only as a guide, precision is not essential and the soil need not be dried before the assessment sieving. The soil should be regarded as belonging to the appropriate finest-grained category as defined above.

The preparation of undisturbed samples requires a different approach, which is touched in Section 3.5 and will be covered in greater detail in Volume 2.

Table 1.9.   MASS OF SOIL REQUIRED FOR ASSESSMENT SIEVING

Largest size of particle present in substantial proportion (more than 10%)	*Minimum mass required*
Less than 75 mm and exceeding 20 mm	15 kg
Less than 20 mm and exceeding 2 mm	2 kg
2 mm and smaller	100 g

1.5.2   Selection for Test

The most important requirement for any test sample is that it be fully representative of the material from which it is taken. Taking samples on site is beyond the scope of this book, so the sample received in the laboratory has to be treated as being representative of the original material *in situ*. However the mass of soil available for each test should not be less than that given in Table 1.10 for the relevant soil category. These minimum masses are

Table 1.10. MASS OF SOIL REQUIRED FOR EACH TEST

Type of test	Section	Clause in BS1377:1990	*Minimum mass according to soil group*		
			Fine-grained	Medium grained	Coarse grained
Moisture content (oven drying)	2.5.2	Part 2:3.2	50 g	350 g	4 kg
Moisture content (sand bath)	2.5.3	–	50 g	350 g	3 kg
Saturation moisture content of chalk	2.5.4	3.3	lump, 300 to 500 ml		
Liquid limit (penetrometer)	2.6.4	4.3	500 g	1 kg	2 kg
Liquid limit (one-point penetrometer)	2.6.5	4.4	100 g	200 g	400 g
Liquid limit (Casagrande)	2.6.6	4.5	500 g	1 kg	2 kg
Liquid limit (one-point (Casagrande)	2.6.7	4.6	150 g	250 g	500 g
Plastic limit	2.6.8	5.3	50 g	100 g	200 g
Volumetric shrinkage (TRRL)	2.7.2	6.3	500 g	1 kg	2 kg
Volumetric shrinkage (ASTM)	2.7.3	6.4	100 g	200 g	400 g
Linear shrinkage	2.7.4	6.5	500 g	800 g	1.5 kg
Particle density (small pyknometer)	3.6.2	8.3	100 g	100 g	100 g
Particle density (gas jar)	3.6.4	8.2	300 g	600 g	600 g
Particle density (large pyknometer)	3.6.5	8.4	1.5 kg	2 kg	4 kg
Particle size distribution (sieving)	4.6.1 to 4.6.7	9.2, 9.3	150 g	2.5 kg	17 kg
Particle size distribution (pipette sedimentation)	4.8.2	9.4	100 g	–	–
Particle size distribution (hydrometer sedimentation)	4.8.3	9.5	250 g	–	–
pH value	5.5.2–4	Part 3:9.5	150 g	600 g	3.5 kg
Sulphate content	5.6.2–6	5.8	150 g	600 g	3.5 kg
Organic matter content	5.7.2,3	3.4	150 g	600 g	3.5 kg
Carbon content	5.8.1–5	6.3, 6.4	150 g	600 g	3.5 kg
Chloride content	5.9.2–5	7.2, 7.3	150 g	600 g	3.5 kg
Total dissolved solids (water)	5.10.2	8.3	(about 500 ml)		
Loss on ignition	5.10.3	4.3	150 g	600 g	3.5 kg
Compaction test (light, 1 litre mould)*	6.5.3	Part 4:3.3		25 kg (10 kg)	
Compaction test (light, CBR mould)*	6.5.5	3.4		80 kg (50 kg)	
Compaction test (heavy, 1 litre mould)*	6.5.4	3.5		25 kg (10 kg)	
Compaction test (heavy, CBR mould)*	6.5.5	3.6		80 kg (50 kg)	
Compaction test (vibration)*	6.5.9	3.7		80 kg (50 kg)	
Maximum density (sands)	3.7.2,3	4.2	6 kg	—	—
Maximum density (gravelly soil)	3.7.4	4.3	16 kg	16 kg	30 kg
Minimum density (sands)	3.7.5	4.4	2 kg	—	—
Minimum density (gravelly soil)	3.7.6	4.5	16 kg	16 kg	30 kg
Moisture condition value	6.6.3	5.4	3 kg	3 kg	6 kg
MCV/moisture content	6.6.4	5.5	6 kg	6 kg	12 kg
MCV rapid assessment	6.6.5	5.6	3 kg	3 kg	6 kg
Chalk crushing value	6.7.2	6.4	—	2 kg	4 kg

* The masses given in brackets apply only when the soil is not susceptible to crushing during compaction.

based on Table 1.6 of BS1377: Part 1:1990. The total mass of soil required to allow for multiple test determinations, or for more than one test on each sample, should be determined by multiplication or addition of the relevant masses.

For testing purposes a sample smaller than the received sample is usually required (see Table 1.10). The actual quantity of material depends upon the type of test and the nature of the soil. Coarse-grained soils require a larger sample, to be properly representative, than do fine-grained soils. The standard procedure for obtaining a small representative sample from a large quantity of granular soil is given in Section 1.5.5. This principle should always be followed. Taking a single random scoopful of material from a bagful or heap does not provide a true representative sample of the whole.

Even though the correct procedure has been followed, a check should be made by visual inspection that the selected test sample is indeed truly representative. For instance, if medium to coarse gravel is present, it can be observed whether the test sample contains an excess or deficiency of the larger particle sizes compared with the original material.

With cohesive soils (clays and silts) the riffling procedure is not possible. To obtain a representative disturbed sample of this type of material, a number of small samples should be selected from several locations and mixed together, rather than taking all the samples from one place. If the soil is laminated or non-homogeneous in any way, careful visual observation is needed to ensure that the test sample is properly representative, and it may be necessary to consider the sample as being of more than one type of material.

1.5.3   Drying

Before a large disturbed sample of cohesionless soil can be properly subdivided, it must first be partially dried until it is in a state in which it can be crumbled. This should normally be done by air drying — that is, by leaving the soil spread out on trays in the laboratory, with free access to air, for a period of up to 3 days. The trays may be placed in a warm area, such as over an oven, but the temperature should not exceed 50°C.

Air drying does not remove all the moisture in the soil. If it is necessary to know the dry mass of soil (as defined in Section 2.3.1), this can be calculated from the air-dried mass and the air-dried moisture content which is determined by taking representative specimens as described in Section 2.5.2, Moisture Content Test.

Cohesive soils should not be dried (even by air-drying at laboratory temperature) before testing, unless it has been ascertained beforehand that drying has no significant effect on the test results. Drying, even partial drying by exposure to air at room temperature, can cause changes to take place in the physical behaviour of some soils. Some of these changes are not reversible when the soil is re-mixed with water. These effects are particularly significant in organic soils and in many tropical residual soils. (See Section 5 of Geological Society Engineering Group Report, 1990). These changes can result in changes in the index properties, and in the measured particle size distribution due to aggregations of fine particles forming strongly-bonded larger particles. Tests should therefore be carried out on the soil in its natural state, or the state in which it is received at the laboratory. If necessary the dry mass of a sample can be determined from the measured moisture content of a representative portion.

Other soils (whether cohesive or cohesionless) which should not be subjected to drying include those containing organic matter; soils for pH and certain chemical tests; and all tropical residual soils.

1.5.4   Disaggregation of Particles

Aggregations of particles must be broken down before cohesionless soil samples are prepared for test, but for most purposes crushing of individual particles must be avoided. This process is best done in a mortar, with a rubber pestle (Fig. 1.33). If a large quantity of material is to be prepared, it should be done in batches, to avoid the mortar being overloaded with soil.

If a quantity of material passing a certain size of sieve is required, the mortar should be emptied into the sieve at intervals and the material retained on the sieve returned to the mortar for further treatment. This avoids interference with the breakdown process due to the presence of fine material.

Disaggregation of particles may be considered to be complete when only individual particles are retained when sieved through a 2 mm sieve.

For some tests such as chemical tests, for which it is permissible to crush individual particles, a conventional stone pestle may be used with the mortar for grinding down to the required degree of fineness.

Fig. 1.33   *Sample preparation equipment*

This procedure is not generally suitable for soils containing cohesive material, for which disaggregation by immersion in water us usually appropriate.

1.5.5   Riffling and Quartering

Two methods may be used for subdividing a large sample of granular soil to obtain a representative test sample. These are:

(1)   Riffle box method
(2)   Cone and quartering method
Either process may be referred to as 'riffling'.

RIFFLE BOX

Three suggested sizes of riffle box which will cover most requirements are given in Section 1.2.5 and are shown in Fig. 1.25. The material to be divided must first be thoroughly mixed, then poured evenly into the riffle box from a scoop or shovel. It should be distributed along all or most of the slots, not confined to two or three slots near the middle. Each receiving container should then receive an identical sample, one of which is rejected and the other remixed and poured back into the riffle-box after an empty receiver has been inserted. This process is repeated as many times as necessary until the required reduction has been achieved, the material from alternate sides after successive pourings being retained.

CONE AND QUARTER

This procedure is slower and requires more effort than the use of the riffle box, but is equally reliable if the mixing is done thoroughly so that segregation of particle sizes does not occur.

The initial material is mixed on a tray, and formed into a circular conical heap. Any coarse particles around the base of the cone should be evenly distributed. With a cruciform sample splitter, or a straight edge, the heap is divided into four equal portions as shown in Fig. 1.34. Two diagonally opposite portions (A and C) are separated out and mixed thoroughly together and formed into a smaller heap for quartering as before. The portions B and D are rejected and returned to the original sample container. The above process is repeated as many times as necessary until a small enough representative sample is obtained.

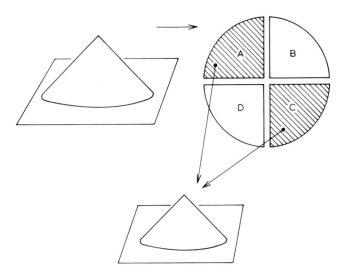

Fig. 1.34   Cone-and-quartering

1.6   SAFETY IN THE LABORATORY

1.6.1   Basic Rules

Safety in the laboratory implies the avoidance of accidents. Accident prevention is largely a matter of common-sense, but it does require the observance of simple precautions, together with tidiness, cleanliness, vigilance, careful working and cautious movements at all times. Short-cuts and potentially dangerous expedients should never be used. The effects of an accident can be harmful or disastrous to others besides those directly involved.

The most important requirements can be summarised as follows:

(1)   Study and observe the laboratory's own rules and procedures.
(2)   Before handling electrical or mechanical apparatus, refer to the manufacturer' instructions — and observe them. (This is particularly important with a new or unfamiliar piece of equipment.)
(3)   Chemicals should be handled only after receiving instruction in their use and potential hazards.
(4)   If in doubt, or if the safety of a device or procedure looks at all dubious, do not hesitate to consult the person in charge.

The precautions which follow under various headings are given in summary form as a general guide to safe practice and conduct. These points may be amplified by the laboratory's own rules and practice.

1.6.2   General Conduct and Handling

Keep laboratory benches and working areas tidy.

Waste materials and soil should be disposed of in proper containers, which should be emptied regularly.

Cleanliness in a soil laboratory is difficult to maintain, but accumulations of dust and soil trimmings and old samples should be minimised by regular tidying up and cleaning.

NEVER RUN in a laboratory building (except in dire emergency).

A laboratory is not the place for practical jokes, which can easily misfire, with disastrous consequences.

Always wear appropriate protective clothing (see Section 1.6.11).

Winchester bottles and the like should not be lifted by the neck.

Never attempt to lift any item which is too heavy or too bulky to handle unaided. Obtain assistance.

Before picking up a heavy or awkward item, make sure that there is a clear and safe space in which to put it down.

Large or heavy objects should be transported any distance on a suitable trolley. This applies particularly to glassware, such as a tray loaded with glass weighing bottles.

A trolley should be pulled from in front, not pushed from behind, when passing by doorways.

Hot objects should be placed on a suitable heat-proof mat, not directly on to the bench.

Use oven gloves and tongs for handling hot items.

1.6.3 Fire

The local fire rules and drill must be learnt and understood.

Smoking is one of the greatest fire hazards, and should be prohibited in a laboratory.

Paper-work and wastepaper containers should be kept away from naked flames and hotplates.

Some vapours from chemical reactions are highly inflammable. The same applies to the use of certain adhesives which give off inflammable vapours. Smoking should certainly be banned in areas where these are used.

Glass reagent bottles and other containers should not be left in direct sunlight, because of the lens effect, which can cause localised intense heat.

Electrical overheating of motors and heaters, and electric sparking, are other potential fire hazards.

1.6.4 Electricity

Connection of electric items to a conventional socket outlet requires the usual precaution that the wires be correctly connected to the plug. The old British and current International colour codes are as follows:

Terminal	International Standard	Obsolete British Standard
Live	Brown	Red
Neutral	Blue	Black
Earth	Green and yellow	Green

Ensure that the capacity of the fuse fitted is adequate, but not excessive, for the maximum current rating of the equipment.

Many electric motors require connection to a 3-phase supply by a competent electrician.

Suitable leads and cables should be used, and should be secured neatly on the bench. Keep cables well away from sources of heat such as hot-plates.

Never touch electrical apparatus or switches with wet hands.

Switch off before removing a plug from its socket, and see that the switch is off before inserting a plug.

Avoid the use of multiple outlet adaptors on a single socket.

If in doubt, or if any modification is necessary, obtain the services of a competent electrician.

Check the condition of leads and cables regularly. If the outside covering is worn or damaged, the cable should be replaced.

1.6.5 Gas

Check flexible gas tubing regularly. Worn or punctured tubing should be replaced — do not attempt to make repairs.

If a leak is suspected, extinguish any naked flames, open the windows, and notify the Gas Board immediately. Do not attempt any emergency repairs.

Never search for a leak with a naked flame.

Learn the correct way to use a bunsen-burner or other gas appliances.

1.6.6 Drains

Sinks and gullies at which wet sieving (Section 4.6.4) is carried out, or where washing of equipment with adhering soil takes place, should be fitted with removable silt traps. The traps must be cleaned out at regular intervals, normally weekly, but every day if large numbers of wet sievings are being done. If not trapped in this way, silt and clay can very quickly build up in gullies and cause blockage to drains.

Waste soil and sample remains should never be disposed of by washing down a drain. Proper receptacles should be provided for this purpose.

1.6.7 Mercury

Handling of mercury can be hazardous unless the appropriate precautionary measures are taken. Mercury vapour is poisonous, and if sufficient liquid mercury is exposed in an enclosed room at normal temperature (18–22°C), the concentration of mercury vapour can rise to more than 100 times greater than the accepted maximum allowable concentration of $100 \mu g/m^3$. However, normal ventilation will prevent the concentration rising as high as that, but it is important to control the surface area of mercury exposed to the air. Further information on the hazards of mercury will be found in Technical Data Note 21 by HM Factories Inspectorate (1976).

PREVENTATIVE MEASURES

To avoid the scattering of mercury and build-up of vapour during handling, the following routine precautions should always be taken so far as they are applicable:

(1)   Good general ventilation (preferably by mechanical means) should be provided.
(2)   Storage should be in airtight or water-sealed shatterproof vessels.
(3)   Containers of mercury should be kept on trays (plastics types are very suitable) which will catch and retain any spillage which may result from the breakage of glass containers or from other causes.
(4)   Mercury should be handled over trays sloping into a gulley leading to a water-sealed trap, or over a perforated tray above a bath containing a sufficient depth of water.
(5)   Mercury should be handled in an enclosed system if possible, or in a fume-cupboard or an enclosure connected to an exhaust system which can maintain an air speed of about 0.5 m/s over all openings.
(6)   The exhaust from vacuum pumps should be vented outside the workroom at a safe height.
(7)   Care should be taken to ensure that no mercury can lodge on any heated surfaces, such as local lighting units, radiators.
(8)   Contact with the skin should be avoided, protective clothing (laboratory coat, rubber gloves) being used where appropriate.

A method for the determination of mercury vapour in the air is given in booklet No. 13 in the series 'Methods for the Detection of Toxic Substances in Air' published by HMSO.

SPILLAGE

If any spillage is not immediately cleaned up, it will, during normal traffic over the area concerned, be broken up into tiny droplets offering a large surface area to the air. The droplets will lodge in any rough or irregular surface and in any cracks, and will be exceedingly difficult to remove.

This danger is not always appreciated until an area has already become contaminated, and cleaning will then be of vital importance.

Spillages should first be cleaned up as far as practicable by mechanical means, such as by vacuum probe (a flexible tube connected to a conical vacuum flask which is connected to a vacuum line). The affected areas should then be treated with a wash composed of equal parts of slaked lime and flowers of sulphur mixed with sufficient water to form a thin paste. This yellow wash, which has the appearance of distemper, should be liberally applied and allowed to dry on the floors, the lower parts of the walls, work-benches and any other contaminated surfaces. Twenty-four hours later the wash should be removed with clean water and the surfaces again allowed to dry.

This method of cleaning ensures that any droplets of mercury which cannot be mechanically removed are converted into mercury sulphide, so eliminating the danger of vaporisation. It also ensures by visual observation of the removal of the yellow wash that the cleaning is effectively performed.

1.6.8   Chemicals

Most chemicals are potentially hazardous and poisonous, and must be handled with caution. Wash hands immediately after handling them. Good technique, orderliness and cleanliness are essential.

Chemicals which emanate fumes should be used only if a properly ventilated fume-hood is available. If there is one — use it.

Do not sniff gases given off by chemical substances or evolved by a reaction.

Do not look into a test tube or flask while heating or mixing chemicals.

Always add acid to water — NEVER water to an acid.

Do not use the mouth to suck corrosive or toxic or volatile liquids into a pipette.

Any spillages should be cleared up immediately. Acids or alkalis must first be neutralised.

Acid or alkali (including ammonia) on the skin should be washed off immediately with plenty of clean water.

Unlabelled containers of chemicals are useless and potentially dangerous. They should be disposed of, but by the proper means.

Food and drink should be kept out of chemical testing areas.

Chemicals likely to be used in a soil laboratory which are particularly hazardous are as follows.

ACIDS AND ALKALIS

Neutralise spilled acids with sodium carbonate.

Neutralise spilled caustic alkalis with ammonium chloride.

Strong solutions of these substances should be kept readily available for emergency use.

AMMONIA

Ammonia affects the skin and causes painful burns. Drops on the skin should be washed off immediately with plenty of clean water. Its effect on the eyes is immediate and disastrous, unless the eye is held open while being washed with clean water for at least 15 min.

Always open a bottle of ammonia in a fume-cupboard behind glass. Alternatively wear an eye-shield as a protection against the inevitable slight spray. Special care is needed with 0.880 g/ml ammonia; 25% ammonia (density 0.91 g/ml) is less dangerous.

MERCURY

See Section 1.6.7.

NITRIC ACID

Strong nitric acid can ignite organic material, and spillages must not be mopped up with a rag or sawdust unless first neutralised.

SODIUM HYDROXIDE

Sodium hydroxide is a strong caustic alkali and should be handled with great care. Emergency precautions are as for ammonia.

SULPHURIC ACID

The addition of water to strong sulphuric acid generates heat rapidly and causes sputtering. Always add acid to water. NEVER water to acid.

1.6.9 Noise

Exceptionally noisy procedures, such as the use of a vibrating hammer, should be kept away from the main working area and carried out in an acoustically insulated room. Ear-muffs should be worn when operating very noisy equipment such as a vibrating hammer.

Intermittent noise of high intensity is not the only possible source of irritation and discomfort in a laboratory. A persistent noise at a much lower level, such as that from a continuously running vacuum pump, can be very troublesome in a laboratory working area. Items such as this should be housed elsewhere, or enclosed in an acoustic surround (making provision for air and exhaust inlet and outlet).

1.6.10 Miscellaneous

GLASSWARE

Any glassware which is damaged in any way, or even scratched, should be disposed of immediately.

If breakage occurs, gather up all fragments of glass immediately and place them in a stout box. (Someone else will have to dispose of the refuse.) Ensure that no glass splinters are left in sinks. A piece of Plasticine is effective in picking up small fragments for disposal.

Take care when inserting bungs in cylinders, or when pushing glass tubes through bungs. Alcohol may be used as a lubricating medium.

Use protective gloves when cutting glass tubes and the like. Smooth off-cut ends in a flame. Glass cutting and shaping should be done only by someone who has had sufficient experience.

All glass vessels should be properly supported when in use.

Do not store glass containers in strong sunlight, or on a high shelf, or adjacent to heaters or hot water supplies.

TOOLS

Do not use a metal file unless it is properly fitted with a handle.

Use the correct size spanner on nuts and bolts, and preferably not an adjustable spanner.

Always wear eye protection when using a grinding wheel.

Machinery with exposed gears, belt drives and other moving parts should be fitted with guards.

VACUUM

Examine glassware which is to be evacuated (filter flasks, desiccators, etc.). If the slightest flaw is seen, do not use it but dispose of it.

Always use a protective screen over glassware subjected to vacuum.

Filtration flasks must be of thick-walled glass and designed for use under vacuum. Never use an ordinary flask and two-hole stopper for filtering by suction.

Open vacuum valves slowly.

Lubricate glass vacuum cocks with silicone vacuum grease, and do not force them if stuck.

Ensure that rubber bungs are large enough to avoid being sucked in.

Do not stop the vacuum pump until all connections have been opened to atmosphere.

1.6.11 Protective Clothing

Normal laboratory practice requires the wearing of a white laboratory coat at all times. Additional protection is necessary for certain operations, as referred to above. Items of protective clothing which should be available in the laboratory are as follows:

Overalls.
Boiler suit.
Eye-shield.
Dust mask.
Ear muffs.
Hard hat.
Gumboots.
Toe-tector shoes or boots.
Rubber gloves.
Heat-insulated gloves.
Industrial gloves.

1.7 CALIBRATION

1.7.1 General

Calibration is the process of checking the graduations of a measuring instrument in order to verify its accuracy, by comparing it against a similar instrument of known accuracy, and if appropriate, graduating it correctly. All measuring instruments must be properly calibrated, and re-calibrated at regular intervals, if accuracy of test results is to be maintained. No instrument can have an accuracy greater than the accuracy with which it was calibrated.

To ensure that consistency is maintained between different laboratories, instruments used for test measurements should be calibrated so that their calibrations are traceable to national standards of measurement through an unbroken chain of calibrations. This is an essential requirement for laboratory accreditation by NAMAS (National Measurement Accreditation Service) which is a division of the National Physical Laboratory in the U.K. Traceable calibration can be achieved either by calibrating laboratory working instruments against certificated reference standards held by the laboratory (in-house calibration) or by entrusting calibration of working instruments to an organisation which is accredited by NAMAS for such calibration (external calibration).

Whenever an instrument has been damaged, repaired, adjusted, overhauled, overloaded or dismantled, or when a change in its accuracy is suspected, it should be re-calibrated before further use.

1.7.2   Reference Standards

Reference standards of measurement which are held in the laboratory for calibration of working instruments should be used for no other purpose. When not in use they should be stored separately from working instruments in a suitable environment, preferably under lock and key.

Reference standards appropriate to the tests described in this volume which may be held by a soil laboratory are listed below, together with the precision category specified in BS1377: Part 1:1990 where appropriate. These reference standards and instruments should have certification provided by a competent body, such as NPL, National Weights & Measures Laboratory, or a NAMAS accredited calibration laboratory. Certification is required to show traceability to national standards.

Length

Steel rule, 300 mm graduated 0.5 mm
Gauge blocks, 2.5 mm to 100 mm, and long slip, 150 mm (BS4311, Grade 1)
Reference gauge plate with hole, for use with liquid limit apparatus
A dial gauge comparator stand is required.

Mass

Reference weights:
1 mg to 100 g (Grade F1)
200 g to 10 kg (Grade F2)

Temperature

Mercury-in-glass thermometers complying with BS593, or calibrated:
range 0 to 150°C, graduated to 0.5°C
range 0 to 50°C, graduated to 0.2°C

Time

Reference to British Telecom speaking clock

Sieves

Set of master sieves (BS410), aperture sizes 63 μm to 5 mm
All items should be calibrated initially and accompanied by a calibration certificate which is traceable to national standards. Reference standards should be calibrated thereafter at regular intervals by a NAMAS accredited calibration organisation. Maximum intervals between calibrations, as specified in BS1377: Part 1:1990, are given in Table 1.11.

Table 1.11.   CALIBRATION INTERVALS FOR LABORATORY REFERENCE STANDARDS

Reference standard	*Maximum interval between calibrations*
Steel rule	5 years
Gauge blocks	5 years
Balance weights:	First calibration after 2 years, then 5 years
up to 200 g	
exceeding 200 g	5 years
Mercury-in-glass thermometers	5 years (calibrate or replace)
Master sieves	Replace after 50 uses

1.7.3 Calibration of Working Instruments

GENERAL REQUIREMENTS

Measuring instruments used for carrying out tests (working instruments) should be calibrated before being put into use, if they are not already accompanied by a certificate of calibration which is traceable to national standards. Re-calibration should be carried out at regular intervals thereafter, the interval between calibrations depending on the type of instrument, intensity of use and the conditions in which used. Calibration intervals specified in BS1377: Part 1:1990 for the types of instrument referred to in this volume are summarised in Table 1.12.

Details of every calibration should be recorded and retained on file. Wherever practicable, a label should be fixed to the instrument immediately after calibration to show clearly:

(a) The date of calibration.
(b) By whom calibrated.
(c) The date when next calibration is due.

If any instrument is found not to comply with the specified standard of accuracy, or to be faulty in any way, it should be clearly labelled as such and removed from the working area for overhaul.

IN-HOUSE CALIBRATION

In-house calibration of working instruments against the laboratory's own reference standards should be entrusted only to suitable qualified and experienced staff. Written procedures should be followed.

Calibration records should include the following:

(a) Description and identification number of instrument calibrated
(b) method of calibration
(c) equipment used

Table 1.12. CALIBRATION INTERVALS FOR LABORATORY WORKING INSTRUMENTS

	Maximum intervals between:		Refer to Section
Item	routine checks	calibrations	
Steel rules	When new	5 years	*
Vernier calipers	Check zero before every use and condition of jaws	1 year	1.3.3
Dial gauges	Check free movement before every use	1 year	(Volume 2)
Balances	Check zero before every use Verify with known weights weekly	6 months	1.3.4
Balance weights		1 year	1.3.4
Thermometers (mercury in glass) (other)	When new	5 years 6 months	*
Timers	1 month	Check against BT speaking clock annually	*
Vacuum gauge	When new	1 year	1.2.5

* Procedures given below

(d) calibration certificate number of the reference instrument against which the working instrument was calibrated
(e) calibration temperature
(f) calibration data and results
(g) date of calibration
(h) date when next calibration is due
(i) name and signature of person responsible for the calibration

EXTERNAL CALIBRATION

Calibration of working instruments under contract by an external organisation should be carried out at intervals not exceeding those given in Table 1.12 for in-house calibration. The certificate issued by the calibrating organisation should define the route of traceability to national standards, and should include their own name; for whom the item was calibrated; and the calibration location. Otherwise the certificate should provide the information listed as items (a) to (g) and (i) above.

1.7.5 Calibration and Checking of Test Equipment

CALIBRATION OF TEST APPARATUS

Numerous items of test apparatus, additional to the measuring instruments referred to above, need to be calibrated or checked. The items referred to in this volume are given in Table 1.13, together with the maximum intervals between calibrations or checks specified in BS1377: Part 1:1990.

Specifications for test equipment given in BS1377 include permissible manufacturing tolerances on critical dimensions (which may be linear dimensions or mass). Working tolerances allow for changes in these dimensions as a result of wear in use. In general, when working tolerances exceed twice the specified manufacturing tolerances the equipment no longer complies with the Standard. (See BS1377: Part 1:1990, Clause 4.1.3).

CALIBRATION PROCEDURES

Calibration procedures for some measuring instruments and major items of equipment are outlined in the appropriate chapters under the sections referred to in Tables 1.12 and 1.13. Suitable procedures for other items (where indicated by * in Tables 1.12 and 1.13) are given in outline below.

Thermometers

Mercury-in-glass thermometers, if marked as complying with BS593, require no initial calibration for soil testing purposes. They should be replaced after 5 years — which is probably more convenient than re-calibration. Thermometers of this type which are not thus marked can be calibrated against the reference thermometer by immersing both in water over a range of temperatures. For temperatures just over 100°C, a good-quality cooking oil can be used in place of water.

Other types of thermometer, e.g. thermocouple, can be calibrated over this range in a similar manner. Thermocouples used for higher temperatures, such as for verifying temperatures inside a muffle furnace, should be calibrated by a suitably accredited external organisation.

Timers

Laboratory timers can be calibrated over a 24-hour period by reference to British Telecom speaking clock. A good quality stop-watch, if calibrated regularly, may be used as a reference standard for other timers.

If wall clocks are used as timers in tests they should also be calibrated.

Table 1.13. CALIBRATION AND CHECKING INTERVALS FOR LABORATORY TEST APPARATUS

Item	Maximum intervals between:		Refer to Section
	routine checks	*calibrations*	
Drying ovens	1 week	Calibrate against temperature control setting annually	1.2.4
Sieves	Visual check before every use	Perforated plate sieves — measurement of apertures at least twice a year. Woven wire sieves — check against master set at least once every 3 months, depending on use use	4.5.6
Desiccator	Change & dry desiccant daily	—	1.2.5(1)
LL cone apparatus	Check mass when new and every 12 months. Check tip sharpness weekly. Check free movement before every use	Calibrate penetration gauge and auto-timer ever 12 months	2.6.4
LL Casagrande apparatus	Check drop before every use	Measure grooving tool every 50 uses. Check cup and bearing for wear every 3 months	2.6.6
Constant temperature water bath	Check daily	Calibrate against control settings annually	1.2.5(2)
Soil hydrometer	When new	2 years	4.8.4
Sampling pipette	When new	2 years	4.8.4
Moulds and cutting rings	When new	1 month	*
Compaction rammers	When new	3 months	*
Vibrating hammer	Inspect before every use	6 months	6.5.9
Glassware	Initial calibration not necessary if to Grade B of relevant BS, if justifiable. Verify at intervals depending on use.		*
Distilled water	Inspect daily	Chemical & conductance analyses annually	1.2.5(4)
De-ionized water	Check conductance whenever drawn from supply	Calibrate conductance indicator annually. Water analysis annually	1.2.5(4)
Sieve shaker	Inspect before each use	3 months	1.2.5(9)
Bottle shaker	Inspect before each use	1 year	1.2.5(10)
Centrifuge	Inspect before each use	1 year	1.2.5(11)
Muffle furnace	Check monthly	1 year	1.2.5(12)
pH meter	Check before and after each use	1 month	5.5.2
Automatic compactor	Mass of rammer, when new. Check auto counter each use; height of drop monthly	1 year	6.5.8

* Procedures given below.

Steel rules and tapes

These items should be verified before first use by careful alignment against the reference steel rule. Working steel rules should be re-calibrated within 5 years. Flexible tapes should be re-calibrated more frequently, depending on use, including a check on the error on the movable end stop.

Dial gauges and depth gauges

Calibration of dial gauges is described in Volume 2. Depth gauges can be calibrated on a flat surface (e.g. clean plate glass) by using calibrated gauge blocks.

Sample moulds and compaction moulds

Dimensions of moulds used for preparing test specimens should be determined at regular intervals. Internal diameters should be measured using internal vernier calipers on at least two diameters at each end. Several measurements of length should be made, using either vernier calipers or a steel rule graduated to 0.5 mm.

Moulds used for compaction and MCV tests should be measured in the same way. If the sample is weighed in the mould for determination of density, the mould should be weighed before each test.

Compaction rammers

The mass and height of drop of hand compaction rammers should be verified regularly.

Glassware

Volumetric glassware, if of Grade A or Grade B of the relevant British Standard, need not be calibrated if the use of uncalibrated glassware is justifiable. Calibration should be verified at regular intervals depending on use. Volumetric containers can be calibrated by weighing the amount of pure water they hold up to the graduation mark (bottom of meniscus), measuring the water temperature and calculating the volume using tabulated values of the density of water given in BS1797.

ROUTINE CHECKS

At regular intervals between calibrations, test equipment should be inspected and checked to ensure that reliability and accuracy is being maintained. This is additional to routine inspection which should be made as part of every test procedure before starting a test. Suitable intervals for routine checks are included in Tables 1.12 and 1.13.

If as a result of inspection and checking it is suspected that the accuracy or reliability of an item falls below that which is specified, the regular calibration procedure should be followed and corrective action taken if necessary.

BIBLIOGRAPHY

*BS1377: Parts 1, 2, 3, 4:1990, 'Methods of testing soils for civil engineering purposes'. British Standards Institution, London
BS5930:1981, Code of Practice for Site investigations (British Standards Institution, London)
The Concise Oxford English Dictionary of Current English, 5th edition. Clarendon Press, Oxford
HM Factories Inspectorate, Health and Safety Executive (June, 1976). Technical Data, Note 21 (rev.), 'Mercury'. HMSO, London
Methods for the Detection of Toxic Substances in Air. No. 13, 'Mercury and Compounds of Mercury'. HMSO, London
Nixon, I. K. and Child, G. H. (1975). *Civil Engineer's Reference Book*, 3rd edition (ed. L. S. Blake), Chapter 11, 'Site investigation'. Butterworths, London
Safety in Laboratories. University of Oxford committee on Safety
The Shorter Oxford English Dictionary, 3rd edition. Clarendon Press, Oxford
Technical Terms, Symbols and Definitions in Eight Languages. I.S.S.M.F.E., International Society of Soil Mechanics & Foundation Engineering (1981)
Tropical Residual Soils, QJEG, Vol. 23, No. 1, Geological Society Engineering Group Working Party Report (1990)

* These British Standards are quoted frequently in the succeeding chapters, and are not listed elsewhere. Extracts from BS1377:1990 are reproduced by permission of BSI, 2 Park Street, London W1A 2BS, from whom complete copies can be obtained.

Chapter 2

Moisture content and index tests

2.1 INTRODUCTION

2.1.1 Scope

This chapter deals with the moisture content (or water content) of soils, and the way in which the amount of water in soils can influence their behaviour. Measurement of moisture content, both in the natural state and under certain defined test conditions, can provide an extremely useful method of classifying cohesive soils and of assessing their engineering properties. These specified conditions have gained world-wide recognition, and are referred to as the index properties, or consistency limits. The tests to determine them are known as Index Tests or Limit Tests.

2.1.2 Types of test

The moisture content of a soil is the characteristic which is most frequently determined, and applies to all types of soil. The other index properties considered here relate only to cohesive soils, often including silts, and are referred to as their plasticity characteristics.

The tests to be described are for the determination of:

Moisture content
Saturation moisture content of chalk
Liquid limit (abbreviated LL)
Plastic limit (abbreviated PL)
Shrinkage limit (abbreviated SL)
Sticky limit
Linear shrinkage
Puddle characteristics
Swelling capability

The liquid limit, plastic limit and shrinkage limit are known collectively as the Atterberg Limits, after the Swedish scientist Dr A. Atterberg, who first defined them for the classification of agricultural soils in 1911. Originally they were determined by means of simple hand tests using an evaporating dish. The procedures were defined more precisely for engineering purposes by Professor A. Casagrande in 1932. The mechanical device he designed for determining the liquid limit is still known as the Casagrande apparatus, but this has now been largely superseded in many countries by various types of cone penetrometer, which has numerous advantages over that apparatus. The tests for determining the liquid and plastic limits are specified in BS1377: Part 2:1990, and in ASTM D4318-84, and are by far the most widely used of the index tests.

The tests grouped under the heading of Empirical Index Tests (Section 2.8) include simple traditional tests to assess the suitability of clay for use as a puddle material. A simple test to indicate the possible swelling characteristics of clays, and the sticky limit test, are also included.

2.2 DEFINITIONS

MOISTURE CONTENT (w)   The mass of water which can be removed from the soil by heating at 105°C, expressed as a percentage of the dry mass. (Also referred to as WATER CONTENT.)

NATURAL MOISTURE CONTENT   The moisture content of natural undisturbed soil *in situ*.

OVEN DRYING   Drying a soil to constant mass in an oven controlled to within certain limits of temperature, as in the standard moisture content test. STANDARD oven drying is at a temperature of 105–110°C.

AIR DRYING   Allowing a soil to lose most of its moisture by exposure to a warm dry atmosphere

DESICCATION   The process of drying out.

LIQUID LIMIT (w_L)   The moisture content at which soil passes from the plastic to the liquid state, as determined by the liquid limit test.

PLASTIC LIMIT (w_P)   The moisture content at which a soil passes from the plastic state to the solid state, and becomes too dry to be in a plastic condition, as determined by the plastic limit test.

PLASTICITY INDEX (I_P)   The numerical difference between liquid limit and plastic limit.

NON-PLASTIC (NP)   A soil with a plasticity index of zero, or on which the plastic limit cannot be determined by the standard procedure.

RELATIVE CONSISTENCY (C_r)   The ratio of the difference between liquid limit and moisture content to the plasticity index.

LIQUIDITY INDEX (I_L)   The ratio of the difference between moisture content and plastic limit to the plasticity index.

SHRINKAGE LIMIT (w_S)   The moisture content at which a soil on being dried ceases to shrink.

SHRINKAGE RATIO (R_S)   The ratio of the change in volume to the corresponding change in moisture content above the shrinkage limit.

SHRINKAGE RANGE   The difference between the field moisture content, or the moisture content at which a clay is placed, and the shrinkage limit.

LINEAR SHRINKAGE (L_S)   The change in length of a bar sample of soil when dried from about its liquid limit, expressed as a percentage of the initial length.

STICKY LIMIT   The moisture content below which a cohesive soil does not stick to metal tools.

2.3 THEORY

2.3.1 Moisture in Soils

GENERAL PRINCIPLES

Naturally occurring soils nearly always contain water as part of their structure. The moisture content of a soil is assumed to be the amount of water within the pore space between the soil grains which is removable by oven drying at 105–110°C, expressed as a percentage of the mass of dry soil. By 'dry' is meant the result of oven drying at that temperature to constant mass, usually for a period of about 12–24 h. In non-cohesive granular soils this procedure removes all water present.

There are several ways in which water is held in cohesive soils, which contain clay minerals existing as plate-like particles of less than 2 μm across (see Section 7.5). The shape and very small size of these particles, and their chemical composition, enable them to combine with or hold on to water by several complex means. A simplified illustration of the 'zones' of water surrounding a clay particle is obtained by considering the following five categories of water, indicated diagrammatically in Fig. 2.1.

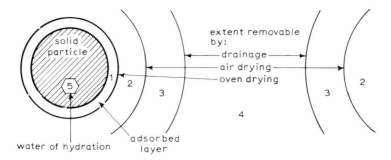

Fig. 2.1   Representation of categories of water surrounding clay particles

(1)   Adsorbed water, held on the surface of the particle by powerful forces of electrical attraction and virtually in a solid state. This layer is of very small thickness, perhaps of the order of 0.005 μm. This water cannot be removed by oven drying at 110°C, and may, therefore, be considered to be part of the solid soil grain.

(2)   Water which is not so tightly held and can be removed by oven drying, but not by air drying (hygroscopic moisture).

(3)   Capillary water, held by surface tension, generally removable by air drying.

(4)   Gravitational water, which can move in the voids between soil grains and is removable by drainage.

(5)   Chemically combined water, in the form of water of hydration within the crystal structure. Except for gypsum, and some tropical clays, this water is not generally removable by oven drying.

For the purpose of routine soil testing, moisture content relates only to the water which is removable by oven drying at 105–110°C. The water of category (1) above is not taken into account in the determination of moisture content and will not be referred to again in that context. The possible presence of water of category (5) is one reason for avoiding oven drying of tropical soils (Section 2.5.2).

Moisture content is usually expressed as a percentage, always on the basis of the *oven-dry* mass of soil. If the mass of water removed by drying at 105°C is denoted by m_w, and the mass of dried soil by m_D, the moisture content, w, is given by the equation:

$$w(\%) = \frac{m_w}{m_D} \times 100$$

SALINE WATER

If the water in the soil is saline, e.g. as in off-shore deposits, the dissolved salts will remain in the soil after drying and can give an incorrect moisture content result. A more appropriate value is the 'fluid content', i.e. the mass of fluid (water plus salts) per unit dry mass of soil. The fluid content, denoted by w_f (%), can be calculated from the measured moisture content ($w\%$) in one of two ways, depending on whether the dissolved salt content is measured by mass per unit mass, or mass per unit volume, of fluid.

(a)   If the salt content is known in terms of mass of salts per gram of fluid (p mg per gram, i.e. p parts per thousand), the fluid content w_f can be calculated from the equation

$$w_f = \frac{1000w}{1000 - p\left(1 + \dfrac{w}{100}\right)}\%$$

(b)   If the salts content is known in terms of mass of salt per litre of fluid (q grams per litre), the density of the fluid, ρ_f (Mg/m^3) also needs to be known. The fluid content w_f can then be calculated from the equation

$$w_f = \frac{1000w}{1000 - \dfrac{q}{\rho_f}\left(1 + \dfrac{w}{100}\right)}\%$$

As an example, assuming sea water contains 35 g per litre dissolved salts, and has a density of 1.024 Mg/m^3, typical values of the fluid content, w_f, corresponding to measured moisture contents, w, are as follows (to 3 significant figures).

$$w = 10 \quad 20 \quad 40 \quad 60 \quad 80 \quad 100 \quad 120 \quad \%$$
$$w_f = 10.4 \quad 20.9 \quad 42.0 \quad 63.5 \quad 85.2 \quad 107 \quad 130 \quad \%$$

Intermediate values can be obtained by interpolation.

2.3.2   The Atterberg Limits

The condition of a clay soil can be altered by changing the moisture content; the softening of clay by the addition of water is a well-known example. For every clay soil there is a range of moisture contents within which the clay is of a plastic consistency, and the Atterberg limits provide a means of measuring and describing the plasticity range in numerical terms.

If sufficient water is mixed with a clay, it can be made into a slurry, which behaves as a viscous liquid. This is known as the 'liquid' state. If the moisture content is gradually reduced by allowing it to dry out slowly, the clay eventually begins to hold together and to offer some resistance to deformation; this is the 'plastic' state. With further loss of water the clay shrinks and the stiffness increases until there is little plasticity left, and the clay becomes brittle; this is the 'semi-solid' state. As drying continues, the clay continues to shrink in proportion to the amount of water lost, until it reaches the minimum volume attainable by this process. Beyond that point further drying results in no further decrease in volume, and this is called the 'solid' state.

These four states, or phases, are shown diagrammatically in Fig. 2.2. The change from one phase to the next is not observable as a precise boundary, but takes place as a gradual transition. Nevertheless three arbitrary but specific boundaries have been established empirically, as indicated in Fig. 2.2, and are universally recognised. The moisture contents at these boundaries are known as the

Liquid limit (LL) (symbol w_L)
Plastic limit (PL) (symbol w_P)   The Attererg limits or consistency limits
Shrinkage limit (SL) (symbol w_S)

The moisture content range between the PL and LL is known as the plasticity index (PI) (symbol I_P), and is a measure of the plasticity of the clay. Cohesionless soils have no plasticity phase, so their PI is zero.

The tests to determine the Atterberg limits are carried out only on the fraction of soil which passes a 425 μm sieve. For soils that contain particles coarser than that size, the particles retained on the 425 μm sieve must be removed as part of the sample preparation procedure.

2.3.3   Consistency of Clays

Moisture content by itself is not sufficient to define the state of consistency of a clay soil. This can be done only by relating its moisture content to its liquid and plastic limits. For

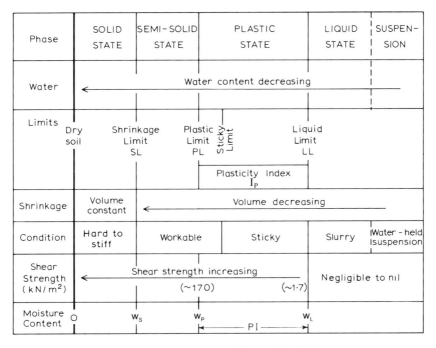

Fig. 2.2   Phases of soil and the Atterberg limits

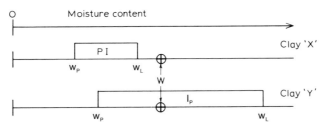

Fig. 2.3   Consistencies of two clays

example, two different clays can have the same moisture content, denoted by w in Fig. 2.3, yet show quite different characteristics. For clay X the moisture content w is greater than the liquid limit, so the clay is in the 'liquid' state (i.e. a slurry). For clay Y the same moisture content lies between the plastic and liquid limit. This clay is in the 'plastic' state, and would be of a firm consistency.

The relationship of the moisture content to the liquid and plastic limits can be expressed numerically in two ways, using parameters known as the relative consistency, denoted by C_r (Terzaghi and Peck, 1948) or the liquidity index, denoted by I_L (Lambe and Whitman, 1969) They are determined as follows:

$$C_r = \frac{w_L - w}{w_L - w_P} = \frac{w_L - w}{I_P}$$

$$I_L = \frac{w - w_P}{w_L - w_P} = \frac{w - w_P}{I_P}$$

Note that $I_L = 1 - C_r$.

Table 2.1. RELATIVE CONSISTENCY AND LIQUIDITY INDEX

Moisture content range w		Relative consistency C_r	Liquidity index I_L
Below PL	$w < w_P$	>1	negative
At PL	$w = w_P$	1	0
Between PL and LI	$w_P < w < w_L$	1 to 0	0 to 1
At LL	$w = w_L$	0	1
Above LL	$w > w_L$	negative	>1

The moisture content used in deriving these relationships should relate to the same fraction of the soil as is used for the Atterberg limits tests, i.e. the fraction passing the 425 μm sieve. If the natural soil contains material retained on a 425 μm sieve, the moisture content w (%) measured on the whole sample should be corrected to allow for the percentage by dry mass, p_a (%), of the portion which passes a 425 μm sieve. The corrected moisture content, w_a (%), is given by the equation

$$w_a = \frac{100w}{p_a}$$

Values of relative consistency and liquidity index throughout the moisture content range depicted in Fig. 2.2 are summarised in Table 2.1. These two parameters must not be confused.

The liquidity index is used more generally than the relative consistency. Between PL and LL it gives an easily visualised indication of the position in the plasticity range at which the soil lies. Below the plastic limit, the value of I_L is negative.

At the liquid limit ($I_L = 1$), a slowly drying clay slurry first begins to show a small but definite shear strength, the value of which is about 1.7 kN/m² (Wood and Wroth, 1976). As the moisture content decreases (and I_L approaches zero), the shear strength increase considerably, and at the plastic limit ($I_L = 0$) the shear strength may be 100 or more times greater than at the liquid limit. (Skempton and Northey, 1953; Wood and Wroth, 1978).

2.3.4 Activity of Clays

The Atterberg limits are related to the combined effects of two intrinsic properties of a clay, namely particle size and mineral composition. For a particular clay mixed with coarser material, Skempton (1953) showed that the plasticity index depended on the clay fraction (the percentage of particles finer than 2 μm), and that the ratio plasticity index/clay fraction was constant. For different clays different ratios were obtained, but the ratio was more or less constant for each clay type. The relationship between PI and clay fraction for a particular clay was termed the colloidal activity, or simply 'activity', where

$$\text{activity} = \frac{I_P}{\text{clay fraction}}$$

To be consistent with the Atterberg limits, the clay fraction used here should be expressed as a percentage of that portion of the soil sample which passes a 425 μm sieve.

On this basis, clays can be classified into four groups as shown in Table 2.2.

Approximate values of the activity of some clay minerals are shown alongside their liquid limits in Table 2.3.

Table 2.2. ACTIVITY OF CLAYS

Description	Activity
Inactive clays	<0.75
Normal clays	0.75–1.25
Active clays	1.25–2
Highly active clays	>2
(e.g. bentonite)	(6 or more)

Table 2.3. TYPICAL RANGES OF INDEX PROPERTIES OF SOME COMMON CLAY MINERALS

Clay mineral	Liquid limit range	Plasticity index range	Activity (approx.)
Kaolinite	40–60	10–25	0.4
Illite	80–120	50–70	0.9
Sodium montmorillonite	700	650	7
Other montmorillonites	300–650	200–550	1.5
Granular soils	20 or less	0	0

2.3.5 Flow Curve

The 'flow curve' for a cohesive soil is derived from the Casagrande liquid limit test (Section 2.6.6). Values of moisture content, w, are plotted against log N, where N is the number of standard blows required to close the groove. The resulting curve is known as the 'flow curve', and is virtually a straight line over the range covered by a typical test. (See Fig. 2.22). A plot of log w against log N gives a straight line, inclined at an angle B to the horizontal axis. The slope of this line, tan B, is, according to the British Standard, equal to 0.092 for most British soils. This is the basis of the table of factors for the one-point liquid limit test (Table 2.6) described in Section 2.6.7. In ASTM D4318 the corresponding value is 0.121.

The equation to the flow line is

$$LL = w\left(\frac{N}{25}\right)^{\tan B}$$

In this equation tan B is an exponent, not a multiplier.

2.3.6 Shrinkage Characteristics

SHRINKAGE LIMIT

When the water content of a fine-grained soil is reduced below the plastic limit, shrinkage of the soil mass continues until the shrinkage limit is reached. At that point the solid particles are in close contact and the water contained in the soil is just sufficient to fill the voids between them. Further reduction of water content cannot bring the particles closer together, so there is no further decrease in volume of the soil mass. Air enters the voids and the soil takes on a lighter colour. Below the shrinkage limit the soil is considered to be a solid in which the particles remain in contact and in an arrangement which gives a dense state of packing.

Clays are more susceptible to shrinkage than silts and sands. In most cohesive soils the shrinkage limit is appreciably below the plastic limit. In silts the two limits are close together and may be difficult to measure.

The shrinkage of a cohesive soil is illustrated graphically in Fig. 2.4, in which the change in volume of the soil mass is plotted against its moisture content as the soil is allowed to

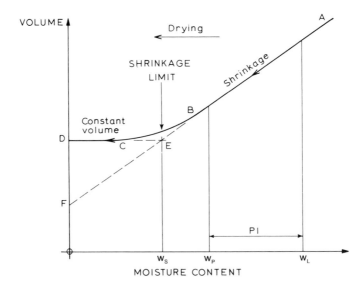

Fig. 2.4   Shrinkage curve for clay soil

dry. The portion AB of the curve is linear, and shows that the decrease in volume is directly proportional to the loss of water. To the left of point C there is no further decrease in volume as the soil dries. The portion BC represents the transition zone between these two conditions. The point E is found from the intersection of the two straight lines AB and DC, and this enables the shrinkage limit value to be determined as shown.

The straight line AB is produced back to meet the volume axis at F, which represents the state of the soil of shrinkage could continue without the formation of air voids — that is, by the solid particles merging together to eventually form a single lump. The volume OF therefore represents the volume of the dry solids in the soil mass. If expressed in ml, it is theoretically equal to the mass of dry soil (g) divided by the particle density of the soil grains.

SHRINKAGE RATIO

The 'shrinkage ratio' is the ratio of the change in volume, expressed as a percentage of the final dry volume, to the change in moisture content, above the shrinkage limit. In Fig. 2.5 the shrinkage ratio is equal to the slope of the line AE. (Points A, E, D, F correspond to those in Fig. 2.4.) The heights of the vertical columns represent diagrammatically the total volumes of soil at various points on the graph, and their subdivisions indicate the volumes of solid, water (V_w), and (below SL) air, contained within the soil.

Volume of dry soil $= V_0$ (ml)
Mass of dry soil $= m_0$ (g)
Mass of water in soil $= \rho_w V_w$
Moisture content $w = \dfrac{\rho_w V_w}{m_0} \times 100\%$

Above the shrinkage limit (w_S) water occupies the whole of the voids between particles. The change in volume on drying from point 1 to point 2 (both above w_S) is therefore equal to the volume of water lost between points 1 and 2.

Change in volume $= V_1 - V_2$

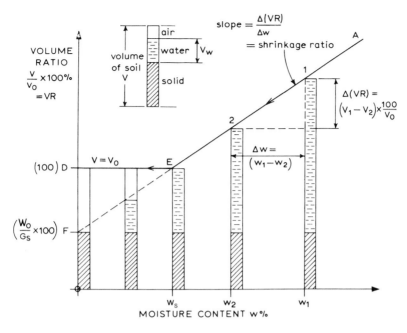

Fig. 2.5   Derivation of shrinkage ratio

Volume change expressed as percentage of final dry volume, denoted by $\Delta(VR)$:

$$\Delta(VR) = \frac{V_1 - V_2}{V_0} \times 100\%$$

Change in moisture content (Δw):

$$\Delta w = \frac{\rho_w V_1}{m_0} - \frac{\rho_w V_2}{m_0} \times 100\%$$

$$= \frac{\rho_w}{m_0}(V_1 - V_2) \times 100\%$$

By definition, shrinkage ratio $R_S = \Delta VR/\Delta w$ (above w_S). Therefore,

$$R_S = \left[\frac{V_1 - V_2}{V_0} \times 100\right] \div \left[\frac{\rho_w}{m_0}(V_1 - V_2) \times 100\right] = \frac{m_0}{\rho_w V_0}$$

In SI units, $\rho_w = 1 \text{ g/cm}^3$. Therefore,

$$R_S = \frac{m_0}{V_0}$$

Thus, the shrinkage ratio is equal to the ratio of the mass (g) to the volume (ml) of the oven-dried pat of soil at the end of the test described in Section 2.7.3.

SHRINKAGE RANGE

The 'shrinkage range' is the difference between the moisture content at which a clay is placed (m_{pl}) and the moisture content at the shrinkage limit:

$$\text{shrinkage range} = m_{pl} - w_S \, (\%)$$

LINEAR SHRINKAGE

The measurement of linear (one-dimensional) shrinkage of fine-grained soils is a different procedure from the measurement of volumetric shrinkage referred to above. The linear shrinkage is found by determining the change in length of a semi-cylindrical bar sample of soil when it dries out, starting from near the liquid limit.

If the original length when made up at about the LL is denoted by L_0, and the dried length by L_D, the change in length is equal to $L_0 - L_D$, and the linear shrinkage, L_S, is given by:

$$L_S = \frac{L_0 - L_D}{L_0} \times 100\%$$

In addition to indicating the amount of shrinkage, this test can provide an approximate estimate of the plasticity index for soils in which the liquid and plastic limits are difficult to determine. Examples are soils of low clay content, or soils from which it is difficult to obtain reproducible results, such as those with a high mica content. In these instances the linear shrinkage test can give more consistent results. An empirical relationship between plasticity index and linear shrinkage was given in BS1377:1967 as

$$I_P = 2.13 L_S$$

This relationship was based on experience with some British soils, and is not necessarily valid elsewhere. If a similar relationship is required for other soils it should be obtained from relevant test data.

2.4 APPLICATIONS

2.4.1 Moisture Content

The reasons for carrying out moisture content tests on soils fall into three categories:

(1) To determine the moisture content of the soil *in situ*, using undisturbed or disturbed samples.
(2) To determine the plasticity and shrinkage limits of fine-grained soils, for which moisture content is used as the index.
(3) To measure the moisture content of samples used for laboratory testing, usually both before and after test. This is normally done on all test samples as a routine procedure.

2.4.2 Classification

The liquid and plastic limits provide the most useful way of identifying and classifying the fine-grained cohesive soils. Particle size tests provide quantitative data on the range of sizes of particles and the amount of clay present, but say nothing about the type of clay. Clay particles are too small to be examined visually (except under an electron microscope), but the Atterberg limits enable clay soils to be classified physically, and the probable type of clay minerals to be assessed.

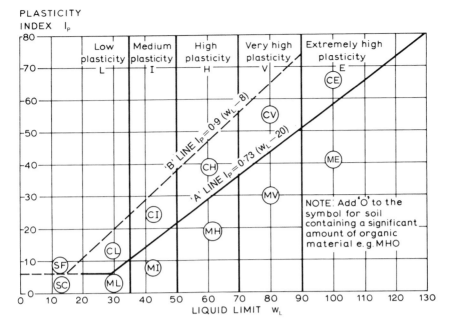

Fig. 2.6 Plasticity chart

Classification is usually accomplished by means of the plasticity chart (also referred to as the A-line chart). This is a graphical plot of the liquid limit (I_L) as ordinate against plasticity index (I_P) as abscissa. The standard chart is shown in Fig. 2.6.

When the values of w_L and I_P for inorganic clays are plotted on this chart, most of the points lie just above the line marked 'A-line', and in a narrow band parallel to it. The A-line is defined by the relationship

$$I_P = 0.73\ (w_L - 20)$$

This line was derived from experimental evidence and does not represent a well-defined boundary between soil types, but it does form a useful reference datum.

The dashed line labelled 'B-line' is a tentative upper limit for all soils, which has also been drawn from experimental data. It is defined by the relationship

$$I_P = 0.9\ (w_L - 8)$$

In British practice this chart is divided into five zones, giving the following categories for clays:

(1)   Clays of low plasticity (CL), less than 35 liquid limit.
(2)   Clays of medium plasticity (CI), liquid limit from 35 to 50.
(3)   Clays of high plasticity (CH), liquid limit from 50 to 70.
(4)   Clays of very high plasticity (CV), liquid limit from 70 to 90.
(5)   Clays of extremely high plasticity (CE), liquid limit exceeding 90.

The letters in parentheses are the standard symbols by which each group is known.

Silts when plotted on this chart generally fall below the A-line. They are divided into five categories similar to those for clays. The group symbols are ML, MI, MH, MV, ME.

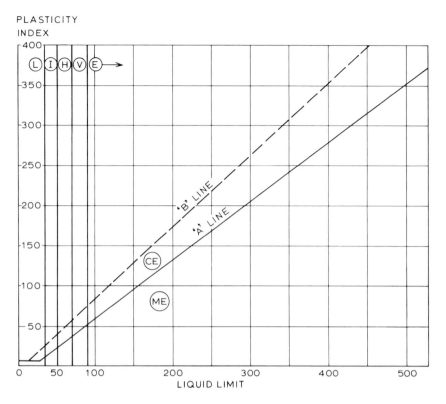

Fig. 2.7   Extended plasticity chart

Clays containing appreciable amounts of organic matter also plot below the A-line. Organic clays and silts include the letter 'O' after the group symbol (for example, CHO, MIO).

In the USA, the plasticity chart is divided into three groups:

Low plasticity (CL), liquid limit less than 30.
Intermediate plasticity (CI), liquid limit from 30 to 50.
High plasticity (CH), liquid limit exceeding 50.

An extended version of the A-line chart, embracing liquid limits up to 500, is given in Fig. 2.7.

Liquid limit values for some of the commoner clay minerals are typically within the ranges given in Table 2.3. Approximate values of activity (I_P/clay fraction) are also included.

2.4.3   Engineering Properties

PLASTICITY

The Atterberg limits may be used to correlate soil strata occurring in different areas of a site, or to investigate in detail the variations of soil properties which occur within a limit zone. Results of limits tests can also be applied to the selection of soils for use as compacted fill in various types of earthworks construction.

In general, clays of high plasticity are likely to have a lower permeability, to be more compressible and to consolidate over a longer period of time under load than clays of low plasticity. High-plasticity clays are more difficult to compact when used as fill material.

While the liquid and plastic limits indicate the *type* of clay in a cohesive soil, the *condition* of the clay is dependent upon its moisture content in relation to those limits, as expressed by the liquidity index. The moisture content used for determining liquidity index should be related to the fraction of soil passing the 425 μm sieve (see Section 2.3.3). Those engineering properties which govern shear strength and compressibility are largely dependent upon this relationship.

For many straightforward applications it is possible to obtain sufficient understanding of the nature of clay soil from Atterberg limits and moisture content tests, and little else, if the geological history of the soil is also known. Where it is necessary to obtain additional information by further testing, the results obtained from limits tests carried out as a first stage facilitate the selection of samples for more complex tests later.

SHRINKAGE AND STICKINESS

The shrinkage limit of most British soils lies within the range 10–15% moisture content. For some tropical and lateric soils the range may be much wider, and for these soils the shrinkage limit is perhaps more useful as a criterion for classification.

The shrinkage limit and shrinkage ratio of a clay soil, considered in relation to the field moisture content, can indicate whether it is likely to shrink on drying (such as by exposure to atmosphere, movement of ground water or loss of moisture taken up by vegetation or trees); and if so, by how much. These values are particularly useful in connection with the placing of puddle clay in reservoir embankments or canal linings. To prevent excessive cracking if some drying out of the clay is likely to occur, the shrinkage range can be limited to, say, 20% by controlling the placement moisture content in relation to the shrinkage limit.

The sticky limit can provide useful information to a contractor prior to an earthworks contract. If the sticky limit is lower than the natural moisture content of a clay to be excavated, or is lower than the moisture content at which it is to be placed, the handling of the clay may prove to be difficult, owing to its tendency to stick to construction tools and equipment.

MOISTURE CONTENT OF CHALK

Chalk is a very variable material with engineering properties ranging from those of rock to those of soil. When natural soil is excavated some of the material breaks down into fines ('putty chalk') and some remains as relatively intact lumps.

In assessing the properties of chalk to be placed as fill it is necessary to determine

(a) the proportion of the fines content, by sieving (Chapter 4);
(b) the moisture content of the fines, using the procedure given in Section 2.5.2;
(c) the 'saturation moisture content' of the intact lumps, i.e. the moisture content when all the voids are filled with water. This value is significant because it usually reflects the condition of the natural undisturbed chalk. (Ingoldby and Parsons, 1977).

The saturation moisture content is used together with the chalk crushing value (CCV) (Section 6.7) to assess the stability and other properties of freshly-placed chalk fill.

2.5 MOISTURE CONTENT TESTS

2.5.1 Test Methods

The standard method for determining moisture content of soils is the oven-drying method, and this is the procedure recommended for a soils laboratory. A subsidiary method using a heated sand-bath is also described for use on site where an oven is not available or would be impracticable. Another site method used extensively for moisture control of fill is the carbide method using a 'Speedy' moisture tester. This is not included here, but detailed instructions are provided by the manufacturer with the equipment.

Another method, in which methylated spirit is mixed with the soil and ignited, was given in BS1377:1975, where it was referred to as the Alcohol Method. This procedure is not recommended as a laboratory method, because of the risk of fire. It is not described here, and it does not appear in BS1377:1990.

Table 2.4. MOISTURE CONTENT CONTAINERS

Type of soil	Size of container diameter (mm)	height (mm)	Capacity	Typical mass (approx., g)
Clay and silt	65	20	75 g	12
Medium-grained	90	20	150 g	18
	100	100	500 g	60
Stony	250 × 250 (tray)	65	4 kg	300

2.5.2 Oven Drying (BS1377: Part 2:1990:3.2, and ASTM D2216)

A drying temperature of 105–110°C is specified as the standard procedure, and this should be used as a general rule. However, this temperature may be too high for certain types of soil. For peats and soils containing organic matter a drying temperature of 60°C is to be preferred, to prevent oxidation of the organic content. For soils containing gypsum the water of crystallisation may be lost at temperatures above 100°C, so a temperature not exceeding 80°C should be used. The presence of gypsum can be confirmed by heating a small quantity of soil on a metal plate (e.g. an electric hotplate). Grains of gypsum will turn white within a few minutes, but most other mineral grains will remain unaltered (Shearman, 1979). Some soils from tropical regions might also require drying temperatures not exceeding 80°C. These lower temperatures may need longer drying periods, and check weighings should be made at 2 to 4 hour intervals.

Throughout this book the apparatus and procedure detailed in this section are referred to as the 'standard moisture content apparatus and procedure'.

APPARATUS

(1) Thermostatically controlled drying oven, capable of maintaining a temperature of 105–110°C, and adjustable to a lower temperature (down to 60°C) if necessary. (See general notes on Ovens in Chapter 1, Section 1.2.4.)
(2) Desiccator container or cabinet, containing anhydrous silica gel.
(3) Balance, of a capacity and accuracy appropriate to the size of specimens to be tested. (See under under 'Selection of sample', below.) An automatic top-pan balance reading to 0.01 g is suitable for all except coarse-grained soils.
(4) Numbered glass weighing bottles or non-corrodible containers. For work requiring high accuracy, glass weighing bottles with ground-glass stoppers (both of which are numbered) are preferable. For routine work aluminium containers (usually referred to as moisture tins) or trays are used . Containers with close-fitting lids should be used for cohesive soils. Container and lid should each be stamped with the reference number. Suitable approximate sizes are given in Table 2.4.

Typical weighing bottles and containers are shown in Fig. 1.13.

All containers must be washed clean, dried thoroughly before use and weighed carefully. The mass of glass weighing bottles can be permanently recorded, although they should be check-weighed periodically. Aluminium containers should be weighed before each use.
(5) Trimming knife, spatula, scoop; other small tools as appropriate.
(6) Moisture content printed forms, or printed forms for other tests with provision for moisture content test data.

PROCEDURAL STAGES

(1) Weigh container.

(2)   Select soil sample.
(3)   Weigh wet.   Sequence: Balance
(4)   Dry in oven.   Oven
(5)   Cool in desiccator.   Desiccator
(6)   Weigh dry.   Balance
(7)   Calculate.
(8)   Report results.

The layout of balance, oven and desiccator should allow for convenient operation of this repetitive sequence.

TEST PROCEDURE

(1) *Weighing container*

If necessary, clean and dry the container and lid. Make sure that both are marked with the same number or reference letter. Weigh to the appropriate degree of accuracy, usually to within 0.02% of the mass of specimen to be used, as follows:

Mass of specimen up to:	50 g	500 g	4 kg
Weigh to nearest:	0.01 g	0.1 g	1 g

The correct use of balances is described in Section 1.3.4.
Enter the mass in the space provided on the test form.

(2) *Selection of sample*

The test sample must be selected so that it is properly representative of the soil sample from which it is taken. The nature of the soil material must also be taken into account. Normal practice should be to make three, or at least two, separate moisture content determinations and to average the results. However, if only a very small quantity of soil is available, it is better to use all the material for one measurement.

For the measurement of natural moisture content, the approximate mass of specimen required for each main soil type, and the accuracy of weighing, are as follows:

Homogeneous clays and silts	30 g, weigh to 0.01 g
Medium-grained soils	300 g, weigh to 0.1 g
Coarse-grained (stony) soils	3 kg, weigh to 1 g

In laminated clays where it is required to measure the moisture content of individual thin layers, it may be necessary to use much less than 30 g, and greater accuracy in weighing may then be desirable. With very stony soils it is sometimes preferable to measure the moisture content of the finer fraction only (for instance, the fraction passing a 6.3 mm sieve), and to report it as such.

The test specimen should be taken from near the middle of a mass of soil — not from the outside, where the soil may have partially dried by exposure to the atmosphere. The specimens should be crumbled and placed loosely in the containers so that the soil will dry right through.

Large samples should be placed in trays, and the soil broken up to expose a large surface area, before being placed in the oven. Replace the lids immediately to prevent loss of moisture before weighing.

Observe and record the type of sample, the soil type, and its condition, at this stage. Include comments on any apparent inadequacy of sealing of the sample in its container, possible drying out or segregation of water. Always record any unusual features which relate to the condition of the sample being tested.

(3) *Wet weighing*

Each container with wet soil should be weighed as soon as practicable after taking the specimen. If weighing is likely to be delayed, the lid must be fitted tightly and the container left in a cool place.

The container and soil is weighed to the appropriate degree of accuracy referred to in stage (1), and the mass is entered in the space on the test sheet.

(4)   Oven drying

Remove the lid from the container and place both on a shelf in the oven.

The thermostat control should already be set to maintain the required temperature, normally 105–110°C unless a lower temperature is desirable for the reasons referred to above. To avoid overheating of the soil, it is essential that air be able to circulate around the container, which should not be placed on the floor of the oven or close to the heating elements. The proper use of ovens is outlined in Section 1.2.4.

Drying in the oven should continue until the specimen has dried out to reach a constant mass. (See Section 1.2.4 and stage (6), below.) In practice drying overnight is usually sufficient for small samples.

Large samples, especially of wet soil, should not be dried at the same time as small samples of cohesive soil because the latter could absorb some moisture. If practicable a separate oven should be used for drying very large or wet samples.

(5)   Cooling in desiccator

Remove the specimen container and lid from the oven and place both separately in a desiccator cabinet until cool. Weighing must not be done while hot, because warm air currents can lead to inaccuracies with a sensitive balance. Cohesive soil must not be allowed to cool in the open, because dry soil can absorb moisture from the atmosphere.

(6)   Dry weighing

When cool, replace the lid on the container and weigh. Enter on the moisture content sheet as the mass of container plus dry soil.

If it is uncertain whether or not the soil has been completely dried, repeat stages (4), (5) and (6) at intervals of 2–4 h until a constant mass is observed. Constant mass is achieved when the difference between successive weighings is less than 0.1% of the dry mass of soil used. Repeated weighings should always be made if an oven temperature lower than 105°C has to be used.

(7)   Calculation

The moisture content of a soil is expressed as a percentage of its *dry* mass:

$$\text{moisture content } w = \frac{\text{loss of moisture}}{\text{dry mass}} \times 100\%$$

Let m_1 = mass of container; m_2 = mass of container and wet soil; m_3 = mass of container and dry soil. Then

$$w = \frac{m_2 - m_3}{m_3 - m_1} \times 100\%$$

The value of w for each individual moisture content specimen is calculated to the nearest 0.01%. If two or three separate measurements have been made on one sample, the average value is then calculated.

The specimen should not be disposed of until the moisture contents have been calculated

Moisture Content

Location:	*Bracknell*		Loc. No. *2456*	
Sample Nos.	*17/7*			
Relevant test	*Natural Moisture*	Operator *C. B. A.*	Date started *3. 1. 78*	

Sample No. & Ref.			*17/7* *A*	*B*	*C*
Container No.			*6*	*17*	*32*
Wet soil & Container	g	m_2	*52·68*	*61·39*	*58·42*
Dry soil & Container	g	m_3	*47·17*	*54·31*	*52·01*
Container	g	m_1	*15.53*	*16.22*	*15.75*
Dry soil	g	$m_3 - m_1$	*31·64*	*38·09*	*36·26*
Moisture loss	g	$m_2 - m_3$	*5.51*	*7·08*	*6·41*
MOISTURE CONTENT	%	w	*17.41*	*18.59*	*17·68*
AVERAGE MOISTURE	%			*17.89*	

$$\text{Moisture content} = w = \frac{m_2 - m_3}{m_3 - m_1} \times 100\%$$

Example (Container no. 6)

$$w = \frac{52.68 - 47.17}{47.17 - 15.53} \times 100$$

$$= \frac{5.51}{31.64} \times 100 = 17.41\%$$

$$\text{Average value} = \frac{17.42 + 18.59 + 17.68}{3} = 17.89$$

Reported moisture content: 18%

Fig. 2.8   Moisture content test results and calculations

and verified as being sensible values. If there is any doubt, check the dry weighings, or repeat the whole test if necessary.

(8)   Results

The final moisture content w is reported to two significant figures. That is:

If less than 10%, report to nearest 0.1% (e.g. 7.6%).
If 10% or greater, report to nearest 1% (e.g. 13%).
If an exact 0.5%, report to the nearest even number. For example,

13.50% would be reported as 14%;
16.50 would be reported as 16%;

but

16.51% would be reported as 17%;
16.49% would be reported as 16%.

Remarks on the results sheet should include whether the test was to determine natural moisture content or whether it was done in connection with another test; and the oven temperature, if different from the standard value.

Typical calculations are given in Fig. 2.8.

(9)   *Correction for coarse material*

When the moisture content of a cohesive soil is to be compared with the liquid limit and plastic limit (e.g. for calculation of liquidity index), and the soil contains particles retained on a 425 μm sieve, the calculated moisture content should be corrected to obtain the moisture content of the fraction passing the 425 μm sieve. A simple way of doing this is to place the dried moisture content specimens together in a weighed container, and weigh them. Transfer the specimens to a 425 μm sieve and wash on the sieve (as described in Section 4.6.4) until the washing water runs clear. Dry and weigh the particles retained on the 425 μm sieve.

Let m_4 = total mass of dry soil (g)
    m_5 = dry mass retained on the 425 μm sieve (g).

The percentage by dry mass of the material passing the 425 μm sieve, p_a, is given by the equation

$$p_a = \frac{m_4 - m_5}{m_4} 100\%$$

The corrected moisture content, w_a, is then calculated from the equation

$$w_a = \frac{100w}{p_a} \%$$

where w (%) is the measured moisture content.

2.5.3   Sand-bath Method

This method is intended as a site test for moisture content, where a drying oven is not available. It can be used in a main laboratory as a rapid method for granular soils. This method is not included in BS1377:1990.

   This procedure should not be used for soils containing gypsum, calcareous matter or organic matter.

APPARATUS

(1)   Sand-bath containing clean sand to a depth of at least 25 mm.
(2)   Moisture content containers for fine soils, as used for oven drying.
(3)   For coarser soils heat-resistant trays 200–250 mm square and 50–70 mm deep, the size depending on the quantity of soil required for test.
(4)   Heating equipment, such as a bottled gas burner or paraffin pressure stove, or electric hot-plate if mains electricity is available.
(5)   Scoop, spatula, appropriate small tools.

PROCEDURAL STAGES

(1)   Weigh container or tray.
(2)   Select soil sample.
(3)   Weigh wet.
(4)   Dry on sand-bath.
(5)   Cool.
(6)   Weigh dry.
(7)   Calculate.
(8)   Report results.

Stages (1)–(3)

Similar to the oven drying method.

If a tray is used as the specimen container, it should first be cleaned and weighed, the wet soil should be spread out evenly on it and the trayful then weighed immediately.

(4) *Drying on sand-bath*

Place the sample container or tray on the sand-bath, and heat on the stove. Do not overheat. Small pieces of white paper mixed with the soil will act as an indicator and turn brown if overheated. The soil should be frequently turned with a spatula during heating, to assist evaporation of moisture.

The period of drying will vary with the type of soil, size of sample and prevailing conditions. Check weighings should be made to determine the minimum drying time necessary. If the loss in mass after heating for a further period of 15 min does not exceed the following, the soil may be considered to be dry:

Fine-grained soils	0.1 g
Medium-grained soils	0.5 g
Coarse-grained soils	5 g

(5) *Cooling*

Since this is a less precise test than the oven-drying method, cooling in a desiccator is not essential, but a container should have the lid fitted when cooling. A tray can be dry-weighed as soon as it is cool enough to handle.

Stages (6) and (7)

As for oven drying.

(8) *Results*

The moisture content is reported to the nearest whole number, and the method used is reported as the sand-bath method.

2.5.4 Saturation Moisture Content of Chalk (BS1377: Part 2:1990:3.3)

This procedure is used for the determination of the potential moisture content of intact chalk lumps when fully saturated. It is based on the determination of the dry density of chalk lumps, from which the saturation moisture content is calculated on the assumption that the density of the chalk solids is 2.70 Mg/m^3.

APPARATUS AND PROCEDURE

The apparatus and procedures for the determination of the density of chalk lumps by immersion in water, and for the determination of their moisture content, are as described in Section 3.5.5. At least three lumps of chalk should be used, each of a volume of about 300 to 500 ml, and the results averaged.

CALCULATIONS

Calculate the volume V_s (ml) of each lump of chalk from the equation

$$V_s = m_w - m_g - \left(\frac{m_w - m_s}{\rho_p} \right)$$

where m_w is the mass of the chalk lump and wax coating (g); m_g is the apparent mass of the waxed lump when suspended in water (g); m_s is the mass of the chalk lump (g); ρ_p is the density of the paraffin wax (g/cm^3).

Calculate the bulk density of each lump, ρ(Mg/m^3) from the equation

$$\rho = \frac{m_s}{V_s}$$

Calculate the dry density of each lump ρ_{DI} (Mg/m^3) from the equation

$$\rho_{DI} = \frac{100\rho}{100 + w}$$

where w is the moisture content of the lump (%).

Calculate the mean dry density of the lumps tested, ρ_D (Mg/m^3), and use the following equation to calculate the saturation moisture content, w_s (%).

$$w_s = 100\left(\frac{1}{\rho_D} - \frac{1}{2.7}\right)$$

This assumes that the particle density of the chalk is 2.7 Mg/m^3, which is typical.

Report the saturation moisture content to two significant figures.

2.6 LIQUID AND PLASTIC LIMIT TESTS

2.6.1 Types of Test

The following tests for the determination of the Atterberg limits are given in BS1377: Part 2:1990:

Clause 4.3 Liquid limit (cone penetrometer).
Clause 4.4 Liquid limit (one-point cone method).
Clause 4.5 Liquid limit (Casagrande apparatus).
Clause 4.6 Liquid limit (one-point Casagrande method).
Clause 5.3 Plastic limit.

These tests are described in this section.

The method devised by Casagrande (1932), using the apparatus known by his name, was for 40 years the universally recognised standard method for the determination of the liquid limit of clay soils. This has now been superseded in Britain by the cone penetrometer test, although the Casagrande procedure is retained in the British Standard as a subsidiary method. The plastic limit test, in which the soil is rolled into a thread until it crumbles, remains unaltered in principle, although some details of procedure have been changed.

2.6.2 Comparison of Methods

Results obtained by the cone method have been proved to be more consistent, and less liable to experimental and personal errors, than those obtained by the Casagrande method (Sherwood and Ryley, 1968). For liquid limit values of up to 100% there is little difference between the results obtained by each method. Above 100% the cone method tends to give slightly lower values (Littleton and Farmilo, 1977). The relationship between results obtained from the Casagrande and cone methods, based on evidence available in 1978, is shown graphically in Fig. 2.9.

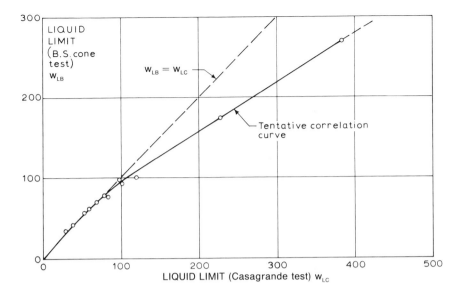

Fig. 2.9   Correlation of liquid limit results from two test methods

Both tests are carried out on remoulded soil, the fraction passing a 425 μm sieve being used. The cone method may not be any quicker to perform, but it is fundamentally more satisfactory, because the mechanics of the test depend directly on the static shear strength of the soil. The Casagrande procedure, on the other hand, introduces a dynamic component which is not related to shear strength in the same way for all soils. The definition of the liquid limit is dependent on the point at which the soil begins to acquire a recognisable shear strength (about $1.7 \, kN/m^2$), so a test which is based on this property is to be preferred to any other empirical method.

The one-point methods are useful as 'rapid' test procedures, or when only a very small amount of soil is available and when a result of lesser accuracy is acceptable.

2.6.3   Sample Preparation

EFFECT OF DRYING

Wherever possible the soil used for Atterberg limits tests should not be dried before testing. Drying, even air drying at laboratory temperature, can cause irreversible changes in the physical behaviour of some soils, especially tropical residual soils, which can result in dramatic changes in their plasticity properties. (Geological Society Engineering Group Report, 1990). (See Section 1.5.3).

The effect of drying may not be very great for most British sedimentary soils. Nevertheless drying, pestling and dry sieving as was the customary procedure should be avoided, and BS1377:1990 requires that the soil should not be allowed to become dry for the definitive methods of testing. Use of soil in its natural state wherever possible, or a wet preparation procedure for removal of particles retained on the 425 μm sieve, are specified instead.

The method of preparation of the soil should always be reported with the test results.

USE OF NATURAL SOIL (BS1377: Part 2:1990:4.2.3)

When the soil consists of clay and silt with little or no material retained on a 425 μm sieve, it can be prepared for testing from its natural state. This method should be used whenever practicable.

Fig. 2.10   Mixing water into soil for liquid limit test

Take a representative sample of about 500 g of soil and chop into small pieces, or shred with cheese-grater. Mix with distilled or de-ionised water on a glass plate, using two palette knives (see Fig. 2.10). During this process remove any coarse particles by hand or with tweezers. Mix the water thoroughly into the soil until a thick homogeneous paste is formed, and the paste has absorbed all the water with no surplus water visible. The mixing time should be at least 10 minutes, with vigorous working of the palette knives. A longer mixing period (sometimes up to 45 minutes) may be needed for some soils which do not readily absorb water.

Place the mixed soil in an airtight container, such as a sealed polythene bag, and leave to mature for 24 hours. A shorter maturing time may be acceptable for low plasticity clays, and very silty soils could be tested immediately after mixing. If in doubt, comparative trial tests should be performed. In a laboratory with a continuous work load it is good practice to be consistent and allow 24 hours maturing for all soils.

The mixed and matured material is then ready for the tests described in sections 2.6.4 to 2.6.8.

WET PREPARATION (BS1377: Part 2:1990:4.2.4)

If the soil includes particles retained on a 425 μm sieve which it is not practicable to remove by hand, the following procedure should be used. This is the procedure specified in BS1377:1990, and enables the coarse particles to be removed without the need for drying, pestling and dry sieving.

Take a representative sample of the soil at its natural moisture content to give at least 350 g of material passing the 425 μm sieve. This quantity allows for a liquid limit and a plastic limit test. Chop into small pieces, or shred with a cheese-grater, and place in a weighed beaker. Weigh, and determine the mass of soil, m (g) by difference.

Take a similar representative sample and determine its moisture content, w (%) as described in Section 2.5.2. The dry mass of soil in the test sample, m_D (g), can then be calculated from the equation

$$m_D = \frac{100m}{100 + w}$$

Add enough distilled water or de-ionised water to the beaker to just submerge the soil. Break down the soil pieces and stir until the mixture forms a slurry.

Nest a 425 μm sieve on a receiver, under a guard sieve (e.g. 2 mm) if appropriate. Pour the slurry through the sieve or sieves, and wash with distilled or de-ionised water, collecting all the washings in the receiver. Use the minimum amount of water necessary, but continue washing until the water passing the 425 μm sieve runs virtually clear. Transfer all the washings passing the sieve to a suitable beaker without losing any soil particles.

Collect the washed material retained on the sieves, dry in the oven and determine the dry mass, m_R (g).

Allow the soil particles in the beaker to settle for several hours, or overnight. If there is a layer of clear water above the suspension, this may be carefully poured or siphoned off, without losing any soil particles. However if the soil contains water-soluble salts which might influence its properties, do not remove any water except by evaporation.

Stand the container in a warm place (e.g. on top of an oven), or in a current of warm air, so that it can partially dry. Protect from dust. Stir the soil/water mixture frequently to prevent local over-drying. Alternatively, excess water may be removed by filtration (using vacuum or pressure). When the mixture forms a stiff paste (such that the penetration of the cone penetrometer would not exceed 15 mm) the soil is ready for mixing on the glass plate as described above. No additional curing time is required, and the material is ready for the tests described in Sections 2.6.4 to 2.6.8.

Calculate the percentage, by dry mass, of soil in the original sample passing the 425 μm sieve (p_a) from the equation

$$p_a = \frac{m_D - m_R}{m_D} \times 100\%$$

DRY PREPARATION METHOD (ASTM D421)

This method of preparation is not recognised in BS1377:1990 for the definitive plasticity tests, but is specified for the corresponding tests in ASTM standards. If the dry preparation method is unavoidable, the procedures is as follows.

(1) Allow the soil sample to air dry at room temperature, or in an oven at a temperature not exceeding 50°C (see Section 1.5.3). The soil should be brought to a state in which it can be crumbled.
(2) Break down aggregations of particles in a mortar using a rubber pestle, but avoid crushing individual particles. Sieve the sample on a 2 mm sieve, and use the rubber pestle to break down aggregations of soil particles so that only individual particles are retained (see Section 1.5.4).
(3) If necessary, take suitable representative samples for determination of the air-dry moisture content, following the procedure given in Section 2.5.2.
(4) Mix the sieved soil thoroughly, and sub-divide by riffling or quartering (Section 1.5.5) to produce a representative sample containing about 250 g finer than a 425 μm sieve if the liquid limit and plastic limit tests are to be performed.
(5) Sieve the sample on a 425 μm sieve. The fraction passing the 425 μm sieve is used for the plasticity tests.
(6) Place the soil on a glass mixing plate, or in an evaporating dish, and add distilled or de-ionised water in increments. Thoroughly mix each increment of water into the soil by kneading and mashing with two spatulas (see Fig. 2.10). Continue adding water and mixing until a thick homogeneous paste is formed. The minimum mixing time is 10 minutes, but up to 45 minutes may be required with some soils to obtain a uniform consistency with no un-absorbed water.
(7) Place the mixed soil in an airtight container, such as a sealed polythene bag, and leave for 24 hours to mature before proceeding with the test. The soil is then ready for the tests described in Sections 2.6.4 to 2.6.8.

(8)   If the percentages of material retained on the 2 mm and 425 μm sieves are to be
      determined, the soil should be weighed before and after riffling, and the particles
      retained on the sieves should be weighed. Corresponding dry masses can be calculated
      by using the measured moisture content. (Percentage calculations follow the same
      principle as described for sieving in Section 4.6).
(9)   Report the method of preparation with the test results.

IMPORTANCE OF MIXING

Whichever procedure is used for determining the liquid limit and preparing the soil, an
essential factor is the thorough mixing of the soil with water, followed by proper maturing.
There is no short-cut for these processes, which demand effort, patience and skill on the
part of the operator.

WATER

In carrying out these tests, tap-water must not be used for mixing the soil paste. Always
use distilled water, or de-ionised water. Otherwise there is the possibility that ion exchange
will take place between the soil and impurities in the water, which could affect the plasticity
of the soil.

2.6.4   Liquid Limit — Cone Penetrometer Method (BS1377: Part 2:1990:4.3)

This is the British Standard definitive method for determining the liquid limit of soils. It is
based on the measurement of penetration into the soil of a standardised cone of specified
mass. At the liquid limit the cone pentration is 20 mm. The method was developed at TRRL
from various cone tests which have been in use in other countries, and was adopted by the
BSI with a few modifications. It requires the same apparatus as is used for bituminous
materials testing, (BS2000: Part 49), but fitted with a special cone.

APPARATUS

(1)   Penetrometer apparatus complying with the requirements of BS2000, Part 49, and
      illustrated in Fig. 2.11.
(2)   Cone for the penetrometer, the main features of which are as follows (see also Fig. 2.12.):
            stainless steel or duralumin
            smooth, polished surface
            length approximately 35 mm
            cone angle 30°
            sharp point
            mass of cone and sliding shaft 80 g ± 0.1 g
(3)   Sharpness gauge for cone, consisting of a small steel plate 1.75 mm ± 0.1 mm thick with
      a 1.5 mm ± 0.02 mm diameter hole accurately drilled and reamed.
(4)   Flat glass plate, about 500 mm square and 10 mm thick, with bevelled edges and
      rounded corners.
(5)   Metal cups, of brass or aluminium alloy, 55 m diameter and 40 mm deep. The rim
      must be parallel to the base, which must be flat.
(6)   Wash bottle containing distilled or de-ionised water.
(7)   Metal straight-edge, about 100 mm long
(8)   Palette knives or spatulas:
            two 200 mm long × 30 mm
            one 150 mm long × 25 mm
            one 100 mm long × 20 mm
            one square-ended, 150 mm long × 25 mm (preferably of rubber or plastics)
(9)   Moisture content apparatus (Section 2.5.2).

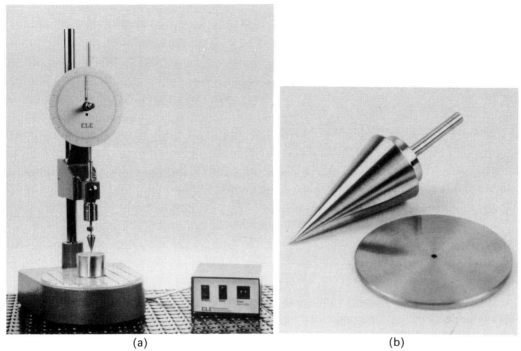

Fig. 2.11   *Apparatus for cone penetrometer liquid test:* (a) *Cone penetrometer with automatic timing device,* (b) *cone and gauge plate*

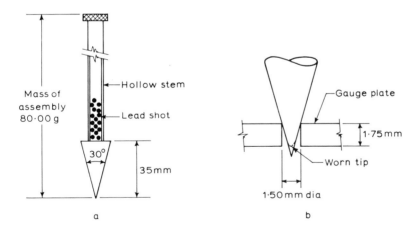

Fig. 2.12   *Details of cone for penetrometer*

PROCEDURAL STAGES

(1)   Select, prepare and mature the sample
(2)   Check apparatus.
(3)   Mix and work.
(4)   Place in cup.
(5)   Adjust cone.
(6)   Adjust dial gauge
(7)   Measure cone penetration.
(8)   Repeat penetration.

(9)   Measure moisture content.
(10)   Remix with extra water.
(11)   Calculate and plot graph.
(12)   Report result.

PROCEDURE

(1)   *Selection and preparation of sample*

Prepare and mature the test sample as described in Section 2.6.3. Use the natural material if possible; if not use the wet preparation method. Place and about 300 g of the prepared soil paste on the glass plate.

If the plastic limit test is also to be done, set aside a small portion in a sealed bag or container before adding too much water, and while the soil is still firm (see Section 2.6.8).

(2)   *Checking apparatus*

The cone penetrometer apparatus must comply with Clauses 4.3.2.3 and 4.3.2.4 of BS1377: Part 2:1990. The main points to check are:

(a)   The cone designed specially for testing soils must be fitted.
(b)   Mass of cone and stem 80 ± 0.1 g. This is *most important*. The stem is hollow, so that lead-shot can be inserted to bring the cone and stem assembly to the specified mass.
(c)   Sharpness of the cone point can be checked by pushing the tip into the hole of the sharpness gauge plate. If the point cannot be felt when brushed lightly with the tip of the finger, the cone should be replaced (see Fig. 2.12(b)).
(d)   The cone must fall freely when the release button is pushed, and the sliding shaft must be clean and dry.
(e)   The penetration dial indicator should be calibrated by inserting gauge blocks between the stem of the indicator and the top of the cone sliding shaft. Alternatively, calibrated vernier calipers could be used.
(f)   The apparatus must stand on a firm level bench.
(g)   If the apparatus is fitted with an automatic timing device, this should automatically lock the cone shaft assembly 5 seconds after pressing the button which releases it. This time interval should be verified against a reference timer.

(3)   *Mixing and working*

Mix the soil paste on the glass plate with the spatulas for at least 10 min. Some soils, especially heavy clays, may need a longer mixing time, up to 4 min. If necessary add more distilled or de-ionised water to give a cone penetration of about 15 mm, and mix well in. It is essential to obtain a uniform distribution of water throughout the sample.

Keep the soil together near the middle of the glass plate, to minimise drying out due to exposure to air. Cover with a damp cloth or polythene when not mixing.

Thorough mixing and kneading is hard work, but it is the most important feature of the Atterberg tests (see Section 2.6.3), and must never be overlooked.

(4)   *Placing in cup*

(a) Press the soil paste against the side of the cup, to avoid trapping air. (b) Press more paste well into the bottom of the cup, without creating an air-pocket. (c) Fill the middle and press well down. The small spatula is convenient for these operations. (d) The top surface is finally smoothed off level with the rim using the straight-edge.

(5)   *Adjustment of cone*

Lock the cone and shaft unit near the upper end of its travel and lower the supporting assembly carefully so that the tip of the cone is within a few millimetres of the surface of

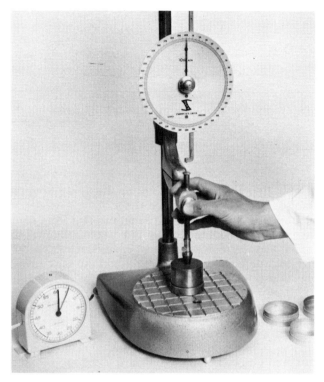

Fig. 2.13   Cone penetrometer test immediately after penetration

the soil in the cup. Hold the cone, press the release button and adjust the height of the cone so that the tip just touches the soil surface. A small sideways movement of the cup should just mark the surface. Avoid touching the sliding shaft, especially if clay is adhering to the fingers. Contamination of the shaft can cause resistance to sliding, and jamming in the sleeve.

(6)   Adjustment of dial gauge

Lower the stem of the dial gauge to make contact with the top of the cone shaft. Record the reading of the dial gauge to the nearest 0.1 mm (R_1). Alternatively, if the pointer is mounted on a friction sleeve, adjust the pointer to read zero (i.e. $R_1 = 0$).

(7)   Measuring cone penetration

Allow the cone to fall by pressing the button, which must be held in the pressed position for 5 s, timed with a seconds timer or watch (see Fig. 2.13). If an auto-timer is used it is necessary only to press the button and release it immediately. Automatic re-locking of the stem is indicated by a click. The apparatus must remain steady and must not be jerked.

After 5 s release the button so as to lock the cone in place. Lower the dial gauge stem to make contact with the top of the cone shaft, without allowing the pointer sleeve to rotate relative to the stem adjustment knob. Record the dial reading to the nearest 0.1 mm (R_2). Record the difference between R_1 and R_2 as the cone penetration. If the pointer was initially set to read zero, the reading R_2 gives the cone penetration directly.

(8)   Repeat penetration

Lift out the cone and clean it carefully. Avoid touching the sliding stem. Add a little more wet soil to the cup, without entrapping air, smooth off, and repeat stages (5), (6) and (7).

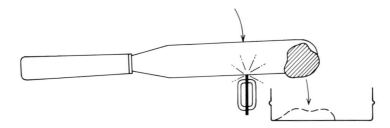

Fig. 2.14   Transferring moisture sample to container

If the second cone penetration differs from the first by less than 0.5 mm, the average value is recorded, and the moisture content is measured (stage 9).

If the second penetration is between 0.5 and 1 mm different from the first, a third test is carried out, and provided the overall range does not exceed 1 mm, the average of the three penetrations is recorded and the moisture content is measured (stage 9).

If the overall range exceeds 1 mm, the soil is removed from the cup and remixed, and the test is repeated from stage (4).

(9)   *Moisture content measurement*

Take a moisture content sample of about 10 g from the area penetrated by the cone, using the tip of a small spatula. This is placed in a numbered moisture content container, which is weighed, oven dried and weighed as in the standard moisture content procedure (Section 2.5.2). The moisture content sample should not be 'smeared' into the container, but dropped cleanly in by tapping the spatula on another spatula held close to the container, as shown in Fig. 2.14.

(10)   *Remixing*

The soil remaining in the cup is remixed with the rest of the sample on the glass plate together with a little more distilled water, until a uniform softer consistency is obtained. The cup is scraped out with the square-ended spatula, wiped clean and dried, and stages (4)–(9) are repeated at least three more times (making four in all), with further increments of distilled water.

A range of penetration values from about 15 mm to 25 mm should be covered, fairly uniformly distributed.

(11)   *Calculation and plotting*

The moisture content of the soil from each penetration reading is calculated from the wet and dry weighings as in the moisture content test.

Each cone penetration (mm) is plotted as ordinate, against the corresponding moisture content (%) as abscissa, both to linear scales, on a graph as shown in Fig. 2.15, which also shows typical data. The best straight line fitting these points is drawn.

(12)   *Results*

From the graph the moisture content corresponding to a cone penetration of 20 mm is read off to the nearest 0.1%. The result is reported to the nearest whole number as the liquid limit (cone test).

The percentage of material passing the 425 μm sieve is reported to the nearest 1%, together with the method of sample preparation.

The plastic limit and plasticity index are usually reported with the liquid limit.

Atterberg Limits - Cone Test

Location	BRACKNELL				Loc. No.	2456
Soil description	Fim blue-grey silty CLAY				Sample No,	6/5
Sample type	Undisturbed	Operator	A.B.S.		Date started	4.1.78

PLASTIC LIMIT

Test number		1	2	3	4	5	Average
Container no.		C1	G5	B7			
Wet soil & container	g	14.99	15.06	17.62			
Dry soil & container	g	13.48	13.60	15.58			
Container	g	7.94	7.99	7.97			
Dry soil	g	5.54	5.61	7.61			
Moisture loss	g	1.51	1.46	2.04			
MOISTURE CONTENT	g	27.26	26.02	26.81			26.70

LIQUID LIMIT

Test number		1		2		3		4	
Cone penetration	mm	15.5	15.1	19.0	19.0	22.0	21.8	25.4	25.2
Average penetration	mm	15.3		19.0		21.9		25.3	
Container no.		20		56		59		62	
Wet soil & container	g	46.78		57.20		63.60		71.72	
Dry soil & container	g	32.51		38.31		41.64		45.78	
Container	g	8.31		8.35		8.26		8.29	
Dry soil	g	24.20		29.96		33.38		37.49	
Moisture loss	g	14.27		18.89		21.96		25.94	
MOISTURE CONTENT	%	58.97		63.05		65.79		69.19	

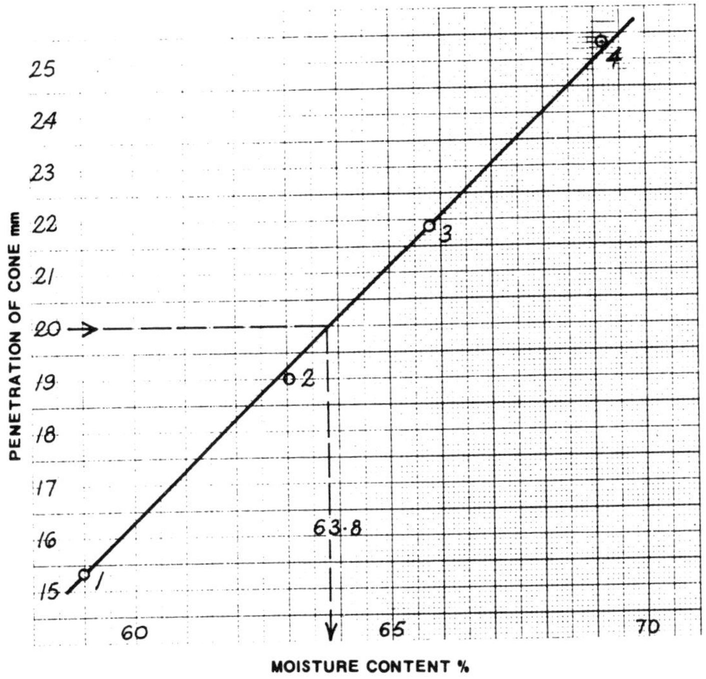

PREPARATION

As received ✓
~~Air dried~~
~~Oven dried~~
~~Pestled~~
~~Passed through~~
~~425 μm sieve~~

RESULTS

LL	64
PL	27
PI	37

Fig. 2.15   Liquid limit (cone test) and plastic limit results and graph

Table 2.5. SUGGESTED FACTORS FOR CONE PENTRATION ONE-POINT LIQUID LIMIT TEST (from Clayton and Jukes, 1978)

Penetration (mm)	Soil of high plasticity	Soil of intermediate plasticity	Soil of low plasticity
15	1.098	1.094	1.057
16	1.075	1.076	1.052
17	1.055	1.058	1.042
18	1.036	1.039	1.030
19	1.018	1.020	1.015
20	1.001	1.001	1.000
21	0.984	0.984	0.984
22	0.967	0.968	0.971
23	0.949	0.954	0.961
24	0.929	0.943	0.955
25	0.909	0.934	0.954
Measured moisture content range	above 50%	35% to 50%	below 35%

2.6.5   Liquid Limit — One-point Cone Penetrometer Method (BS1377: Part 2:1990:4.4)

This method was suggested by Clayton and Jukes (1978) as a possible less elaborate routine method of assessing the liquid limit of a soil than the four-point cone penetrometer test described in Section 2.6.4. It is based on a statistical analysis of experimental data. If the liquid limit is likely to exceed about 120% this test may not be valid and the four-point test should be performed instead, using the appropriate amount of soil.

The apparatus is the same as that described in Section 2.6.4. The procedure is identical to stages (1)–(9) of that section, except that a smaller quantity (about 100 g) of soil is required. Proper maturing, and thorough mixing with water, are just as important as in the standard test. The moisture content of the soil should be adjusted so that a cone penetration of between 15 mm and 25 mm is obtained.

After measuring two or more consecutive values of cone penetration as in steps (7) and (8) of Section 2.6.4, use most of the soil in the cup, including the zone penetrated by the cone, for the determination of the moisture content. Otherwise the procedure is as in step (9) of Section 2.6.4. Express the moisture content to the nearest 0.1%, and then multiply by a factor obtained from Table 2.5 to obtain the liquid limit of the soil. The factor to be used for a given penetration depends upon whether the soil is of low, medium or high plasticity. These categories are explained in Section 2.4.2, but for this purpose high plasticity refers to all soils having a liquid limit exceeding 50%.

The measured moisture content indicates to which group the soil belongs, as indicated at the bottom of Table 2.5. The factor to be used is read off from the appropriate column opposite the measured penetration. Factors for soils of low plasticity are the least reliable because many of the samples used for deriving these factors contained chalk.

The liquid limit (w_L) calculated in this way is reported to the nearest whole number, and the method is reported as the one-point cone penetrometer test. The percentage passing the 425 μm sieve, and the method of sample preparation, are also reported.

The plastic limit test may also be carried out on the same sample.

2.6.6   Liquid Limit — Casagrande Method (BS1377: Part 2:1990:4.5 and ASTM D4318)

This test procedure has been retained in the British Standard as a subsidiary method, but

Fig. 2.16   Casagrande liquid limit apparatus and tools

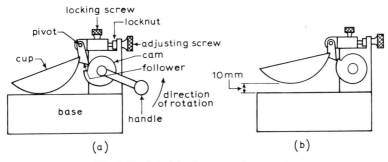

Fig. 2.17   Principle of Casagrande apparatus

the cone penetrometer method is now the standard preferred method. Results obtainable from the two types of test are discussed in Section 2.6.2.

The ASTM test designation D4318 replaced the previous designations D423 and D424 in 1984.

The liquid limit devices specified in BS1377 and ASTM D4318 are similar in principle, but there are some differences in detail. The ASTM devise has a rubber base of a different specification from the BS device. The rubber is harder, and test results from the two types of device may not be compatible. The test procedures defined by the two standards are very similar, except for certain details which are indicated.

APPARATUS

(1)   A mechanical device (the Casagrande apparatus), shown in Fig. 2.16. The principle of its action is shown in Fig. 2.17. The cup must fall freely, without too much side-play.

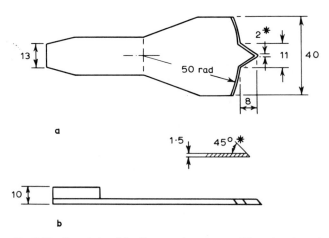

Fig. 2.18   Grooving tool for Casagrande apparatus (dimensions in mm)

The height to which the cup is lifted must be exactly 10 mm above the base. This can be checked with the spacer gauge (a steel block 10 mm thick, or a block on the handle of the grooving tool; Fig. 2.18), which should just pass between cup and base when the cup is at its maximum height. The adjusting screw provides a simple means of adjustment. The lock-nut must be tightened after adjustment, and the maximum height rechecked with the gauge. For the BS apparatus the material and construction of the base must conform to clause 4.5.2.3 of BS1377: Part 2:1990, which makes reference to BS903: Part A8:1963, BS903: Part A26:1969, and BS1154:1978.

Details of the requirements for the base of the ASTM apparatus are given in ASTM D4318, clause 6.1.1. The height of drop of the ASTM cup is measured to the point at which the cup makes contact with the base.

The apparatus may be fitted with a drive motor designed to turn the cam at the correct speed. Vibration from the motor must not be transmitted to the apparatus itself.

(2)   Grooving tool as illustrated in Fig. 2.18. It must be kept clean and dry, and the dimensions of the cutting profile must be checked periodically. The V-groove profile wears with use, and the essential dimensions must not differ by more than 0.25 mm from those specified. When necessary the tool should be reground to the correct profile.
(3)   Flat glass plate, with bevelled edges and rounded corners, 10 mm thick and about 500 mm square.
(4)   Wash bottle containing distilled or de-ionised water.
(5)   Two palette knives (spatulas), blades 200 mm long and 30 mm wide.
(6)   Spatula with blade about 150 mm and 25 mm wide.
(7)   Moisture content apparatus (Section 2.5.2).

PROCEDURAL STAGES

(1)   Select, prepare and mature sample.
(2)   Adjust apparatus.
(3)   Remix.
(4)   Place in bowl.
(5)   Cut groove.
(6)   Apply blows.
(7)   Repeat run.
(8)   Measure moisture content.
(9)   Repeat tests.
(10) Calculate.
(11) Report results.

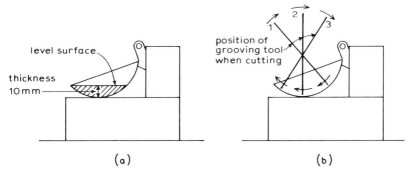

Fig. 2.19   Soil placed in Casagrande bowl, and use of grooving tool

PROCEDURE

Stage (1)

As for the cone penetrometer test (Section 2.6.4).

(2)   *Adjustment of apparatus*

The Casagrande apparatus must be clean, and the bowl must be dry and oil-free. Check that the bowl moves freely but without too much side-play, and that the drop is 10 mm; adjust if necessary. If a blow counter is fitted, turn it to zero. Check that the grooving tool is clean and dry, and conforms to the correct profile. (See items (1) and (2) under 'Apparatus', above.)

The machine should be placed on a firm, solid part of the bench, so that it will not wobble. The position should also be convenient for turning the handle steadily and at the correct speed (two turns per second). Practice against a seconds timer with the cup empty, to get accustomed to the correct rhythm.

(3)   *Remixing*

As for the cone penetrometer test (Section 2.6.4).

The same remarks regarding the vital importance of thorough mixing apply to this test.

(4)   *Placing in bowl*

See that the bowl of the apparatus is resting on the base, and is not supported by the cam. Place a portion of the mixed soil in the cup, pressing it from the middle outwards to prevent trapping any air. The smaller spatula is best for this. The surface of the soil paste should be smoothed off level and parallel to the base, giving a depth at the greatest thickness of 10 mm (see Fig. 2.19a).

(5)   *Cutting groove*

Cut a groove through the sample from back to front, dividing it into two equal halves. Starting near the hinge, draw the grooving tool from the hinge towards the front in a continuous movement with a circular motion, keeping the tool normal to the surface of the cup (see Fig. 2.19b). The chamfered edge of the tool faces the direction of movement. The tip should lightly scrape the inside of the bowl, but do not press hard. The completed groove is shown in Fig. 2.20.

It is sometimes difficult to cut a smooth groove in soils of low plasticity without tearing the soil. The ASTM procedure recommends cutting the groove with several strokes of the grooving tool, or alternatively forming a groove with a spatula and using the tool to trim

Fig. 2.20   Groove before applying bumps

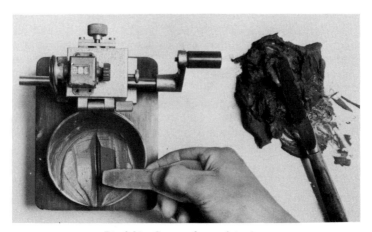

Fig. 2.21   Groove after applying bumps

to the required dimensions. Sliding of the soil pat on the surface of the cup must be avoided, otherwise the soil will slide back again during the test, instead of flowing as it should.

The curved grooving tool previously referred to as the 'ASTM grooving tool' is no longer called for in the ASTM standard.

If a smooth groove cannot be cut with the standard tool, this fact should be recorded. If a groove is formed by any other means, the method should be reported.

(6)   *Application of blows*

Turn the crank handle of the machine at a steady rate of two revolutions per second, so that the bowl is lifted and dropped. Use a seconds timer if necessary to obtain the correct speed. If a revolution counter is not fitted, count the number of bumps, counting aloud if necessary. Continue turning until the groove is closed along a distance of 13 mm. The back end of the standard grooving tool serves as a length gauge. The groove is closed when the two parts of the soil come into contact at the bottom of the groove (Fig. 2.21). Record the number of blows required to reach this condition. If there is a gap between two points of contact, continue until there is a length of continuous contact of 13 mm, and record the number of blows. If this exceeds 50 blows, remove the soil, mix in a little more water, and repeat stages (4)–(6).

The ASTM procedure requires verification that both sides of the groove have deformed with approximately the same shape, and that premature or irregular closing has not been caused by an air bubble. If this is suspected, a repeat run as described in (7) below should be made.

Closing of the groove should occur because of plastic flow of the soil, and not as a result of sliding on the surface of the bowl. If sliding occurs, this should be recorded and the result discarded. The test is repeated until flow does occur. If sliding persists after adding and mixing more water, the test is not applicable and it should be reported that the liquid limit could not be obtained.

(7) Repeat run

Add a little more soil from the mixture on the glass plate, about as much as was removed in forming the groove. Mix with the soil in the cup. Repeat stages (4)–(6) until two consecutive runs give the same number of blows for closure. Record the number immediately.

Ensure that the soil does not dry out between repeat tests, because the number of blows will increase with decreasing moisture content.

(This step is called for in the ASTM procedure only if the groove has closed in an irregular manner in step (6).)

(8) Moisture content measurement

Take a small quantity (about 10 g) of soil from the zone of the groove where the two portions have flowed together, using the small spatula. Place it in a moisture content container, replace the lid and determine the moisture content as in the moisture content test (Section 2.5.2). (See stage (9) of Section 2.6.4.)

(9) Repeat tests

Remove the soil from the bowl and mix it in with the remaining soil paste on the glass plate. Add a little more water, and mix in thoroughly. Clean the bowl of the apparatus and wipe it dry. Reset the revolution counter, if fitted, to zero. Clean and dry the grooving tool and spatula.

Repeat stages (4)–(8) at least three more times (four determinations in all), adding a little more water each time. The moisture contents should be such that the number of blows is roughly evenly spaced over the range from about 50 to 10. There should preferably be at least two blow-counts each side of 25. If necessary, do an additional repeat test. The test must start at the drier condition (about 50 blows) and proceed towards the wetter condition (10 blows).

Whenever the soil is left on the glass plate, it should be covered with a damp cloth to prevent drying out.

The ASTM procedure requires a minimum of 3 determinations covering the range 35 to 15 bumps.

(10) Calculations

Calculate the moisture content for each blow count, as in the moisture content test. Using a semi-logarithmic chart, plot the moisture content as ordinate (linear scale) against the corresponding number of blows as abscissa (logarithmic scale). Draw the best straight line fitting the plotted points. This is called the 'flow curve' (Fig. 2.22).

Draw the ordinate representing 25 blows, and where it intersects the flow curve draw the horizontal line to the moisture content axis. Read off this value of moisture content and record it on the horizontal line to the nearest 0.1%, as in Fig. 2.22.

Atterberg Limits Tests

Location	North Bromwich				Loc. No.	3210
Soil description	Grey clay with brown mottling				Sample No.	4/9
Sample type	Jar		Operator	F. B. J.	Date started	15. 6. 78

PLASTIC LIMIT

Test Number		1	2	3	4	5	Average
Container no.							
Wet soil & container	g						
Dry soil & container	g						
Container	g						
Dry soil	g						
Moisture loss	g						
MOISTURE CONTENT	%						

LIQUID LIMIT

Test Number		1	2	3	4	5	6
Number of bumps		49	40	29	22	16	
Container no.		64	95	57	74	82	
Wet soil & container (m_2)	g	19·47	19·77	21·01	22·72	19·26	
Dry soil & container (m_3)	g	14·78	14·82	15·39	16·31	14·38	
Container (m_1)	g	8·51	8·43	8·38	8·49	8·64	
Dry soil $(m_3 - m_1)$	g	6·27	6·39	7·01	7·82	5·74	
Moisture loss $(m_2 - m_3)$	g	4·69	4·95	5·62	6·41	4·88	
MOISTURE CONTENT W	%	74·8	77·5	80·2	81·9	85·1	

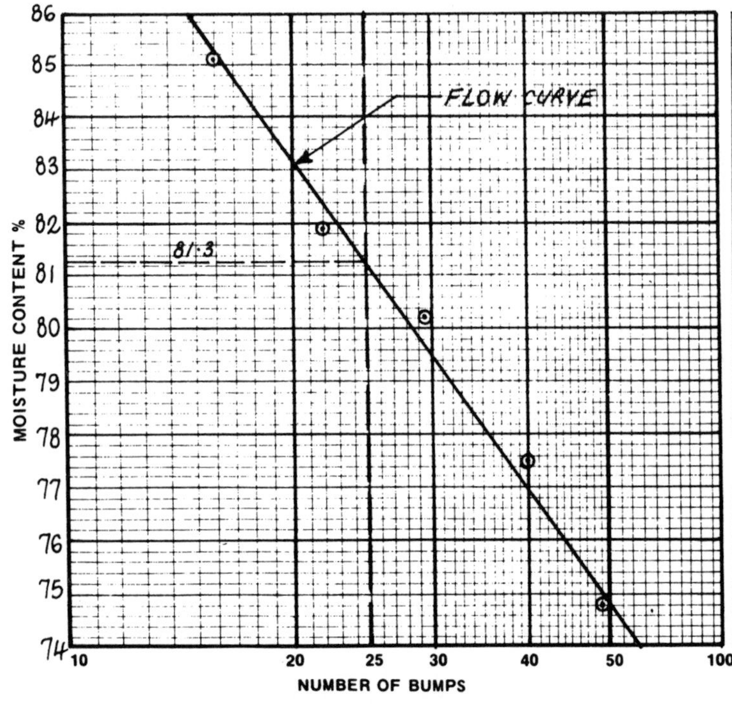

PREPARATION

(As received)

Air dried

Oven dried

Pestled

Passed through 425 μm sieve

RESULTS

LL	81
PL	29
PI	52

Fig. 2.22   Liquid limit (Casagrande test) results and graph

(11) *Results*

The moisture content read off from the flow curve is reported to the nearest whole number as the liquid limit (w_L) of the soil.

The method of test (using the Casagrande apparatus) is reported. Also report the percentage of material passing the 425 μm sieve (if the sample was sieved), and the method of sample preparation.

The plastic limit (w_P) is usually reported at the same time.

2.6.7 Liquid Limit — Casagrande One-point Method (BS1377:Part 2:1990:4.6 and ASTM D4318: 13, 14, 15)

This method provides a quick means of determining the liquid limit of a soil, because only one moisture content measurement is needed. However, the result is likely to be less reliable than the four-point Casagrande method or the cone penetrometer method. If the liquid limit is likely to exceed about 120% this test may not be valid and the four-point test should be performed instead, using the appropriate amount of soil.

It is a useful method when only a very limited quantity of soil is available.

APPARATUS

Similar to that listed in Section 2.6.6.

PROCEDURAL STAGES

(1)   Select, prepare and mature sample.
(2)   Adjust apparatus.
(3)   Mix with water.
(4)   Place in bowl.
(5)   Cut groove.
(6)   Apply blows.
(7)   Repeat run.
(8)   Moisture content.
(9)   Calculate.
(10) Report results.

PROCEDURE

(1) *Selection, preparation and maturing of sample*

About 50 g of soil is required for the test; otherwise the procedure is as described in Section 2.6.6.

Even though a small sample is used, the mixing process is just as important as for the four-point test. The period of maturing should not be omitted for clay soils.

Stages (2)–(5)

As in Section 2.6.6.

(6) *Application of blows*

This is carried out as in Section 2.6.6 but the moisture content should be such that the number of blows for closure of the groove is between 15 and 35.

(7) *Repeat run*

Add a little more soil to the bowl, mix in and make an immediate repeat run. This should

Table 2.6. FACTORS FOR CASAGRANDE ONE-POINT LIQUID LIMIT TEST (From BS1377:1990)

Number of blows	Factor	Number of blows	Factor	Number of blows	Factor	Number of blows	Factor
15	0.95	21	0.98	26	1.00	31	1.02
16	0.96	22	0.99	27	1.01	32	1.02
17	0.96	23	0.99	28	1.01	33	1.02
18	0.97	24	0.99	29	1.01	34	1.03
19	0.97	25	1.00	30	1.02	35	1.03
20	0.98						

give the same number of blows for closure. If not, repeat until two consecutive runs do give the same number.

(8) *Moisture content measurement*

Since only one moisture content determination is necessary, most of the soil in the bowl after the second run can be used for this purpose. It is placed in a moisture container and treated as in the moisture content test (Section 2.5.2). Clean out and dry the bowl.

(9) *Calculation*

The moisture content of the soil from the bowl is calculated in the usual way, and expressed to the nearest 0.1%. This moisture content is multiplied by a factor, given in Table 2.6, corresponding to the number of blows, to give the liquid limit. The theoretical basis of these factors is discussed in Section 2.3.5. The ASTM Standard gives factors for the range 20 to 30 blows which differ but little from those given in Table 2.6.

The moisture content of the soil should be as close as possible to the liquid limit, so that the number of blows required is close to 25. This will minimise the error due to variation of the slope of the flow line from the mean value used for the derivation of the conversion factors.

(10) *Report results*

The liquid limit (w_L) calculated as above is reported to the nearest whole number. The method is reported as the one-point method using the Casagrande apparatus. The percentage passing the $425 \, \mu$m sieve, and the method of sample preparation.

The plastic limit test may also be carried out on the same sample.

2.6.8 Plastic Limit (BS1377: Part 2:1990:5.3 and ASTM D4318)

This test is to determine the lowest moisture content at which the soil is plastic. It can be carried out only on soils with some cohesion, on the fraction passing a $425 \, \mu$m sieve. The test may be carried out either on soil in its natural state or on soil prepared by the wet preparation method, both as described in Section 2.6.3. The test is usually carried out in conjunction with the liquid limit test.

APPARATUS

(1)   The most important piece of apparatus for this test is the hand of the operator, which should be clean and free from grease.

(2)   Glass plate and small tools used for the liquid limit test.

(3)   A separate glass plate reserved for rolling of threads. This should be smooth and free from scratches, and about 300 mm square and 10 mm thick. The surface condition of the plate can affect the behaviour of rolled threads, and the use of unscratched glass reduces the likelihood of discrepancies. An alternative is to reserve one side of the mixing plate for thread rolling, and avoid mixing the soil on this area. Silica particles in soil will inevitably scratch glass during the mixing process.
(4)   Two palette knives or spatulas
(5)   A short length (say 100 mm) of 3 mm diameter metal rod.
(6)   Standard moisture content apparatus (Section 2.5.2).

PROCEDURE STAGES

(1)   Prepare sample.
(2)   Roll into ball.
(3)   Roll into threads until crumbling occurs.
(4)   Measure moisture content.
(5)   Repeat tests.
(6)   Calculate.
(7)   Report results.

PROCEDURE

(1)   *Preparation of sample*

Prepare and mature the test sample by one of the methods described in Section 2.6.3. Take about 20 g of the prepared soil paste and spread it on the glass mixing plate so that it can partially dry. Mix occasionally to avoid local drying out. It is convenient to set aside this sample just before carrying out the liquid limit test.

(2)   *Rolling into a ball*

When the soil is plastic enough, it is well kneaded and then shaped into a ball. Mould the ball between the fingers and roll between the palms of the hands so that the warmth of the hands slowly dries it. When slight cracks begin to appear on the surface, divide the ball into two portions each of about 10 g. Further divide each into four equal parts, but keep each set of four parts together.

(3)   *Rolling into threads*

One of the parts is kneaded by the fingers to equalise the distribution of moisture, and then formed into a thread about 6 mm diameter, using the first finger and thumb of each hand. The thread must be intact and homogeneous. Using a steady pressure, roll the thread between the fingers of one hand and the surface of the glass plate (see Fig. 2.23). The pressure should reduce the diameter of the thread from 6 mm to about 3 mm after between five and ten back-and-forth movements of the hand. Some heavy clays may need more than this because this type of soil tends to become harder near the plastic limit. It is important to maintain a uniform rolling pressure throughout; do not reduce pressure as the thread approaches 3 mm diameter.

Dry the soil further by moulding between the fingers again, not by continued rolling which gives a dried crust. Form it into a thread and roll out again as before. Repeat this procedure until the thread crumbles when it has been rolled to 3 mm diameter. The metal rod serves as a reference for gauging this diameter. Crumbling of the thread occurs in different ways with different types of soil. These include falling apart in small pieces; breaking into a number of short pieces tapered towards the ends; longitudinal splitting from the ends towards the middle and then falling apart. The essential requirement is to reach the crumbling condition, not to finish with spaghetti-like threads.

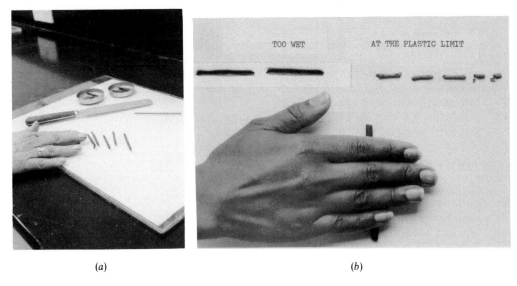

(a)   (b)

Fig. 23   Plastic limit test: (a) Apparatus, (b) soil thread before and after rolling (Courtesy LTG Laboratories, London)

Crumbling must be the result of the decreasing moisture content only, and not due to mechanical breakdown caused by execesive pressure, or oblique rolling or detachment of an excessive length beyond the width of the hand.

The first crumbling point is the plastic limit. It may be possible to gather the pieces together after crumbling, to reform a thread and to continue rolling under pressure, but this should not be done.

(4)   Moisture content measurement

As soon as the crumbling stage is reached, gather the crumbled threads and place them into a weighed moisture content container. Replace the lid immediately.

(5)   Repeat tests

Repeat stage (3) for the other three pieces of soil, and place in the same container. Weigh the container and soil as soon as possible, dry in the oven overnight, cool and weigh dry, as in the standard moisture content procedure (Section 2.5.2).

Repeat stages (3)–(5) on the other set of four portions of the soil, using a second moisture content container.

The ASTM method requires tests on enough portions, each of 1.5 g to 2.0 g, to give at least 6 g of soil in each container.

(6)   Calcultions

Calculate the moisture content of the soil in each of the two containers. Take the average of the two results. If they differ by more than 0.5% moisture content, the test should be repeated.

Typical test data are included in Fig. 2.15.

(7)   Results

The average moisture content referred to above is expressed to the nearest whole number and reported as the plastic limit (w_L) of the soil. The method of preparation of the soil is reported, and so is the percentage of material passing the 425 μm sieve if it was sieved. The result is usually reported on the same sheet as the liquid limit test.

(8) *Plasticity Index* (Clause 5.4 in BS1377: Part 2:1990)

The difference between the liquid limit and plastic limit is calculated to give the plasticity index (I_P) of the soil:

$$I_P = w_L - w_P$$

This value is also reported to the nearest whole number. If it is not possible to perform the plastic limit test, the soil is reported as non-plastic (NP). This also applies if the plastic limit is equal to or greater than the liquid limit; the latter can occur in some soils with a high mica content (Tubey and Webster, 1978).

2.7 SHRINKAGE TESTS

2.7.1 Types of Test

Tests to measure two aspects of the shrinkage properties of soils discussed in Section 2.3.6, are considered here. These are the shrinkage limit and linear shrinkage. Two tests for the determination of the shrinkage limit, and a test for determining linear shrinkage, are given in BS1377: Part 2:1990 and are described below. Both methods for the measurement of shrinkage limit depend upon the accurate measurement of the volume of a soil specimen as it dries out, by immersing it in mercury.

The first (the BS definitive method) was developed by the Transport & Road Research Laboratory (TRRL) for use with standard 38 mm or 51 mm diameter undisturbed samples. It is the more convenient, and potentially more accurate, method provided that corrosion effects at the small electrical contact are avoided.

The second method is the same as that given in ASTM Standards. It requires simple apparatus, but involves the use of exposed mercury. Adequate ventilation is essential, together with precautions to guard against spillage and availability of treatment facilities if spillage does occur (see Chapter 1, Section 1.6.7).

The linear shrinkage test is easy to perform and indicates only the amount of shrinkage.

2.7.2 Shrinkage Limit — TRRL Method (BS1377: Part 2:1990:6.3)

This test is the definitive method for the measurement of volumetric shrinkage in BS1377:1990

APPARATUS

(1) Shrinkage limit apparatus, the essential features of which are an immersion tank for holding mercury; soil specimen cage; micrometer measuring device readable to 0.01 mm; electric circuit with dry cell battery, platinum contact, indicator lamp and test switch; levelling screws and level indicator bubble.
(2) Mercury, sufficient to fill the cell of the apparatus to the required level.
(3) Mould of known volume, for preparation of remoulded specimen.
(4) Trimming and measuring tools, for preparation and measurement of undisturbed specimens.
(5) Balance, 200 g capacity, 0.01 g accuracy.
(6) Tongs, small brush, watch glass and other small tools.
(7) Moisture content apparatus (Section 2.5.2).
(8) Large tray containing a small depth of water, to retain any spilled mercury.
(9) Rubber gloves

Fig. 2.24   Shrinkage limit test — TRRL apparatus

PROCEDURAL STAGES

(1)   Prepare specimen.
(2)   Measure and weigh.
(3)   Adjust apparatus.
(4)   Measure volume and mass.
(5)   Allow partial drying.
(6)   Repeat volume measurements.
(7)   Oven dry and weigh.
(8)   Calculate and plot.
(9)   Report results.

TEST PROCEDURE

(1)   *Preparation of specimen*

A cylindrical specimen with a height equal to twice the diameter is convenient for this test. The size will depend upon the dimensions of the mercury container, but 38 mm diameter and 75 mm long is a typical specimen size.

An undisturbed specimen may be obtained by carefully pushing a thin-walled cutting tube into the main sample, if it is in a relatively large container. A sample in a 100 mm diameter sampling tube should be jacked out and the 38 mm diameter cutting tube pushed in as it emerges. The sample is then pushed out of the cutting tube and trimmed to length. Alternatively, an undisturbed specimen may be prepared by hand trimming, or with the use of a soil lathe.

For a remoulded specimen, the soil is mixed with water if necessary to a stiff plastic consistency (wetter than the plastic limit), and compacted without entrapping pockets of air into a cutting tube or split mould. It is then extruded or removed, and trimmed to length. The soil should not be dried.

(2) Initial measurements

The volume of the specimen (V_1 cm^3) may be determined initially from its linear measurements, as described in Chapter 3, Section 3.5.2. This is useful as a check on the volume-measuring apparatus, but is not essential.

The specimen should be weighed to an accuracy of 0.01 g and this mass recorded as the initial mass.

(3) Adjustment of apparatus

Determine the internal diameter of the apparatus, D, to the nearest 0.1 mm, using internal gauging calipers measured with a vernier caliper.

Set the apparatus on the tray on a firm bench in a fume cupboard or near an air extractor fan. Use the base-adjusting screws to level it, checking with the built-in bubble. Add mercury carefully to the required depth, and recheck for level (the mercury is probably almost as heavy as the apparatus which contains it). The apparatus should remain undisturbed in the same position until the tests in hand are completed, which may cover a period of several days.

Check that the batteries are fresh and that the indicator lamp lights up when the circuit is closed. A test switch is provided for this purpose. Lower the cage into the mercury tank until it is completely immersed. Rotate the cage two or three times to remove trapped air bubbles. Adjust the micrometer until the platinum contact just touches the surface of the mercury. This is indicated by the bulb lighting up. Record the reading of the micrometer as the zero reading, M_0. Raise the cage out of the tank and raise the micrometer adjustment.

(4) Measurements of volume and mass

Place the specimen in the cage, and lower into the mercury until the whole is completely immersed. Air bubbles must be dislodged by rotating or agitating the cage. Reset the micrometer so that electrical contact with the mercury surface is just made, and the bulb just lights up. Record the micrometer reading (M).

Raise the cage, remove the specimen and carefully brush off any droplets of mercury back into the tank. Weigh the specimen immediately, to 0.01 g (W).

Use rubber gloves when handling mercury and wash hands afterwards.

(5) Partial drying

Place the specimen on a watch-glass and leave it standing exposed to the air at room temperature for about 2 h so that it can partially dry.

(6) Repeat measurements

Repeat stages (4) and (5) so as to obtain a series of readings of volume and corresponding mass. Continue until three successive readings show no change in volume with reducing mass. At this stage the soil will appear lighter in colour. This may require several days, but the process must not be hurried. If the specimen dries too quickly, cracks may develop. If necessary, cover the specimen overnight with a damp cloth under a piece of polythene sheet so that it does not lose too much moisture between readings. If fine shrinkage cracks develop, it is possible that mercury will not be able to penetrate them fully, which will result in a false reading of volume. Other cracks may become filled with mercury which may remain entrapped, giving a false mass.

(7)   Oven drying and weighing

When it is clear that no further shrinkage is taking place, dry the specimen overnight, or to constant mass, in an oven at 105–110°C. The oven and its surroundings must be adequately ventilated so as to get rid of any mercury vapour (Section 1.6.7). Weigh to 0.01 g after cooling (m_D).

(8)   Measurement of dry volume

Place the dried specimen in the cage, lower into the mercury and record the micrometer reading as in step (4). This reading is denoted M_d.

(9)   Calculation and plotting

The volume of the specimen at any stage is calculated from the height to which the mercury is displaced in the tank. Let D=diamter of tank, M_0=zero reading of micrometer (mm) without specimen, M=micrometer reading (mm) at any intermediate stage, V=volume at that stage (cm^3), m=mass at that stage (g), m_d=dry mass of soil. Then

$$V = (M - M_0) \times \frac{\pi D^2}{4} \times \frac{1}{1000} \, \text{cm}^3$$

For the dried specimen

$$V_d = (M_d - M_0) \times \frac{\pi D^2}{4} \times \frac{1}{1000} \, \text{cm}^3$$

If the area of cross-section of the tank is 50 cm^2 (i.e. a diameter of about 80 mm), the volume is given by

$$V = (M - M_0) \times 5 \, \text{cm}^3$$

Some older cells were fitted with a micrometer calibrated in inches, and gave a direct reading of volume in cubic inches. Multiply by 16.39 to convert to cubic centimetres.

Calculate the unit volume (U) per 100 g of dry soil at each stage from the equation

$$U = \frac{100 \times V}{m_d} \, \text{cm}^3$$

Calculate the moisture content (w) of the specimen at each stage from the expression

$$w = \frac{m - m_D}{m_d} \times 100\%$$

Plot a graph of U (as ordinate) against w. A typical curve, together with test data and calculations, is shown in Fig. 2.25. The unit volume U_d in the fully dry state is plotted at zero moisture content, where

$$U_d = \frac{100 \times V_d}{m_d} \, \text{cm}^3$$

Shrinkage Limit

T. R. R. L. method

Location:	Hemel Hempstead					Loc. No.	4402	
Soil description	Light brown silty clay					Sample No.	5/2	
Sample type	Jar		Operator	G. H. W.		Date started	8. 11. 78	
Time	Zero reading R_0 units	Volume reading R units	$M - M_0$ $= V'$	$5 \times V'$ $= V$ cm^3	Wet weight of Sample g	Moisture Content %	Volume of 100 g dry soil $= 100 \frac{V}{m_d}$ cm^3	
	3.42	20.66	17.24	86.20	149.65	35.8	78.2	
	2.18	18.95	16.77	83.85	147.12	33.5	76.1	
	6.21	21.95	15.74	78.70	142.27	29.1	71.4	
	6.27	21.17	14.90	74.50	137.64	24.9	67.6	
	6.30	19.23	13.93	69.65	131.14	19.0	63.2	
	6.36	29.91	13.55	67.75	122.43	11.1	61.5	
	5.42	18.93	13.51	67.55	117.58	6.7	61.3	
	6.03	19.43	13.40	67.00		0	60.8	
			Final dry weight m_d		110.20			

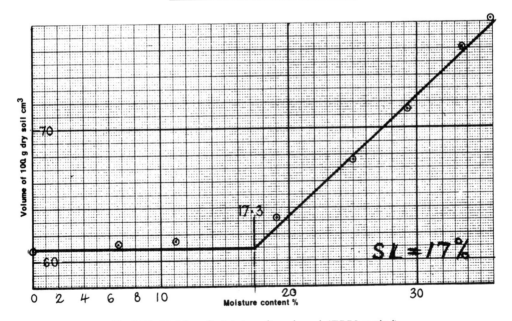

Fig. 2.25   Shrinkage limit test results and graph (TRRL method)

Calculate the shrinkage ratio, R_s, from the equation

$$R_s = \frac{m_d}{V_d}$$

and report to two significant figures.

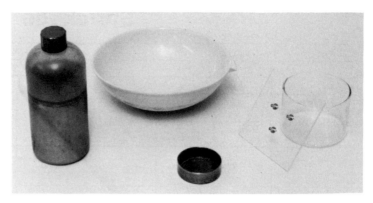

Fig. 2.26   Shrinkage limit test apparatus–alternative method

For a decrease in moisture content from any value w (%) to the shrinkage limit w_s (%), the corresponding volumetric shrinkage of the soil, V_s (cm^3) can be calculated from the equation

$$V_s = \frac{w - w_s}{R_s}$$

(10)   *Results*

A smooth curve may be drawn through the plotted points. But the graph consists essentially of two straight portions. Draw the horizontal abscissa through U_d, and produce the inclined straight line of best fit back so that these two lines intersect at E, as in Fig. 2.25.

The moisture content corresponding to this point of intersection is read off, and is reported as the shrinkage limit (w_s) to the nearest 1%.

Report also the shrinkage ratio (R_s), the initial moisture content and density of the test specimen, the method of specimen preparation and the percentage of material passing the 425 μm sieve.

It is preferable to carry out two tests simultaneously, and to report the average of the two results.

2.7.3   Shrinkage Limit — Alternative Method (BS1377: Part 2:1990:6.4, and ASTM D427)

This test is referred to in BS1377 as the subsidiary method. It is performed on remoulded soil, starting at a moisture content wetter than the liquid limit. It requires only one measurement of volume of the soil specimen, which is done by weighing displaced mercury. Shrinkage limit is calculated directly from initial moisture content, dried mass and volume. In this procedure a shrinkage curve is not produced, but intermediate volume measurements could be made if a curve is required. This would be necessary if the test were used as an alternative method for measuring the shrinkage limit of an undisturbed soil sample.

APPARATUS

(1)   Shrinkage dish, porcelain, about 42 mm diameter and 12 mm deep.
(2)   Glass cup, about 57 mm diameter and 38 mm deep, with rim ground flat.
(3)   Prong plate, glass or clear acrylic, fitted with three non-corrodible prongs as shown in Fig. 2.26. Dimensions are not critical but the plate should be large enough to completely cover the glass cup.
(4)   Glass plate, large enough to cover the shrinkage dish.

(5)   Two porcelain evaporating dishes 150 mm diameter.
(6)   Measuring cylinder 25 ml.
(7)   Mercury, rather more than will fill the glass cup.
(8)   Straight edge, spatula, small tools.
(9)   Balance, 200 g capacity, reading to 0.01 g.
(10) Moisture content apparatus (Section 2.5.2).
(11) Large tray containing a small amount of water, to retain any spilled mercury.
(12) Petroleum jelly (vaseline).

PROCEDURAL STAGES

(1)   Prepare specimen.
(2)   Measure and prepare apparatus.
(3)   Form soil pat.
(4)   Weigh.
(5)   Dry.
(6)   Weigh dry.
(7)   Measure volume.
(8)   Calculate.
(9)   Report result.

PROCEDURE

(1)   *Preparation of specimen*

About 30 g of soil passing the 425 μm sieve prepared from natural soil or by using the wet sieving process described in Section 2.6.3, is required. Place the soil in an evaporating dish and thoroughly mix with distilled water to make into a readily workable paste. Air bubbles must not be included. The moisture content should be somewhat greater than the liquid limit. The consistency should be such as to require about 10 blows of the Casagrande liquid limit apparatus to close the groove or to give about 25–28 mm penetration of the cone penetrometer.

(2)   *Preparation of apparatus*

Clean and dry the shrinkage dish, and weigh it to 0.01 g (m_2). Its internal volume is determined by measuring the volume of mercury held. Place the dish in an evaporating dish and fill it to overflowing with mercury. The evaporating dish will catch the overflow. Press the small glass plate firmly over the top of the shrinkage dish so that excess mercury is displaced, but avoid trapping any air. Remove the glass plate carefully and transfer the mercury to the 25 ml measuring cylinder. Record the volume of mercury in ml, which is the volume of the shrinkage dish (V_1(.

Coat the inside of the shrinkage dish with a thin layer of petroleum jelly. This is to prevent soil sticking to the dish.

(3)   *Forming soil pat*

Add the mixed soil paste to the shrinkage dish so as to about one-third fill it. Avoid trapping air. Tap the dish on the bench surface to cause the soil to flow to the edges of the dish. This should also release any small air bubbles present. The bench should be padded with a few layers of blotting paper or similar material.

Add a second amount of soil, about the same as the first, and repeat the tapping operation until all entrapped air has been released. Add more soil, and continue tapping, so that the dish is completely filled with an excess standing out. Strike off the excess with a straight edge, and clean off adhering soil from the outside.

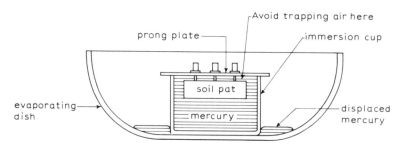

Fig. 2.27   Immersion of sample in shrinkage limit test

(4)   Weighing

Immediately after the above, weigh the soil and dish to 0.01 g (m_3). Calculate the mass of wet soil (m_1) from

$$m_1 = m_3 - m_2$$

(5)   Drying

Leave the soil in the dish to dry in the air for a few hours, or overnight, until its colour changes from dark to light. Place it in oven at 105–110°C, and dry to constant mass.

If the shrinkage curve during drying is required, make a series of volume measurements at suitable intervals before drying in the oven. Leave the soil in the shrinkage dish exposed to warm air, and when it has shrunk away from the dish and can be safely handled, determine its volume and mass as in step (7) below. Place the soil pat on a flat surface to dry further, and repeat the measurements until the colour changes from dark to light. Then dry in the oven.

(6)   Weighing dry

Cool in a desiccator, and weigh the dry soil and container to 0.01 g (W_3). Calculate the mass of dry soil m_d from

$$m_d = m_4 - m_2$$

(7)   Measurement of volume

Remove the dried soil-pat carefully from the shrinkage dish. It should be intact if it was given long enough dry in air before transferring to the oven.

Place the glass cup in a clean evaporating dish standing on the large tray. Fill the cup to overflowing with mercury, and remove the excess by pressing the glass prong plate firmly on top of the cup. Avoid trapping air under the glass plate. Carefully remove the prong plate, and brush off any mercury drops adhering to the glass cup. Place the cup into another clean evaporating dish without spilling any mercury.

Place the soil-pat on the surface of the mercury (it will float). Press the three prongs of the prong plate carefully on the sample so as to force it under the mercury (Fig. 2.27). Avoid trapping any air. Press the plate firmly on to the dish. Displaced mercury will be held in the evaporating dish. Brush off any droplets of mercury adhering to the cup into the dish. Transfer all the displaced mercury to the measuring cylinder and record its volume (V_d). This is equal to the volume of the dry soil-pat.

(8)   Calculations

Calculate the moisture content of the initial wet soil-pat, w_1, from the equation

$$w_1 = \left(\frac{m_1 - m_d}{m_d} \right) \times 100\%$$

SHRINKAGE LIMIT

Shrinkage Dish No.			3
Mass of dish + wet soil	m_3	g	35.84
Mass of dish + dry soil	m_4	g	26.31
Mass of dish	m_2	g	12.73
Mass of water	$m_3 - m_4$	g	9.53
Mass of dry soil	$m_4 - m_2 = m_d$	g	13.58
Mass of wet soil	$m_3 - m_2 = m_1$	g	23.11
Moisture content $\dfrac{m_3 - m_4}{m_d} - 100$	w_1	%	70.2
Volume of dish	V_1	cm³	14.81
Volume of dry soil	V_d	cm³	7.23
Volume change $(V_1 - V_d)$	ΔV	ml	7.58
Unit Volume Change $\dfrac{\Delta V}{m_d} \times 100$	ΔU	%	55.8
Shrinkage Limit $(w_1 - \Delta U)$	w_s	%	14.4
Shrinkage Ratio $\dfrac{m_d}{V_d}$	R_s		1.88

Fig. 2.28   Shrinkage limit test results (ASTM method)

The shrinkage limit (w_S) can then be calculated from the equation

$$w_S = w_1 - \left(\frac{V_1 - V_d}{m_d}\right) \times 100\%$$

where V_1 = volume of wet soil pat (ml); V_d = volume of dry soil pat (ml); m_d = mass of dry soil-pat (g).

The shrinkage ratio, R_s, can be calculated from

$$R_S = \frac{m_d}{V_d} \quad \text{(see Section 2.3.6)}$$

Typical results and calculations are given in Fig. 2.28.

If intermediate measurements of volume and mass were made, corresponding values of U and w can be derived, and the shrinkage curve plotted, as described in section 2.7.2. An appropriate value of V_s can also be calculated.

(9) Results

The shrinkage limit is reported to the nearest whole number. The method of test, the method of preparation and the percentage of material passing the 425 μm sieve, should be reported.

The shrinkage ratio is reported to two places of decimals.

2.7.4 Linear Shrinkage (BS1377: Part 2:1990:6.5)

This test gives the percentage linear shrinkage of a soil. It can be used for soil of low plasticity, including silts, as well as for clays.

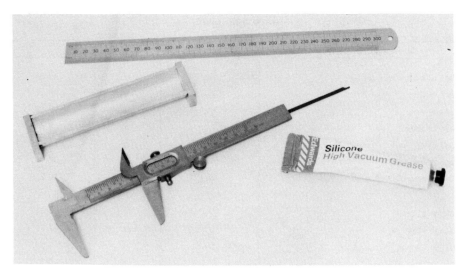

Fig. 2.29   Linear shrinkage apparatus

APPARATUS

(1)   Mould as illustrated in fig. 2.29, of brass or other non-corrodible metal, 140 mm long and 25 mm diameter.
(2)   Flat glass plate (as for the liquid limit test).
(3)   Palette knives.
(4)   Silicone grease or petroleum jelly.
(5)   Vernier calipers measuring up to 150 mm and reading to 0.1 mm, or steel rule graduated to 0.5 mm.
(6)   Moisture content apparatus (Section 2.5.2). Oven drying temperatures of 60–65°C and 105–110°C are required.

PROCEDURAL STAGES

(1)   Prepare mould.
(2)   Prepare soil sample.
(3)   Place in mould.
(4)   Dry.
(5)   Measure length.
(6)   Calculate.
(7)   Report result.

PROCEDURE

(1)   *Preparation of mould*

Clean and dry the mould. Apply a thin film of grease to the inner surfaces to prevent soil from sticking.

(2)   *Preparation of sample*

About 150 g of soil passing the 425 μm sieve, prepared from natural soil or by using the wet sieving process described in Section 2.6.3, is required. This proportion of the original sample passing the 425 μm sieve is recorded.

Place the soil on the glass plate and mix thoroughly with distilled water, as for the liquid limit test. Continue mixing until it becomes a smooth homogeneous paste at about the liquid limit. This is not critical, but it may be checked by using the cone penetrometer, which should give a penetration of about 20 mm.

(3)   Placing in mould

Place the paste in the mould, avoiding the trapping of air as far as possible, so that the mould is slightly overfilled. Tap it gently on the bench to remove any air pockets. Level off along the top edge of the mould with a palette knife or straight-edge. Wipe off any soil adhering to the rim of the mould.

(4)   Drying

Leave the mould exposed to the air but in a draught-free position, so that the soil can dry slowly. When the soil has shrunk away from the walls of the mould it can be transferred to an oven set at 60–65°C. When shrinkage has virtually ceased, increase the drying temperature to 105–110°C to complete the drying.

(5)   Measurement of length

Allow the mould and soil to cool in a desiccator. Measure the length of the bar of soil with the vernier calipers, making two or three readings and taking the average (L_D).

If the specimen has curved during drying, remove it carefully from the mould and measure the lengths of the top and bottom surfaces. Take the mean of these two lengths as the dry length L_D.

If the specimen has fractured in one place, the two portions can be fitted together before measuring the length. If it has cracked badly, and the length is difficult to measure, repeat the test, using a slower drying rate or leaving the mould longer in air before transferring to the oven.

(6)   Calculations

Calculate the linear shrinkage (L_S) as a percentage of the original length of the specimen from the equation

$$L_S = \left(1 - \frac{L_D}{L_0}\right) \times 100\%$$

where L_0 = original length (140 mm if a standard mould is used); L_D = length of dry specimen.

For some soils the plasticity index, I_P, might be found to be proportional to L_S. This can be useful for silty soils for which the plastic limit is difficult to determine.

(7)   Results

The linear shrinkage of the soil is reported to the nearest whole number. The percentage of soil passing the 425 μm sieve is also reported, together with the method of specimen preparation.

2.8   EMPIRICAL INDEX TESTS

2.8.1   Types of Test

Six simple empirical tests on clays are described in this section. They are known as

Puddle-clay index tests: pinch test
                                     tenacity test
                                     elongation test
                                     soaking test
Free swell test
Sticky limit test

The first four are traditional tests for assessing the suitability of a clay for use as puddle-clay in applications such as earth dams, reservoirs and canal bank linings. They were referred to by Bishop (1946) and Glossop (1946), and were adapted for laboratory use in 1954 by I. K. Nixon. While they do not provide precise numerical results, they do give an indication of the properties of the clay which may be used as a guide to the selection of more elaborate tests. These tests may also be useful in situations where no laboratory testing facilities are available.

The free swell test is taken from a procedure described by Gibbs and Holtz (1956). Its purpose is to indicate the possible expansive characteristics of a clay, whether in its natural state *in situ* or as a fill after being compacted.

The sticky limit test is that referred to by Terzaghi and Peck (1948).

Very little apparatus is required for these tests, other than standard oven-drying equipment for moisture content measurements.

2.8.2   Puddle Tests

PREPARATION OF SAMPLE

The clay is tested at the moisture content considered to be suitable for puddling. If this is not know, first determine the liquid and plastic limit of the natural (undried) soil. Calculate the moisture content required to give a liquidity index (I_L) of about 0.4 (see Section 2.3.3). Use either air-dried soil or undried soil at a known moisture content, and take a quantity equivalent to a dry mass of about 1 kg. If undried soil is used, calculate the amount of additional water needed to bring the moisture content up to the required value. Add the water, mix it well in and allow to mature in an airtight container for a few hours or preferably overnight.

Whether the clay is tested 'as received' or after remoulding, the actual moisture content should be measured on a small representative sample immediately before carrying out these tests.

PINCH TEST

(1)   Knead the soil well in the hands and form a ball of clay about 75 mm diameter. It must not be fissured at this stage.
(2)   Squeeze the ball flat on a glass plate or between the hands (see Fig. 2.30) until it forms a disc 25 mm thick.
(3)   Sketch and describe any cracks, if formed. If there are no cracks, state 'no cracks', and the clay has passed the test.

If the test is not possible because the sample is too friable, record that fact.

TENACITY TEST

(1)   Roll a cylinder 300 mm long and 25 mm diameter from the clay sample.
(2)   Hold it up vertically from one end so that 200 mm is unsupported, for a period of 15 s (Fig. 2.31).
(3)   If the clay supports its own weight, record this fact; it has passed the test. If necking or stretching occurs, record the details. Also record whether any cracks are visible.

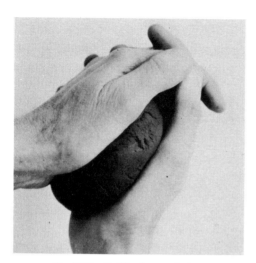

Fig. 2.30 *Puddle clay — pinch test*

Fig. 2.31 *Puddle clay — tenacity test*

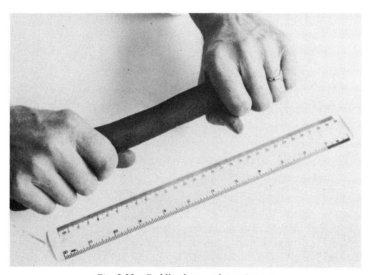

Fig. 2.32 *Puddle clay — elongation test*

ELONGATION TEST

(1) Roll another cylinder 300 mm long and 25 mm diameter.
(2) Grip each end firmly in the hands, leaving 100 mm unsupported (Fig. 2.32).
(3) Holding the cylinder horizontal, stretch it gradually by pulling on both hands until it breaks.
(4) Record the length of neck formed at failure and the type of break. The longer the neck, the more suitable the clay (see Fig. 2.33). If a break occurs after little or no stretching, this should be recorded.

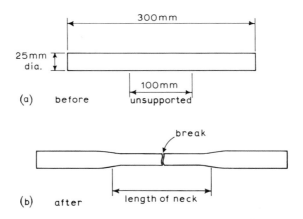

Fig. 2.33   *Puddle clay — extension of elongation test specimen*

Fig. 2.34   *Puddle clay — ball soaking test on two specimens*

SOAKING TEST

(1)   Make up a ball of clay 50 mm diameter, without fissures.
(2)   Place it in a 600 ml beaker and cover it with water. Note the time of immersion.
(3)   Record the state of the sample at the following inetervals after immersion: $\frac{1}{2}$, 1, 2, 4, 8, 24 h; 2, 4 days. Include details of cracking, flaking, breaking away of pieces and final disintegration. If no change takes place, state 'nil'. A suitable clay should not disintegrate (see Fig. 2.34).

Alternatively, if several clays are to be tested, record the times after immersion at which the following events occur: cracking; spalling; flaking; splitting; severe splitting; collapse or disintegration. If these events are marked at equal intervals along the X-axis of a graph (starting with 'unchanged' at zero), and if time is plotted to a suitable scale on the Y-axis, the behaviour of a clay can be represented graphically. Several different clays can be compared at a glance by this method.

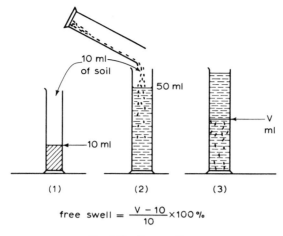

$$\text{free swell} = \frac{V - 10}{10} \times 100\,\%$$

Fig. 2.35   *Free swell test*

REPORT RESULTS

The above four tests are usually carried out together on one sample of clay. Recorded details, and sketches where appropriate, are reported as a set of results on one sheet for a particular clay. After each test should be stated 'suitable' or 'unsuitable', depending on the results.

The moisture content at which the tests were done, and the Atterberg limits, should also be reported.

2.8.3 Free Swell Test (Gibbs and Holtz, 1956)

CHARACTERISTICS

Free swell is defined as the increase in volume of the soil from a loose dry powder form when it is poured into water, expressed as a percentage of the original volume.

Soils with free swell values less than 50% are not likely to show expansive properties. Values of 100% or more are associated with clays which could swell considerably when wetted, especially under light loadings. High-swelling soils such as bentonite might have free swell values up to 2000%.

PREPARATION OF SAMPLE

About 50 g of soil is oven dried, and passed through a 425 μm sieve. Place the dried soil loosely in a dry 25 ml cylinder up to the 10 ml mark. The powder should not be compacted or shaken down.

PROCEDURE

(1) Place 50 ml of distilled water in a 50 ml glass measuring cylinder.
(2) Pour the dry soil powder slowly and steadily into the water. The word used in the original text is 'drizzle', and this is a very apt description.
(3) Allow the main part of the solid particles to come to rest; this will take from a few minutes to half an hour. The finest particles may remain in suspension for much longer but these can be ignored. (See Fig. 2.35.)
(4) Read off and record the volume of settled solids (*V* ml).

CALCULATION AND REPORT

'Free swell' is defined as the change in volume of the dry soil expressed as a percentage of its original volume. It is calculated from the equation

$$\text{free swell} = \frac{V - 10}{10} \times 100\%$$

if the original dry loose volume was 10 ml.

The result is reported to the nearest whole number.

2.8.4 Sticky Limit

This is a simple test on a clay to determine the lowest moisture content at which the clay adheres to metal tools. This method is based on a procedure outlined by Terzaghi and Peck.

Use a pat of clay which has been matured at a moisture content within the plastic range, such that it is 'sticky'—that is, the clay sticks to a clean dry spatula blade or a plated grooving tool. Allow the clay to dry gradually by exposure to the atmosphere, and at intervals draw the tool lightly over the surface of the clay-pat. When the tool no longer picks up any clay, measure the moisture content. Add a little water to the clay so that it becomes sticky again, and repeat the process once or twice more.

If the measured moisture contents are in reasonable agreement (an overall range of 2%) calculate the average moisture content to the nearest 1% and report it as the sticky limit of the clay.

BIBLIOGRAPHY

American Society for Testing and Materials 1990. Annual Book of ASTM Standards, Vol. 04.08, Soil and Rock; Building Stones.

Atterberg, A. (1911). 'Über die physikalische Bodenuntersuchung und über die Plastizität der Tone'. *Internationale Mitteilungen für Bodenkunde*, Vol. 1.

Bishop, A. W. (1946). 'The leakage of a clay core wall'. *Trans. Inst. Water Engineers*, Vol. 51, pp. 97–116

BS903: 'Methods of testing vulcanized rubber'. Part A8:1963, 'Determination of rebound resilience'. British Standards Institution, London

BS903: Part A26:1969, 'Determination of hardness'. British Standards Institution, London

BS1154, 'Specification for natural rubber compounds'. British Standards Institution, London

BS2000, 'Methods of test for petroleum and its products', Part 49, 'Penetration of bituminous materials', British Standards Institution, London

Casagrande, A. (1932). 'Research on the Atterberg limits of soils'. *Public Roads*, Vol. 13, No. 8

Casagrande, A. (1947). 'Classification and identification of soils'. *Proc. Am. Soc. Civ. Eng.*, Vol. 73, No. 6, Part 1

Casagrande, A. (1958). 'Notes on the design of the liquid limit device'. *Géotechnique*, Vol. 8, No. 2

Clayton, C. R. I. and Jukes, A. W. (1978). 'A one point cone penetrometer liquid limit test. *Géotechnique*, Vol. 28, No. 4, p. 469, December 1978

Gibbs, H. J. and Holtz, W. G. (1956). 'Engineering properties of expansive clays'. *Trans. Am. Soc. Civ. Eng.*, Vol. 121, No. 1, Paper 2814

Glossop, R. (1946). Discussion on 'The leakage of a clay core wall'. *Trans. Inst. Water Engineers*, Vol. 51, p. 124

Hansbo, S. (1957). 'A new approach to the determination of the shear strength of clay by the fall-cone test'. *Proc. Royal Swedish Geotechnical Institute, Stockholm*, No. 14

Karlsson, R. (1961). 'Suggested improvements in the liquid limit test, with reference to flow properties of remoulded clays'. *Proc. 5th Int. Conf. S.M. & F.E., Paris, July* 1961, Vol. I

Lambe, T. W. and Whitman, R. V. (1979). *Soil Mechanics*. Wiley, New York

Littleton, I. and Farmilo, M. (1977). 'Some observations on liquid limit values, with reference to penetration and Casagrande tests'. *Ground Engineering*, Vol. 10, No. 4

Norman, L. E. J. (1958). 'A comparison of values of liquid limit determined with apparatus having bases of different hardness'. *Géotechnique*, Vol. 8, No. 2

Norman, L. E. J. (1959). 'The one-point method of determining the liquid limit of a soil'. *Géotechnique*, Vol. 9, No. 1

Nixon, I. K. (1956). Internal communication. Soil Mechanics Limited

Shearman, D. J. (1979). 'A field test for identification of gypsum in soils and sediments. Technical Note. *J. Engineering Geology*, Vol. 12, p. 51.

Sherwood, P. T. (1970). 'The reproducibility of the results of soil classification and compaction tests'. TRRL Report LR 339, Transport and Road Research Laboratory, Crowthorne, Berks.

Sherwood, P. T. and Ryley, D. M. (1968). 'An examination of cone-penetrometer methods for determining the liquid limit of soils'. TRRL Report LR 233, Transport and Road Research Laboratory, Crowthorne, berks.

Skempton, A. W. (1953). 'The colloidal "activity" of clays'. *Proc. 3rd Int. Conf. Soil Mech.*, Zurich, Vol. 1, Session 1/14

Skempton, A. W. and Northey, R. D. (1953). 'The sensitivity of clays'. *Géotechnique*, Vol. 3, No. 1

Terzaghi, K. and Peck, R. B. (1948). *Soil Mechanics in Engineering Practice*. Wiley, New York

Transport and Road Research Laboratory (1952). *Soil Mechanics for Road engineers*, Chapter 3, HMSO, London

Tubey, L. W. and Webster, D. C. (1978). 'The effects of mica on the roadmaking properties of materials'. TRRL Supplementary Report No. 408, Transport and Road Research Laboratory, Crowthorne, berks.

US Department of the Interior, Bureau of Reclamation (1974), *Earth Manual*, 2nd edition. Test designation E.7, Part C. US Government Printing Office, Washington, DC

Wood, D. M. and Wroth, C. P. (1976). 'The correlation of some basic engineering properties of soils'. *Proc. Int. Conf. on Behaviour of Offshore Structures, Trondheim*, Vol. 2

Wood, D. M. and Wroth, C. P. (1978). 'The use of the cone penetrometer to determine the plastic limit of soils'. *Ground Engineering*, Vol. 11, No. 3, April 1978

Chapter 3

Density and particle density

3.1 INTRODUCTION

3.1.1 Scope

In this chapter are described three different laboratory tests for the measurement of density of soil samples, and three for the measurement of the density of soil particles. All these tests are to be found in BS1377: Part 2:1990. In addition, tests for the determination of the 'limiting densities' of granular soils are included, which are given in BS1377: Part 4:1990.

The section on theory (Section 3.3) includes illustrations of the meaning of voids ratio, porosity and degree of saturation, and their relationship to density and moisture content.

The measurement of density *in situ* is beyond the scope of laboratory work and is not included here.

3.1.2 Terminology

DENSITY

The concept of 'density' is well known, and refers to mass per unit volume. Density can be qualified in several different ways, each one of which has a particular application. These terms are discussed in this chapter and defined below.

In British practice the various terms using 'density' relate to the mass per unit volume (or unit mass of the soil in bulk). The accepted SI term (see Appendix) is 'mass density', and is denoted by the symbol ρ. It is expressed as a mass per unit volume (e.g. kg/m^3 or Mg/m^3), which should not be confused with the 'weight density', denoted by γ. Weight density relates to the gravitational force (the weight) acting on the mass, and is expressed as force per unit volume (e.g. N/m^3 or kN/m^3). Weight (force) is equal to mass multiplied by the local acceleration due to gravity, so, if we consider a unit volume,

$$\gamma = \rho \times g$$

The average terrestrial value of g is about $9.81 \, m/s^2$, which for many practical purposes can be written as $(10) \, m/s^2$.

The weight density γ expressed in kilonewtons per cubic metre (kN/m^3), when the mass density ρ is expressed in Mg/m^3, is then given by the equation

$$\gamma = 9.81 \times \rho \quad kN/m^3$$

or approximately

$$\gamma = (10) \times \rho \quad kN/m^3$$

Throughout this book, unless stated otherwise, the term 'density' refers to mass density, ρ. For soils it is expressed in Mg/m^3, which is the same magnitude as g/cm^3. The density of water (denoted by ρ_w) for most practical purposes is equal to $1 \, Mg/m^3$.

116

PARTICLE DENSITY

In BS1377:1990 the term 'particle density' has replaced 'specific gravity' as a measure of the average density of the solid particles which make up a soil mass. Specific gravity is the ratio of the mass of dry particles to the mass of water they displace. This is a dimensionless quatity, i.e. a number without units. In some European countries the specific gravity has been called 'relative density' which in Britain has a different meaning as a descriptive term derived from *in-situ* tests or observations.

The term 'particle density' has been adopted in Britain to comply with international use in ISO Standards. Particle density is denoted by the symbol ρ_s, and values are quoted in Mg/m^3 which is the same unit as used for density. It is also numerically equal to the specific gravity (G_s) which it replaces.

Throughout this book the particle density refers to the density of naturally-occurring soil particles, usually after oven drying at 105°C to 110°C. This is sometimes referred to as the 'apparent particle density' (see Section 3.3.4).

3.2 DEFINITIONS

DENSITY   Mass per unit volume.

BULK DENSITY (ρ) Mass of dry soil, including solid particles, water and air, contained in a unit volume.

DRY DENSITY (ρ_D) Mass of dry soil, after drying at 105°C, contained in a unit volume of undried soil.

DENSITY OF WATER (ρ_w) (see Section 3.3.1).

LIMITING DENSITIES   The dry densities corresponding to the extreme states of packing (loose and dense) at which the grains of soil can be placed. Relevant only for granular soils.

MAXIMUM DENSITY (ρ_{Dmax}) The dry density at the densest practicable state of packing of the particles. (Maximum dry density has a different meaning in connection with the compaction of soil. See Chapter 6.)

MINIMUM DENSITY (ρ_{Dmin}) The dry density at the loosest state of packing of dry particles which can be sustained.

PARTICLE DENSITY (ρ_s) The average mass per unit volume of the solid particles in a sample of soil, where the volume includes any sealed voids contained within solid particles. (In some countries this is known as *relative density*).

VOIDS   The spaces between solid particles of soil. They may contain gas (usually air) or water, or both.

VOIDS RATIO (e) The ratio between the volume of voids (water and air) and the volume of solid particles in a mass of soil.

DENSITY INDEX (I_D) The relationship of the voids ratio e to the limiting voids ratios e_{max}, e_{min} (corresponding to the minimum and maximum densities):

$$I_D = \frac{e_{max} - e}{e_{max} - e_{min}}$$

POROSITY (n) The volume of voids (water and air) expressed as a percentage of the total (bulk) volume of a mass of soil.

DEGREE OF SATURATION (S_r) The volume of water contained in the void space between soil particles expressed as a percentage of the total voids.

$$S_r = \frac{w\rho_s}{e}$$

SATURATION   A soil is fully saturated when all the voids are completely filled with water.

CRITICAL DENSITY   The dry density corresponding to the critical voids ratio of a granular soil.

CRITICAL VOIDS RATIO   The voids ratio of a granular soil at which it will neither expand nor contract when subjected to shear strains.

UNIT MASS   Same as DENSITY.

UNIT WEIGHT (γ)   Weight (force) per unit volume, equal to mass per unit volume multiplied by local acceleration due to gravity.

ABSOLUTE PARTICLE DENSITY   The particle density of the mineral constituents present in the soil. It is measured by pulverising the soil to silt size or finer, so that all impermeable voids in the coarser grains are exposed. (Mainly applicable to rocks.)

APPARENT PARTICLE DENSITY   The particle density of the soil particles as they occur naturally, and referred to simply as particle density. (See definition under PARTICLE DENSITY.)

*BULK PARTICLE DENSITY (SATURATED, SURFACE DRY)   The particle density with the permeable or surface voids of the particles filled with water.

*BULK PARTICLE DENSITY (WET, SURFACE DRY)   The particle density when the permeable voids are not entirely filled with water.

*BULK PARTICLE DENSITY (OVEN DRY)   The minimum particle density of particles, in which the permeable and impermeable voids associated with the particles are included.

* These terms are used mainly for concrete aggregates and not for soils, and are usually expressed in kg/m^3 (1000 kg/m^3 = 1 Mg/m^3).

3.3   THEORY

3.3.1   Mass, Volume and Density

DETERMINATION OF DENSITY

In the laboratory mass is measured in grams. Linear dimensions are measured in millimetres, and a volume calculated from such measurements will be in cubic millimetres (mm^3). It is convenient to convert to cubic centimetres (cm^3) at the outset by dividing by 1000, because 1 cm^3 = 1000 mm^3. Thus, for a rectangular prism, the volume V is given by

$$V = \frac{LBH}{1000} \text{ cm}^3$$

and for a right cylinder

$$V = \frac{\pi D^2 H}{4000} \text{ cm}^3$$

where L, B, H, D relate, respectively, to length, breadth, height and diameter, and are measured in millimetres.

If the mass of a soil sample is in grams, and its volume is V cm^3, the density ρ is given by

$$\rho = \frac{m}{V} \text{ g/cm}^3 \qquad = \frac{m}{V} \text{ Mg/m}^3$$

because 1 Mg = 10^6 g and 1 m^3 = 10^6 cm^3.

The practical unit of density is Mg/m^3, and is obtained directly from laboratory measurements if we convert mm^3 to cm^3.

If the volume V is obtained by direct measurement, such as by water displacement, it is measured in millilitres (ml). For practical purposes 1 ml = 1 cm^3, so the above relationship still applies.

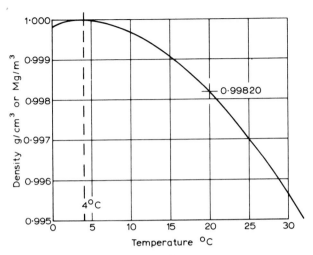

Fig. 3.1   Density of pure water

DENSITY OF WATER

The density of pure water is at its maximum value of 1.000 g/ml (or Mg/m³) at a temperature of 4°C. The density of water at other temperatures may be obtained from the graph in Fig. 3.1, or from Table 4.13 in Section 4.7.3. At 20°C it is equal to 0.998 20 g/ml.

For most soil testing purposes (except the hydrometer sedimentation test; Section 4.8.3), the density of pure water may be taken to be 1.00 g/ml (or Mg/m³).

3.3.2   Voids Ratio and Porosity

The amount of void space within a soil has an important effect on its characteristics. Two expressions are used to provide a measure of the void space — namely 'voids ratio' and 'porosity'. The relationships between density, moisture content, specific gravity, voids ratio and porosity are developed below for soil in four conditions:

(1)   dry
(2)   fully saturated
(3)   partially saturated
(4)   submerged

Equations are summarised at the end of this Section for easy reference, but it is better to understand the principles than to attempt to memorise all the formulae.

DRY SOIL

A dry soil consists of solid particles separated by air spaces (voids), as shown diagrammatically in Fig. 3.2(a). The soil occupies a total volume V and its mass is m. The density of the solid particles is ρ_s.

If we imagine that all the solid particles are fused together into a solid lump they will occupy a volume V_s which is less than V. The difference is the volume of voids V_v, as indicated in Fig. 3.2(b).

The voids ratio, e, is the ratio of volume of voids to volume of solids, or

$$e = \frac{V_v}{V_s} \tag{3.1}$$

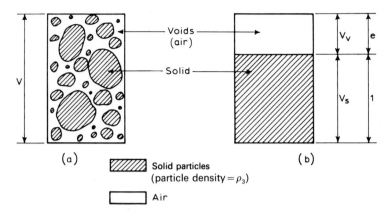

Fig. 3.2   Representation of dry soil

This is a pure number which is usually expressed as a decimal (e.g. 0.35). It can be greater than 1.

Porosity, n, is the ratio of volume of voids to the total volume, or

$$n = \frac{V_v}{V} = \frac{V_v}{V_v + V_s} \tag{3.2}$$

This is often expressed as a percentage, and must be less than 100%.

The relationship between voids ratio and porosity is given by the equations

$$n = \frac{e}{1+e} \quad \text{and} \quad e = \frac{n}{1-n} \tag{3.3}$$

The mass of the volume of soil is equal to the volume of solids multiplied by their density, the voids having zero mass. If the density of the solid particles is ρ_s, then

$$\text{mass of the soil} = V_s \rho_s$$

The volume of this soil is $(V_s + V_v)$.

The density (which in this case is the dry density, ρ_D) is equal to mass divided by volume, or

$$\rho_D = \frac{V_s \rho_s}{V_s + V_v}$$

$$= \frac{\rho_s}{1 + \dfrac{V_v}{V_s}}$$

i.e.

$$\rho_D = \frac{\rho_s}{1+e} \tag{3.4}$$

SATURATED SOIL

In a completely saturated soil the void spaces are completely filled with water, as in Fig. 3.3(a). If we again imagine the solids fused together, the remaining voids will be filled with water as in Fig. 3.3(b).

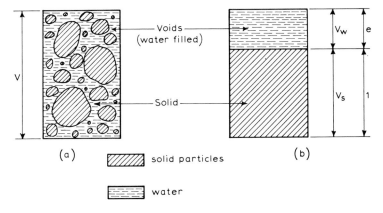

solid particles

water

Fig. 3.3   *Representation of fully saturated soil*

Voids ratio and porosity are defined in the same way as before, i.e.

$$e = \frac{V_w}{V_s} \tag{3.5}$$

$$n = \frac{V_w}{V} = \frac{V_w}{V_w + V_s} \tag{3.6}$$

The total mass is now made up of two parts:

$$\text{solid, } m_s = V_s \rho_s$$

$$\text{water, } m_w = V_w \rho_w$$

The volume is

$$V = V_s + V_w$$

The density is now the saturated density, ρ_{sat}, and is given by

$$\rho_{sat} = \frac{(V_s \rho_s) + (V_w \rho_w)}{V_s + V_w}$$

$$= \frac{\rho_s + \dfrac{V_w}{V_s} \rho_w}{1 + \dfrac{V_w}{V_s}}$$

Therefore

$$\rho_{sat} = \frac{\rho_s + e\rho_w}{1 + e} \tag{3.7}$$

PARTIALLY SATURATED SOIL

Many natural soils contain both air and water in the voids—that is, they are partially saturated. This condition can be represented diagrammatically as in Fig. 3.4. The symbols

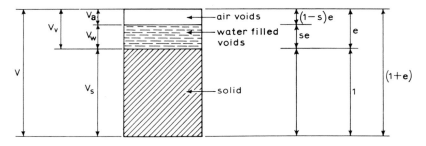

Fig. 3.4   *Representation of partially saturated soil*

on the left of the diagram are those used previously. On the right-hand side the volume of solids, V_s, is reduced to unity, so that the volume of voids is denoted by e. This notation simplifies calculations involving voids ratio changes or degree of saturation. It is much easier to remember this diagram, and to work from the first principles, than to remember these equations.

The percentage of the voids occupied by water is denoted by S_r, so that

$$S_r = \frac{V_w}{V_a + V_w} \times 100\% \tag{3.8}$$

The percentage S_r is known as the degree of saturation. In a fully saturated soil, $S_r = 100\%$ and the volume of air voids, V_a, is zero. In a dry soil, $S_r = 0\%$ and $V_a = V_v$.

The mass of the volume V, represented by $(1+e)$, is made up as follows.

Solid: $1 \times \rho_s$

Water: $e \times \dfrac{S_r \rho_w}{100}$

Air: 0

$\therefore$ Total mass $= \rho_s + \dfrac{S_r e \rho_w}{100}$

The bulk density is denoted by ρ, and

$$\rho = \frac{\text{Mass}}{\text{Volume}} = \frac{\rho_s + \dfrac{S_r e \rho_w}{100}}{1 + e} \tag{3.9}$$

The moisture content, w, is the ratio of mass of water to mass of solid, expressed as a percentage (Section 2.3.1), i.e.

$$w = \frac{S_r e \rho_w}{\rho_s}\% \tag{3.10}$$

Hence

$$S_r = \frac{\rho_s}{\rho_w} \frac{w}{e}\% \tag{3.11}$$

The dry density, denoted by ρ_D, is the mass of dry soil per unit volume, i.e.

$$\rho_D = \frac{\rho_s}{1 + e}$$

which is the same equation as (3.4)

Substituting from (3.10) in equation (3.9),

$$\rho = \frac{\rho_s + w\rho_s}{1+e} = \frac{\rho_s}{1+e}(1+w)$$

and from Equation (3.4).

$$\rho = \rho_D(1+w)$$

Therefore

$$\rho_D = \frac{1}{1+w}\rho$$

If w is expressed as a percentage,

$$\rho_D = \frac{100}{100+w\%}\rho \qquad (3.12)$$

Equation (3.12) should be remembered, because it is often needed for calculating dry density from bulk density and moisture content.

SUBMERGED SOIL

If partially saturated soil is completely submerged and the air in the voids remains entrapped, the apparent (submerged) density is denoted by ρ' and is derived as follows.

Total mass $= \rho_s + \dfrac{S_r e \rho_w}{100}$

volume of water displaced $= (1+e)$

Therefore

mass of water displaced $= (1+e)\rho_w$

By the principle of Archimedes (Section 3.3.3),

Apparent mass $= \left(\rho_s + \dfrac{S_r e \rho_w}{100}\right) - (1+e)\rho_w$

Submerged density $= \dfrac{\text{Apparent mass}}{\text{Volume}}$,

i.e.
$$\rho' = \frac{\rho_s - \left[1+e\left(\dfrac{100-S_r}{100}\right)\right]\rho_w}{1+e} \qquad (3.13)$$

If all the voids are filled with water, $S_r = 100\%$ and the density is the submerged saturated density, ρ'_{sat}, so that

$$\rho'_{sat} = \frac{\rho_s - \rho_w}{1+e} \qquad (3.14)$$

SUMMARY OF EQUATIONS

If we use SI units, the density of water, ρ_w, is $1\,\text{g/cm}^3$ for all practical purposes, and the relationships derived above become

$$\rho_D = \frac{\rho_s}{1+e}$$

$$\rho_{sat} = \frac{\rho_s + e}{1+e}$$

$$\rho = \frac{\rho_s + \dfrac{S_r e}{100}}{1+e}$$

$$\rho' = \frac{\rho_s - 1 - e\left(\dfrac{100 - S_r}{100}\right)}{1+e}$$

$$\rho'_{sat} = \frac{\rho_s - 1}{1+e}$$

However, if sea-water is present, its density (about $1.04 \, \text{Mg/m}^3$) must be inserted for ρ_w in Equations (3.4)–(3.14).

The following equations are worth remembering:

$$S_r = \frac{\rho_s w}{\rho_w e} \, \% \tag{3.11}$$

$$\rho_D = \frac{100 \rho}{100 + w} \tag{3.12}$$

3.3.3   Principle of Archimedes

This principle is explained in textbooks on elementary physics (for example, Abbott, 1969). If a solid body is wholly or partially immersed in a liquid, the upthrust force (or buoyancy force) acting on the body is equal to the weight force of the liquid displaced by the body, and acts vertically upwards through the centre of gravity of the displaced liquid. The apparent mass of a solid body immersed in water is therefore equal to the mass of the body less the mass of water displaced. This is the basis of the water immersion method for measuring density (Section 3.5.5).

The same principle applies to a body immersed in a gas, including the atmosphere, but this effect may be ignored for most practical purposes.

When a body floats in water, the mass of water displaced is equal to the mass of the body, whether it is totally or partially immersed. If it floats wholly immersed, its mean density must be the same as the density of water.

3.3.4   Particle Density

A soil consists of an accumulaton of particles which may be of a single mineral type, such as clean quartz sand, or more usually a mixture of a number of mineral types, each with

Table 3.1. ABSOLUTE PARTICLE DENSITIES OF SOME COMMON MINERALS

Mineral	Composition	Absolute particle density Mg/m^3
Anhydrite	$CaSO_4$	2.9
Barytes	$BaSO_4$	4.5
Calcite, chalk	$CaCO_3$	2.71
Feldspar	$KAlSi_3O_8$, etc	2.6–2.7
Gypsum	$CaSO_4 \cdot 2H_2O$	2.3
Haematite	Fe_2O_3	5.2
Kaolinite	$Al_4Si_4O_{10}(OH)_8$	2.6
Magnetite	Fe_3O_4	5.2
Quartz (silica)	SiO_2	2.65
Peat	Organic	1.0 or less
Diatomaceous earth	'Skeletal' remains of microscopic plants	2.00

a different particle density. For a single mineral type the particle density of the solids comprising a mass of the soil is that of the mineral itself (for example, 2.65 for quartz). But for a soil consisting of a variety of minerals we are concerned only with the mean particle density of the mass as a whole, and this is the sense in which the term is used here. Particle densities of a few common mineral types are listed in Table 3.1.

The particle density of most soils generally lies between 2.60 and 2.80, and for sands which consist wholly or mainly of quartz it is often sufficient to assume that the particle density is 2.65. The presence of particles consisting of other minerals will result in a different value. Clays consist of various minerals, most of which are heavier than quartz, and have higher particle densities, typically 2.68–2.72 for many British soils. The presence of heavy metals in the form of oxides or other compounds can give even higher values. On the other hand, soils which contain appreciable quantities of peat or organic material may have particle densities considerably less than 2.65, and sometimes below 2.0. Soils consisting of particles which contain small cavities (bubbles of gas), such as pumice, also show a low 'apparent' particle density, although the 'absolute' particle density is higher. Tropical soils may have particle densities which are unexpectedly high or low.

Soils containing substantial proportions of heavy or light particles can give erratic values of particle density. A number of repeat tests may be needed to obtain a reliable average value for this type of soil.

Particle density is related to the density of water at 4°C, but most laboratory tests are carried out at an ambient temperature of about 20°C. However, the difference in the density of water between 4°C and 20°C is less than $0.003 \, g/cm^3$ (i.e. within 0.3%), so for practical purposes this discrepancy can be neglected.

There are several different kinds of particle density in use, especially in American practice, and they are defined in Section 3.2. The 'apparent particle density' is generally used in Britain for soils and is referred to throughout this book as 'particle density', and is denoted by ρ_s

The 'absolute' particle density gives the highest particle density value which can be obtained for a given soil. The 'apparent' particle density may be equal to, but it usually slightly smaller than, the 'absolute' particle density. The other particle density terms are listed in order of magnitude of their values, the 'bulk' particle density (oven-dry) giving the lowest value, because it includes the volume of all voids associated with the particles. The 'absolute' particle density can be used as a guide for the identification of rock minerals, and is the parameter used in Table 3.1. The three 'bulk' particle densities apply mainly to concrete aggregates, and will not be referred to again.

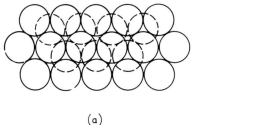 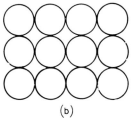

(a)   (b)

Fig. 3.5   States of packing of equal spheres

Table 3.2.   THEORETICAL LIMITING VOIDS RATIOS AND POROSITES FOR EQUAL SPHERES

	Densest state	*Loosest state*
Voids ratio e	0.35	0.91
Porosity n	26%	48%
Symbols used	e_{min}, n_{min}	e_{max}, n_{max}
	ρ_{max}	ρ_{min}

3.3.5   Limiting Densities

STATES OF PACKING

The limiting densities, as applied to sands, are the dry densities corresponding to the two extreme states of packing (loosest and densest) at which the particles can be placed in nature. The maximum density represents the densest packing of particles without crushing the grains, and corresponds to minimum porosity. The minimum density represents the loosest packing, and corresponds to maximum porosity.

If we consider a collection of equal spheres, their extreme states of packing can be represented diagrammatically in two dimensions as shown in Fig. 3.5. They can be densely packed as at (a), or loosely packed as at (b). In each case the spheres are in face-to-face contact. The densities achieved depend upon the density of the spheres (i.e. particle density), but voids ratios and porosities do not, and can be calculated theoretically. These values are given in Table 3.2.

Real soils consist of particles of many sizes, and at their densest packing the voids between large particles contain smaller particles, and the voids between these contain yet smaller particles, and so on (Fig. 3.6a). The Fuller grading curve, referred to in Section 4.4.2, is based on the idealised limit of this concept (Fig. 3.6c). It might be expected that a real soil with a variety of particle sizes would show smaller values of e_{min} and n_{min} than can be obtained with equal spheres, but this is partially offset by the more irregular shape of real sand grains.

In the loosest state it is possible for groups of real soil particles to form 'arch' structures, as represented in Fig. 3.6(b), which can be sustained if left undisturbed. The possible higher values of e_{max} and n_{max} compared with equal spheres is again partially offset by the effect of irregularity of grain shape. For these reasons the voids ratio and porosity limits of many natural sands do not differ greatly from the theoretical values for equal spheres shown in Table 3.2.

In Fig. 3.6, sample (a) is dense, and is stable. Sample (b) is unstable, and the grain structure is likely to collapse under the influence of a sudden shock, vibration or inundation.

The minimum density referred to here must not be confused with the 'bulking' phenomenon of damp sand. It is possible to deposit damp sand at a porosity greater than that achieved by the standard procedures for determining the maximum porosity. This effect is due to a

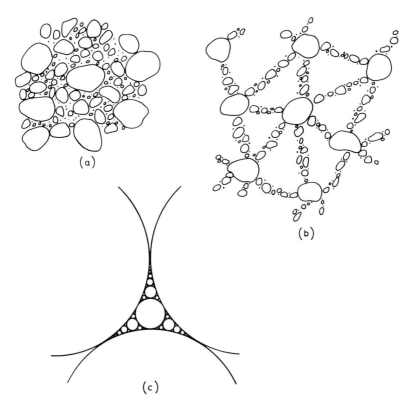

Fig. 3.6   States of packing of soil particles: (a) densely packed, (b) loosely packed and (c) idealised 'Fuller' packing

film of water on the particles preventing direct contact between them, and so increasing the porosity. This is a condition not found in nature, because natural sand deposits have usually been either laid down under water, or wind-blown in dry conditions.

DENSITY INDEX

The voids ratio of a granular soil can be used with reference to the limiting voids ratios (e_{min}, e_{max}) as an index, in much the same way as the moisture content of clays is referred to the Atterberg limits (Chapter 2; Section 2.3.3). The term used here, corresponding to the liquidity index for clays, is 'density index' (I_D), previously referred to as 'relative density', and is the ratio derived as follows. If the voids ratio of the soil is denoted by e, then

$$I_D = \frac{e_{max} - e}{e_{max} - e_{min}}$$

Density index is illustrated diagrammatically in Fig. 3.7, in which values of voids ratio are plotted from left to right. Zero voids ratio relates to a density equal to the density of the individual particles. The range of voids ratios for equal spheres is shown on the top line.

Considering sand (a), the voids ratio in its natural state is e, and is denoted by the point S. Its density index (I_D), as defined by the above equation, is near the middle of the range — that is, sand (a) has a 'medium' density index.

Another material, sand (b), with different values of limiting voids ratios, but at the same natural voids ratio, would have a 'low' density index — that is, it would be at a less dense state of packing than sand (a). It follows that voids ratio (or density), without reference to

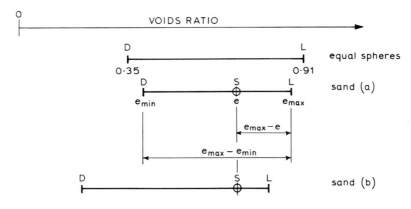

Fig. 3.7   Density index of two sands

the limiting voids ratio (or densities), is not sufficient to define the state of compaction of a sand.

To calculate the density index from dry densities instead of voids ratios, the following expression is used:

$$I_D = \frac{\rho_D - \rho_{D,min}}{\rho_{D,max} - \rho_{D,min}} \times \frac{\rho_{D,max}}{\rho_D}$$

3.4   APPLICATIONS

3.4.1   Density

The *in situ* density of soil is an important property having many applications in soil engineering. In practical problems associated with earthworks and foundations, the weight of the soil itself exerts forces which have to be taken into account in the analysis. It is therefore necessary to know the bulk density of the soil, from which these forces can be calculated. In the analysis of the stability of a slope, such as an embankment or the sides of a cutting, the weight of soil provides the main force, but it is also significant in calculating the bearing capacity and settlement of foundations for other structures. The bulk density of natural soil is usually determined from laboratory measurements on undisturbed samples.

When soil is used as a construction material, such as in embankments or road sub-bases, measurement of dry density provides an important means of quality control. Direct measurement of compacted density *in situ* is beyond the scope of this book. The methods most often used are the sand-replacement method, and the core-cutter method, which are given in BS1377: Part 9:1990. Undisturbed samples of compacted fill can be taken for laboratory measurements of density, moisture content and other properties.

In general, all undisturbed soil samples (whether from natural ground or from compacted soil) used for laboratory tests are measured for the determination of bulk density and moisture content, from which the dry density can also be calculated (Section 3.3.2).

Some typical values of density, voids ratio and porosity for a few soil types are given in table 3.3.

3.4.2   Voids Ratio and Porosity

The main application of voids ratio is in the analysis of data from the oedometer consolidation tests, which is given in Volume 2.

Table 3.3. TYPICAL DENSITIES AND OTHER PROPERTIES OF SOILS

Soil type	Moisture content w (%)	Bulk density, ρ (Mg/m³)	Dry density, ρ_D (Mg/m³)	Voids ratio, e	Porosity, n (%)	Degree of saturation, S (%)
Dry uniform sand, loose	0	1.36	1.36	0.95	49	0
Well-graded sand	7.5	1.95	1.81	0.46	32	43
Soft clay	54	1.67	1.07	1.52	60	99.5
Firm clay	22	1.96	1.61	0.68	41	87
Stiff glacial till	9.5	2.32	2.12	0.27	21	95
Peat	220*	0.98	0.31	3.67	78	85

*Can be up to 10 times greater in bog peat.

Table 3.4. DENSITY INDEX DESCRIPTION

	Density index range, I_D (%)
Very loose	0–15
Loose	15–35
Medium	35–65
Dense	65–85
Very dense	85–100

Voids ratio, and saturation, are important factors relating to the compaction of soils, which is dealt with in Chapter 6.

Voids ratio is used as an index for assessment of the density index of sands (see Section 3.3.5).

Porosity can be useful in some soils as an indication of permeability characteristics.

3.4.3 Particle Density

It is rarely possible to use particle density as an index for soils classification. But knowledge of the particle density is essential in relation to some other soil tests, especially for calculating porosity and voids ratio, and is particularly important when compaction and consolidation properties are considered. The particle density must also be known for the computation of particle size analysis from a sedimentation procedure (Chapter 4; Section 4.8).

3.4.4 Limiting Densities

The application of limiting densities relates only to granular soils, and to sands in particular. Limiting densities of sands are not so widely used as are the Atterberg limits of clays. Nevertheless the assessment of density index can give some indication of the possible effects of loading or of disturbance on sand. This factor is particularly important where a sand deposit is likely to be affected by vibration, such as from machinery, or by earthquake shock. In loose sand or silt deposits below water table, vibration or shock could result in liquefaction and collapse of the grain structure. It is therefore desirable to have some means of assessing how loose a 'loose' deposit is, so that appropriate remedial measures (such as compaction by vibration) can be taken if necessary.

Classification of granular soils, based on density index values, as suggested by Lambe and Whitman (1979) is indicated in Table 3.4.

3.5   DENSITY TESTS

3.5.1   Scope

Measurement of the density of soils is sometimes overlooked, but it can be just as important as the measurement of moisture content. It is good practice to measure the density of all undisturbed samples tested in the laboratory, and sometimes it may be the only test, apart from moisture content.

The tests to be described fall into three categories, which are based on three different principles for the measurement of volume. These are

(1)   Linear measurement.
(2)   Water displacement.
(3)   Water immersion (weighing in water).

In many instances the specimen prepared for some other test, such as compressive strength or consolidation, will be used for the measurement of density. Alternatively, an undisturbed sample may be specially prepared for a density test. Sometimes a relatively undisturbed sample will be too friable to handle and prepare for other tests, and a density measurement in the sampling tube or container itself is the only possibility. The first of these methods is applicable only when the sample can be formed into a regular geometric shape, such as a rectangular prism or cylinder. Either of the other methods may be used when the available sample consists of an irregular lump of soil.

3.5.2   Linear Measurement (BS1377: Part 2:1990:7.2, and Part 1:1990:8)

LINEAR MEASUREMENT

The usual types of sample for which the volume can be determined by linear measurement, and the clause reference in BS1377:1990 in which these methods are specified, are as follows.

(1)   Rectilinear sample, hand trimmed from block sample or tube sample (Part 2:7.2).
(2)   Cylindrical sample, of the same size as the sample taken in the tube (Part 1:8.4).
(3)   Cylindrical sample, of a diameter smaller than that of the sampling tube (Part 1:8.4).
(4)   Cylindrical sample, carved from a block sample (Part 1:8.5).
(5)   Circular or square disc sample, prepared from a sample taken in a tube (Part 1:8.6).
(6)   Circular or square disc sample, prepared from a block sample (Part 1:8.7).

Samples of types (2) to (6) are the types used for shear strength and compressibility tests, and their methods of preparation are described in volume 2, Chapter 9. Preparation of samples of type (1), and the determination of the volume of cylindrical samples, are described below.

APPARATUS

(1)   Cutting and trimming tools:
        Sharpened knife with rigid blade.
        Scalpel (eg. a craft tool with interchangeable blades).
        Wire saw of piano wire about 0.4 mm diameter.
        Spiral wire saw.
        Saw, medium to coarse teeth.
        Steel straight-edge trimmer, $300 \times 25 \times 3$ mm with one bevelled edge.
(2)   Engineers' steel rule, graduated to 0.5 mm.
(3)   Steel try-square.
(4)   Vernier calipers, readable to 0.1 mm.
(5)   Flat glass plate, about 300 mm square and 10 mm thick.

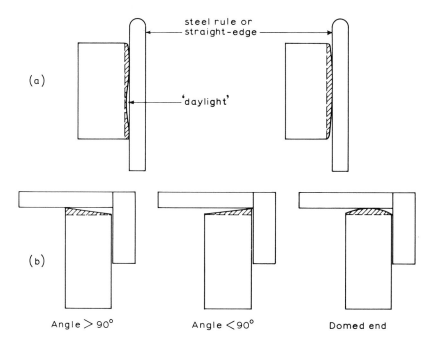

steel rule or
straight-edge

(a)

'daylight'

(b)

Angle > 90°   Angle < 90°   Domed end

Fig. 3.8   Trimming specimens: (a) checks for flatness and (b) checks for squareness

(6)   Mitre box.
(7)   Balance, readable to 0.01 g.
(8)   For large samples that exceed the capacity of the above balance, a balance readable to 0.1% of the sample mass or better.
(9)   Drying oven and moisture content apparatus (Section 2.5.2).

PROCEDURAL STAGES

(1)   Prepare specimen.
(2)   Weigh.
(3)   Measure.
(4)   Calculate.
(5)   Report results.
(6)   Determine dry density if needed.

TEST PROCEDURE

(1)   *Preparation*

A rectangular prism of soil is prepared as follows.

Avoid using the outer face of the original sample, which might have lost some moisture. Cut away at least 10 mm from the outside face, and cut out a rectangular prism of soil a few millimetres larger than the required final dimensions.

Make the ends of the prism flat, plane and parallel, either using the mitre box or by careful trimming and constant checking on the glass plate. Check the surfaces against the light using the steel rule, and trim until they are plane (i.e. no 'daylight'; see Fig. 3.8). Check the corners and angles using the try-square. Use the cutting tools or straight edge trimmer for trimming away surplus soil, *never* the steel rule or try-square.

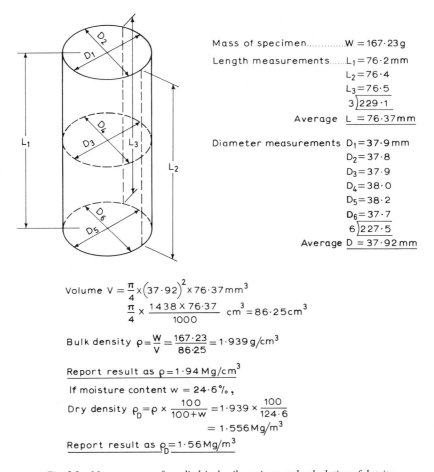

Mass of specimen..............W $= 167 \cdot 23$ g

Length measurements......$L_1 = 76 \cdot 2$ mm

$$L_2 = 76 \cdot 4$$
$$L_3 = 76 \cdot 5$$
$$3 \overline{)229 \cdot 1}$$

Average L $= 76 \cdot 37$ mm

Diameter measurements $D_1 = 37 \cdot 9$ mm

$$D_2 = 37 \cdot 8$$
$$D_3 = 37 \cdot 9$$
$$D_4 = 38 \cdot 0$$
$$D_5 = 38 \cdot 2$$
$$D_6 = 37 \cdot 7$$
$$6 \overline{)227 \cdot 5}$$

Average D $= 37 \cdot 92$ mm

Volume V $= \dfrac{\pi}{4} \times (37 \cdot 92)^2 \times 76 \cdot 37$ mm^3

$$\dfrac{\pi}{4} \times \dfrac{1438 \times 76 \cdot 37}{1000} \text{ cm}^3 = 86 \cdot 25 \text{ cm}^3$$

Bulk density $\rho = \dfrac{W}{V} = \dfrac{167 \cdot 23}{86 \cdot 25} = 1 \cdot 939$ g/cm^3

Report result as $\rho = 1 \cdot 94$ Mg/cm^3

If moisture content w $= 24 \cdot 6\%$,

Dry density $\rho_D = \rho \times \dfrac{100}{100+w} = 1 \cdot 939 \times \dfrac{100}{124 \cdot 6}$

$$= 1 \cdot 556 \text{ Mg/m}^3$$

Report result as $\rho_D = 1 \cdot 56$ Mg/m^3

Fig. 3.9   Measurements of a cylindrical soil specimen and calculation of density

(2)   *Weighing*

Weigh the trimmed specimen to the nearest 0.01 g, or to within 0.1% of its mass for a large specimen (see Section 1.2.3).

(3)   *Measurements*

Measure the length of each face of a rectangular prism along the edges and near the mid-face, using vernier calipers. Record the readings to 0.1 mm.

Measure a cylindrical specimen as follows. Take three measurements of length, using vernier calipers, equally spaced around the cylindrical surface (L_1 to L_3, Fig. 3.9). Take two measurements of diameter, using the calipers, at each end and at mid-length (D_1 to D_6). Read the vernier to 0.1 mm and record each measurement.

When using vernier calipers, first check the zero reading and record it. Close the jaws carefully on to the specimen so that they make contact without bedding in. The calipers must be perpendicular to the surfaces being measured, not on the skew. Read the vernier, record it, and subtract the zero reading to obtain the length being measured.

The specimen should not be exposed to the air for any longer than necessary, and should not be unduly handled, as otherwise it will lose moisture and may shrink. Wear thin plastics gloves to minimise moisture loss due to the warmth of the hands.

(4) *Calculation*

Calculate the mean value (mm) of each dimension from the several measurements, to the nearest 0.1 mm. For a rectangular block, if L=mean length (mm), B=mean breadth (mm) and H=mean height (mm), the volume V is given by

$$V = LBH \quad \text{mm}^3$$

$$= \frac{LBH}{1000} \quad \text{cm}^3$$

For a cylinder of diameter mean D and length L,

$$V = \frac{\pi D^2 L}{4000} \quad \text{cm}^3$$

If the mass of the specimen is m grams, the bulk density ρ is given by the equation $\rho = m/V \, \text{Mg/m}^3$.

An example calculation for a cylindrical specimen is given in Fig. 3.9.

(5) *Result*

The result is reported to the nearest 0.01 Mg/m^3.

(6) *Dry density*

If the dry density is also required, and no other test is to be done, the whole specimen can be broken up and placed in the oven to dry, in order to obtain the dry mass from which the moisture content is calculated. Make sure that no fragments of soil are lost.

The moisture content may be determined instead by using small representative portions of the specimen, or by using trimmings placed directly into a moisture content container when the spacimen is being prepared. The trimmings should be taken from immediately adjacent to the test specimen, not from the outer face of the original sample.

If the measured moisture content is w%, the dry density ρ_D is calculated from the equation

$$\rho_D = \frac{100}{100+w} \rho \quad \text{Mg/m}^3$$

3.5.3   Sample in Tube

If the density of the sample has to be measured in the sampling tube, the following outline procedure is used. The details refer to a U100 tube sample, but the same principle can be used for tube samples of other sizes if weighings and measurements are made to the appropriate degree of accuracy.

(1) Clean the outside of the tube, but do not remove the identification label. Record sample number, and tube number if marked.
(2) Remove end caps and wax or other protective material from each end.
(3) Trim each end of the sample to give flat surface normal to the axis of the tube. Remove all loose material (see Fig. 3.10a).
(4) Measure the overall length of the sampling tube (L_1) to the nearest mm, using a calibrated metre-stick or steel tape.
(5) Using a steel rule and straight-edge, or a depth gauge, measure from the ends of the tube to the trimmed surface of the sample to 0.5 mm, taking three or four readings at each end (Fig. 3.10b). Calculate the mean of each set to the nearest 0.5 mm (L_2 and L_3).

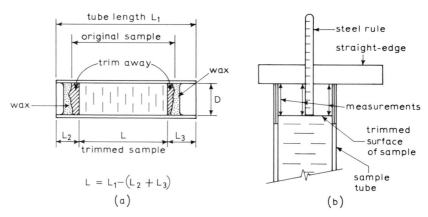

Fig. 3.10   *Measurements of a sample in a tube: (a) trimming ends and (b) measurements for sample length*

(6)   Using the internal-measuring jaws of the vernier calipers, measure the internal diameter of the tube on two perpendicular diameters at each end, to 0.1 mm. Calculate the average to 0.1 mm (D). Make sure that the calipers are opened to the maximum extent possible, to give a diametrical reading.

(7)   Weigh the tube and sample, to an accuracy of 0.1% or better (m_1). A 20 kg capacity balance may be required if the sample is contained in a steel tube. The sample is weighed in the tube as a precautionary measure in case the sample cannot be kept together on extrusion.

(8)   Extrude the sample and collect all material together.

(9)   Weigh the extruded material (m_2) to the nearest 0.1 g if possible.

(10) Clean out the tube, and weigh it (m_3) to 0.1% or better.

(11) Check that $m_1 - m_3 = m_2$.

(12) Calculate sample length L:

$$L = L_1 - (L_2 + L_3)   \text{mm};$$

sample volume V

$$V = \frac{\pi D^2 L}{4000}   \text{cm}^3$$

bulk density

$$\rho = \frac{m_2}{V}   \text{or}   \frac{m_1 - m_3}{V}   \text{Mg/m}^3$$

Report the result to the nearest 0.01 Mg/m^3, and that the density was measured in a U100 tube.

(13) Measure the moisture content $w\%$ of the extruded sample, and calculate dry density ρ_D as in Section 3.5.2, stage (6), if it is required.

3.5.4   Water Displacement (BS1377: Part 2:1990:7.4)

When it is not possible to trim a specimen to a regular cylindrical or prismoidal shape, or when the only available sample is an irregular lump, determination of the volume by simple

Fig. 3.11   *Water displacement apparatus (siphon can)*

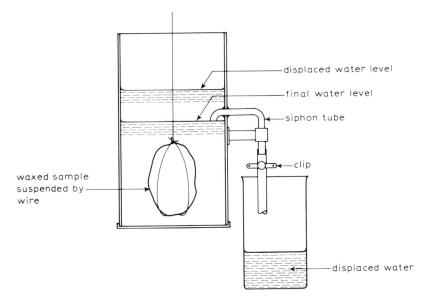

Fig. 3.12   *Principle of water displacement apparatus*

linear measurement is not practicable. The volume is then measured by one of the immersion methods. Water displacement is the simplest method, though less accurate than weighing in water (Section 3.5.5).

APPARATUS

(1)   Water-displacement apparatus, as shown in Fig. 3.11. A cross-section of the apparatus showing the siphon tube is given in Fig. 3.12.

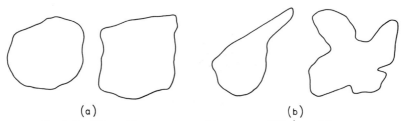

Fig. 3.13   *Trimmed lump specimens: (a) correct and (b) not satisfactory*

(2)   A watertight container to act as a receiver for water siphoned from the above apparatus.
       A convenient size is 250 mm diameter and 250 mm deep. A large beaker may be used
       if appropriate.
(3)   Balance reading to 1 g.
(4)   Paraffin wax and thermostatically controlled bath for melting wax. Alternatively, a
       double pot such as a glue-pot will serve to prevent the wax being overheated.
(5)   Drying oven and moisture-content apparatus (Section 2.5.2).
(6)   Plasticene or putty.

PROCEDURAL STAGES

(1)   Prepare specimen.
(2)   Weigh.
(3)   Fill voids.
(4)   Coat with wax.
(5)   Weigh.
(6)   Prepare displacement apparatus.
(7)   Immerse in water.
(8)   Measure displaced water.
(9)   Measure moisture content.
(10) Calculate.
(11) Report result.

TEST PROCEDURE

(1)   *Preparation of specimen*
Trim the specimen to a convenient size and shape, preferably about 100 mm in each direction.
Avoid making a long, narrow section, and avoid the formation of re-entrant angles (see
Fig. 3.13). Use the largest specimen possible, compatible with the capacity of the apparatus.

(2)   *Weighing*
Weigh the specimen to the nearest 1 g (m_s).

(3)   *Filling the voids*
Fill all the surface air voids with Plasticine or putty, and trim off level with the surface of
the specimen. Only air voids, and not cavities from which stones have been removed, should
be filled. Paraffin wax could be used for this purpose if it is applied carefully (see stage 4),
and if the voids are not large. Weigh the specimen after filling, to the nearest 1 g (m_f).
   The reason for this operation is to ensure that naturally occurring voids are included as
part of the sample volume, and are not penetrated by the protective coating of wax applied
in stage (4).

(4)   *Coating with wax*
Apply a first coat of paraffin wax, by brushing molten wax on to the surface. This should

be done carefully, avoiding trapping air bubbles especially in deep cavities. The wax must be heated to just above its melting point; it must not be overheated, as otherwise shrinkage and cracking will result. When the wax has dried, immerse the whole sample in the wax-bath, remove and allow the wax to set, and repeat the immersion several times.

(5) Weighing

When cool, weigh the waxed sample to the nearest 1 g (m_w). The mass of wax (m) is given by

$$m = m_w - m_f$$

(6) Preparation of apparatus

Place the water-displacement apparatus on a level base, close the clip on the outlet tube, and pour water to a level about 50 mm above the siphon tube. Release the clip on the outlet tube and allow excess water to run to waste. Retighten the clip without moving the can.
 Weigh the empty receiver container to the nearest 1 g (m_1).

(7) Immersion of sample

Place the receiving container below the siphon outlet. Lower the waxed specimen carefully into the apparatus until it is completely immersed, taking care to ensure that no air bubbles are trapped underneath the specimen. This operation is easier if the specimen is suspended by a length of fine wire looped around it.
 Release the clip on the outlet tube, allowing displaced water to siphon into the receiver. Wait until all excess water has been siphoned off, then retighten the clip.

(8) Measurement of displaced water

Weigh the receiver and water to the nearest 1 g (m_2).

(9) Measurement of moisture content

Remove the specimen from the can, dry the surface and break it open. Take representative samples, free from filler and wax, for the determination of the moisture content (w).

(10) Calculations

The volume of the immersed waxed specimen (V_b cm^3) is equal to the volume of water displaced (cm^3), which is equal to its mass in grams:

$$V_b = m_2 - m_1$$

The volume of paraffin wax is equal to m/ρ_P, where ρ_P is the density of the wax. A typical value is 0.91 Mg/m^3, but the actual value should be verified by trimming, measuring and weighing a rectangular prism of wax, following a procedure similar to that described in Section 3.5.2. The volume of specimen, V_s, is given by the equation

$$V_s = V_b - \frac{m}{\rho_P}$$

The bulk density, ρ, of the soil specimen is equal to its mass divided by its volume — i.e.

$$\rho = \frac{m_s}{V_s} \quad \text{Mg/m}^3$$

The dry density, ρ_D is calculated as in Section 3.5.2, stage (6).

(11)   *Results*

The bulk density and dry density are reported to the nearest 0.01 Mg/m^3, and the moisture content to two significant figures. The method is reported as the BS water-displacement method.

ALTERNATIVE PROCEDURE

A simple method can be used if the sample is small.

(1)   Take the smallest glass graduated measuring cylinder into which the waxed specimen will conveniently fit.
(2)   Add water up to a known mark, and record the reading.
(3)   Prepare, wax and weigh the specimen as for stages (1)–(5) of the water displacement method.
(4)   Immerse the specimen in the water in the cylinder, avoiding trapping any air.
(5)   Read and record the new water level in the cylinder. The difference between the two readings is the volume ($V_b \text{cm}^3$) of the waxed specimen.
(6)   Measure moisture content, and calculate bulk density and dry density, as for stages (9)–(11) above. Report as the measuring cylinder water-displacement method.

3.5.5   Immersion in Water (BS1377: Part 2:1990:7.3)

This method requires relatively simple weighing apparatus, and can be used for determining the density of quite large lump samples or compacted samples. It makes use of the principle of Archimedes (Section 3.3.3) for the measurement of volume.

APPARATUS

(1)   Balance, 15 or 20 kg capacity, reading to 1 g. For convenience it is preferable for it to be fitted with a flat pan instead of a scoop pan.
(2)   Suitable cradle and supporting frame for suspending the sample below the balance such as that shown in Fig. 3.14.
(3)   Paraffin wax and wax-bath (see item (4) of Section 3.5.4).
(4)   Watertight container (metal or plastics) such as a dust-bin.
(5)   Drying oven and moisture-content apparatus.
(6)   Plasticene or putty.

PROCEDURAL STAGES

(1)   Prepare apparatus.
(2)   Prepare specimen.
(3)   Weigh.
(4)   Fill voids.
(5)   Coat with wax.
(6)   Weigh.
(7)   Weigh immersed in water.
(8)   Measure moisture content.
(9)   Calculate.
(10) Report result.

TEST PROCEDURE

(1)   *Preparation of apparatus*

(a)   Place the balance on a board or battens projecting over the edge of a firm bench, so that the cradle can be suspended over the middle of the water container standing on

Fig. 3.14   *Apparatus for weighing in water*

the floor. Place a counterbalance weight on the board or battens to ensure that it will not tip over when carrying the specimen. Alternatively, use the apparatus shown in Fig. 3.14.

(b)   Adjust the height of the container and support it below the balance so that the largest specimen to be tested, when placed in the cradle, will be completely immersed, and will not touch the sides or bottom of the container.

(c)   Fill the container with water to about 80 mm below the top.

(d)   With the cradle completely immersed, add a counterweight to the scale pan to bring the balance reading to zero, if it is of the two-pan type; or use the tare adjustment to set the balance reading to zero.

If a counterweight is added it is preferable to use a jar or tin containing sand, water or lead-shot for this purpose, so that it will not be confused with the weights added later. If balance weights are used, they should be clearly identified as counterweights — for example, by standing them on a sheet of thin coloured paper.

Stages (2)–(6)

Similar to stages (1)–(5) in Section 3.5.4, using the balance in the normal manner.

mass of prepared sample $= m_s$
mass of sample and filler $= m_f$
mass of wax-coated sample $= m_w$
mass of wax $= m = m_w - m_f$ (masses in grams)

(7)   *Weighing in Water*

Place the waxed specimen in the cradle and suspend the cradle from the supporting frame so that the sample is completely immersed in the water in the container. Check that no air bubbles have been trapped underneath the specimen. Add more water or adjust the supporting frame, if necessary, to ensure complete immersion of the specimen. Stage (1)(d) should then be repeated with the sample removed, followed by a check-weighing of the immersed specimen.

Measure the apparent mass of the specimen in water to the nearest 1 g (m_g).

(8) *Measurement of moisture content*

Remove the sample from the cradle, dry the surface and break it open. Take representative samples, free from filler and wax, for the determination of moisture content (w).

(9) *Calculations*

From Archimedes' principle, the volume of the immersed waxed specimen, V_g (cm^3), is given by

$$V_g = m_w - m_g$$

The volume of the wax coating is equal to its mass divided by its density—i.e. m/ρ_P. The volume of the specimen itself (V_s) is therefore given by

$$V_s = V_g - \frac{m}{\rho_P} = (m_w - m_g) - \frac{m}{\rho_P}$$

The bulk density of the specimen is given by

$$\rho = \frac{m_s}{V_s} \text{ g/cm}^3 \ (= \text{Mg/m}^3)$$

The dry density is calculated as in Section 3.5.2, Stage (6).

(10) *Results*

The bulk density and dry density are reported to the nearest 0.01 Mg/m^3, and the moisture content to two significant figures. The method is reported as the BS immersion in water method.

3.6 PARTICLE DENSITY TESTS

3.6.1 Scope

Three tests for the measurement of the density of soil particles are described. These have different applications, which may be summerised as follows:

(1)   Density bottle method—for fine-grained soils only.
(2)   Gas jar method—for most soils including those containing gravel-sized partcles.
(3)   Pycnometer method—for use in a site laboratory on medium- and coarse-grained soils.

These methods are given in BS1377: Part 2:1990. The first two are the usual laboratory methods. The third is useful as a simple on-site procedure where full laboratory facilities are not available and when a result of lower accuracy is acceptable.

3.6.2 Density Bottle (Small Pyknometer) Method (BS1377: Part 2:1990:8.3)

This is the traditional method for the accurate measurement of the density of particles heavier than water, or of the density of liquids. It is referred to in BS1377:1990 as the small pyknometer method.

Fig. 3.15   Density bottles and large pyknometer

Fig. 3.16   Density bottles in constant-temperature bath

This method is suitable for soils consisting of particles finer than 2 mm. If the soil contains larger particles they may be ground down to pass a 2 mm sieve before testing, but if these particles contain internal voids a test on broken-down material could give a result different from that obtained on the natural material.

Distilled water is normally used as the density bottle fluid, but if the soil contains soluble salts, an alternative liquid should be used. The usual liquid is kerosene (paraffin) or, alternatively, white spirit. The density of the liquid must be measured separately (Section 3.6.3). The test procedure which follows is the same whether water or another fluid is used, but if the latter, the calculation must allow for the density of the liquid.

APPARATUS

(1)   Density bottles (50 ml) with stoppers, numbered and calibrated (three for each sample). (See Fig. 3.15.) Two density bottles are specified for each test in BS1377:1990, but it is preferable to make three determinations.

(2)   Constant-temperature water-bath, with shelf for holding density bottles, maintained at 25°C ± 0.2°C with thermometer reading to 0.2°C (see Fig. 3.16).

(3)   Vacuum desiccator with protective cage (Fig. 1.16).
(4)   Drying oven and moisture-content apparatus.
(5)   Analytical balance reading to 0.001 g.
(6)   Small riffle-box.
(7)   Source of vacuum, with vacuum gauge calibrated at a pressure of 20 mm of mercry, and vacuum tubing,
(8)   Chattaway spatula 150×3 mm.
(9)   Wash bottle containing de-aired density fluid.
(10)   Rubber-coated tongs, or rubber gloves.

MATERIALS

De-aired distilled water or de-aired kerosene (or white spirit), the specific gravity of which must be known or measured. Obtain air-free distilled water by boiling the water for at least 30 minutes and cooling in a container which can be sealed from the atmosphere during cooling; it must be strong enough to resist the resulting internal reduction of pressure.

   Although dissolved air does not greatly affect the accuracy of the result, there is the danger of air coming out of solution in the form of small bubbles when the temperature is raised to 25°C. A wetting agent may be necessary if the soil contains non-polar materials, eg. coal particles. Add 20% by volume of a very dilute solution of wetting agent to the water in the wash-bottle.

PROCEDURAL STAGES

(1)   Prepare density bottles.
(2)   Prepare test specimen.
(3)   Weigh and dry specimen in bottle (three bottles).
(4)   Add liquid and apply vacuum.
(5)   Stir and repeat until air is removed.
(6)   Top up, transfer to temperature bath.
(7)   Weigh bottle + soil + liquid.
(8)   Weigh bottle + liquid.
(9)   Calculate density values.
(10)   Report results.

TEST PROCEDURE

(1)   *Preparation of density bottles*

Wash each density bottle with stopper. Rinse with acetone or an alcohol–ether mixture and dry by blowing warm air through them. Cool in the desiccator and weigh to the nearest 0.001 g (m_1). Repeat to ensure that the bottle has been dried to constant weight.

   Do not dry density bottles in an oven, otherwise distortion may occur.

   Each density bottle should carry a unique identification number. If this is not engraved by the manufacturer, a laboratory number should be lightly scratched on the neck with a diamond-tipped pencil.

(2)   *Preparation of test specmen*

Obtain a sample of 50–100 g by quartering down the original sample. If gravel-sized particles are present a representative proportion should be included, but these must be ground with pestle and mortar to pass a 2 mm BS sieve. Riffle the sample to obtain a test specimen of about 30 g, oven dry at 105–110°C, and cool in the desiccator.

   If at this temperature the particle density is likely to be changed owing to loss of water of hydration, a drying temperature of 80°C should be used, and the drying time extended if necessary. This fact must be reported.

(3) Placing in bottles

Divide the dried specimen into three approximately equal parts, and place each into a density bottle using tongs or wearing rubber gloves. This should be done directly from the desiccator. Weigh each bottle with soil to 0.001 g (m_2).

(4) Adding liquid and applying vacuum

Add the de-aired liquid (distilled water or kerosene) carefully to each bottle so that the soil is just covered and the bottles are no more than half full. Pour the liquid down the side of the bottle to avoid disturbing the soil and thereby entrapping the air.

Place the bottles (without stoppers) in the vacuum desiccator. Reduce the pressure gradually to about 20 mm mercury and leave under vacuum for at least one hour. Air trapped in the soil will bubble out, but this must not occur too violently, as otherwise small drops of suspension may be lost through the mouth of the bottle. The vacuum should be maintained until no further loss of air can be seen.

(5) Removal of air

Release the vacuum, remove the desiccator lid and carefully stir the soil in the bottle with the Chattaway spatula. Wash any soil particles adhering to the blade back into the bottle with a little of the de-aired liquid. Replace the desiccator lid and apply the vacuum again until no further loss of air can be seen.

This process is repeated until no more air is evolved from the soil. Complete removal of air is essential for obtaining reliable results from this test. If in doubt, leave the bottles for a further period under vacuum, preferably overnight.

(6) Transfer to constant-temperature bath

Remove each bottle from the desiccator and add further air-free liquid until full. Insert the stopper, and place the stoppered bottle in the constant-temperature bath so that it is immersed up to the neck. Leave for 1 h at least, or longer if necessary for it to attain the temperature of the bath.

Surplus liquid will exude through the capillary tube in the stopper. If the liquid is other than water, it should be carefully absorbed on a filter paper to avoid contamination of the water-bath.

If there is a decrease in volume of the liquid, remove the stopper and top up. Repeat if necessary after a futher 1 h in the water-bath until the bottle remains full, up to the top of the stopper.

(7) Weighing

Remove the stoppered bottle from the bath and wipe it dry. Avoid prolonged contact with the hands, which may raise the temperature further. Weigh (bottle + stopper + soil + liquid) to 0.001 g (m_3).

(8) Weighing bottle with liquid

Clean out each bottle, and fill completely with de-aired liquid. Insert the stopper and immerse in the constant-temperature bath as before. Repeat stages (6) and (7) as necessary, and weigh (bottle + stopper + liquid) to 0.001 g (m_4).

(9) Calculation

Calculate the particle density, ρ_s, of the soil in each bottle from the following equation:

$$\rho_s = \frac{\rho_L(m_2 - m_1)}{(m_4 - m_1) - (m_3 - m_2)}$$

where ρ_L = density of the liquid used, at the constant temperature (if distilled water is used, ρ_L may be assumed to be equal to 1.000 g/ml for most purposes); m_1 = mass of density bottle (g); m_2 = mass of bottle + dry soil (g); m_3 = mass of bottle + soil + liquid (g); m_4 = mass of bottle + liquid only (g).

The average of the three values obtained is calculated. If any one value differs from the average value by more than 0.03 Mg/m³, the test should be repeated.

(10)   Results

The average value derived above is reported to the nearest 0.01 Mg/m³ as the particle density of the soil.

The liquid used in the test should be reported with the result.

3.6.3   Calibrations

CALIBRATION OF DENSITY BOTTLE

Each density bottle is dried to constant weight, cooled and weighed to 0.001 g. The bottle is then filled with air-free distilled water. Insert the stopper and place in the constant-temperature bath until it has attained a temperature of 25°C, as for stage (6) in Section 3.6.2. Take the bottle out of the bath, wipe it dry (care being taken not to raise the temperature of the bottle by prolonged contact with the hands) and weigh to 0.001 g. The test is repeated twice more.

The volume of each density bottle is calculated as follows:

m_1 = mass of density bottle with stopper (g)
m_4 = mass of density bottle with stopper full of distilled water at 25°C

The density of water at 25°C is 0.997 04 g/ml. Then

$$V_d = \frac{m_4 - m_1}{0.997\,04}$$

where V_d = volume of density bottle (ml).

DENSITY OF LIQUID

Calibrate three density bottles, using distilled water as described above. Empty and dry them, and allow them to cool.

Fill with liquid, insert stopper, place in temperature-bath, top up as for stage (6) in Section 3.6.2.

Remove, dry and weigh (m_5).

The density, ρ_L, of the liquid is calculated from the equation

$$\rho_L = \frac{m_5 - m_1}{m_4 - m_1} \text{ g/ml}$$

where m_1 = mass of density bottle; m_4 = mass of bottle with stopper filled with distilled water; m_5 = mass of bottle with stopper filled with liquid.

Calculations should be done to at least 5 decimal places. The result should be rounded off to the third decimal place.

The value of ρ_L is used in the equation given in step (9) of Section 3.6.2 for calculating the particle density of soil, when a fluid other than distilled water is used.

Fig. 3.17   Gas jars and mechanical shaker

3.6.4   Gas Jar Method (BS1377: Part 2:1990:8.2)

This test appeared for the first time in the 1975 British Standard. It can be used for soils containing up to 10% by mass of particles retained on a 37.5 mm sieve. Particles larger than this should first be broken down so as to pass a 37.5 mm sieve. (See the comments on breaking down of particles in Section 3.6.2).

APPARATUS

(1)   Two gas jars, 1 litre capacity, each with rubber bung and ground-glass cover-plate.
(2)   Mechanical shaker, giving end-over-end shaking motion at about 50 rev/min (Fig. 3.17).
(3)   Balance, 5 kg capacity, readable to 0.1 g.
(4)   Thermometer readable to 1°C.
(5)   Drying oven and moisture content apparatus.

PROCEDUREAL STAGES

(1)   Prepare specimens.
(2)   Prepare gas jar.
(3)   Weigh jar and sample.
(4)   Add water and shake.
(5)   Top up and fit glass plate.
(6)   Weigh jar, sample, water, plate.
(7)   Weigh jar and water.
(8)   Calculate.
(9)   Report result.

TEST PROCEDURE

(1)   *Preparation of specimens*

Prepare a representative sample of about 1 kg by riffling or quartering the original sample.

Break down any coarse particles in excess of 10% by mass retained on a 37.5 mm sieve, or any particles retained on a 50 mm sieve, to pass the 37.5 mm sieve.

Obtain two test specimens from the prepared sample by riffling. Each should be of about 200 g for fine-grained soil, or 400 g for coarse-grained soil. Dry in the oven at 105° to 110°C, cool in the desiccator and store in airtight containers until required.

(2)   *Preparation of gas jar*

The gas jars and ground-glass plates are cleaned and dried in the oven, cooled and weighed to the nearest 0.2 g (m_1).

(3)   *Weighing*

Place each prepared specimen into a gas jar. Weigh the gas jar, contents and ground-glass plate to 0.2 g (m_2).

(4)   *Adding water and shaking*

About 500 ml of water at room temperature is added to the soil in the jar. The rubber bung is inserted, and the jar is allowed to stand for 4 h. The gas jar is then shaken by hand to put the particles in suspension, and placed in the shaking apparatus and shaken for 20–30 min.

The jar and bung must be clamped tightly enough to prevent loss of water, but not so tightly that the jar might split while being shaken. Adhesive plastics tape wrapped around the jar and supporting frame of the shaker will ensure that the jar is held in place.

(5)   *Topping up and fitting glass plate*

After the required shaking period the jar is taken out of the shaker. Remove the rubber bung carefully to avoid losing fine particles. Any soil particles adhering to the bung must be washed carefully back into the jar with a jet of distilled water. Any froth that has formed is dispersed with the water jet.

Add water to the gas jar to within about 2 mm of the top, with the jar standing on a level surface. After standing for about 30 min to allow the soil to settle, fill the gas jar to the brim with water. Place the ground glass plate on top of the jar without trapping any air. To do this, tilt the jar under a slow stream of water and slide the ground glass plate upwards over the rim.

(6)   *Weighing jar with sample, water and plate*

The outside of the jar and plate are carefully dried with an absorbent cloth or blotting paper, without disturbing the plate. Jar, contents and plate are weighed to 0.2 g (m_3).

(7)   *Weighing jar and water*

The gas jar is emptied and washed out thoroughly. It is filled to the brim with water at room temperature, and the ground-glass plate is slid on without entrapping any air. The jar is dried, and the whole is weighed to 0.2 g (m_4).

Stages (3)–(6) are repeated on the other specimen of the same soil.

(8)   *Calculations*

The particle density, ρ_s, of the soil particles for each determination is calculated from the equation

$$\rho_s = \frac{m_2 - m_1}{(m_4 - m_1) - (m_3 - m_2)} \ \text{Mg/m}^3$$

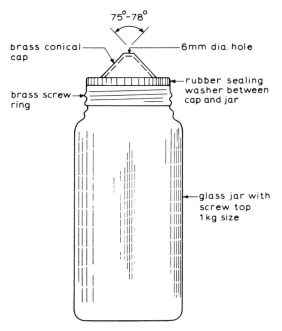

Fig. 3.18   *Pyknometer*

where; m_1 = mass of jar and plate; m_2 = mass of jar, plate and dry soil; m_3 = mass of jar, plate, soil and water; m_4 = mass of jar filled with water, and plate.

(9)   Results

If the two values of particle density differ by 0.03 Mg/m³ or less, the average of the two values is reported to the nearest 0.01 as the particle density of the soil. If the difference is greater than 0.03 Mg/m³, the tests should be repeated.

The method is reported as the British Standard gas jar method.

3.6.5   Large Pyknometer Method (BS1377: Part 2:1990:8.4)

This method is satisfactory for use with non-cohesive soils in a site laboratory with limited facilities. It is not recommended for general laboratory use, because it is less accurate than the gas jar method (Section 3.6.4). It is not suitable for clay soils.

Particles coarser than 20 mm should be broken down to pass a 20 mm sieve before testing. (See the comments on breaking down of particles in Section 3.6.2).

APPARATUS

(1)   Pyknometer, as illustrated in Fig. 3.18. This consists of a screw-top glass jar of about 1 kg capacity with a brass conical cap, screw ring and rubber sealing ring.
(2)   Balance reading to 0.5 g.
(3)   Oven and moisture-content apparatus,
(4)   Glass stirring rod.
(5)   Thermometer, readable to 1°C.

PROCEDURAL STAGES

(1)   Prepare pyknometer.
(2)   Prepare sample.

(3)   Place in pyknometer and weigh.
(4)   Add water and stir.
(5)   Remove air and top up.
(6)   Weigh.
(7)   Fill pyknometer with water.
(8)   Weigh.
(9)   Repeat test.
(10) Calculate.
(11) Report results.

TEST PROCEDURE

(1)   Preparation of Pyknometer

Clean and wash the pyknometer, including the cap assembly, dry and weigh to the nearest $0.5\,g\,(m_1)$.

Check that the rubber sealing ring is in good condition. If it has gone hard, replace it. Screw the securing ring so as to make the seal watertight. Make locating marks adjacent to one another on the conical cap, screw ring and glass jar. When the pyknometer is used, the ring should be tightened to this position every time so that the pyknometer volume remains constant.

If a watertight seal is difficult to make, grind the rim of the pyknometer jar by rubbing it on a sheet of fine carborundum paper laid on a perfectly flat surface, such as the glass plate used for liquid limits. Leakage can give rise to serious errors.

(2)   Preparation of sample

Prepare a representative sample of about 1 kg by riffling or quartering the original sample. Break down any particles larger than 20 mm to pass a 20 mm sieve.

Obtain two test specimens each of about 400 g, from the prepared sample by riffling. Dry in the oven at 105° to 110°C, cool in the desiccator and store in airtight containers until required.

(3)   Placing in Pyknometer

Removing the screw top from each pyknometer and insert the prepared sample direct from the sealed container. Weigh the jar and sample, together with the cap assembly, to $0.5\,g\,(m_2)$.

(4)   Adding water

Add water at about room temperature (within $\pm 2°C$) to about half-fill the jar. Stir thoroughly with the glass rod to remove entrapped air. Replace the cap assembly and fill with water. Check that the locating marks coincide. If necessary the water should be allowed to stand in the laboratory until the temperature is within the stated range.

(5)   Removing air and topping up

Shake the pyknometer, holding one finger over the hole in the conical cap, and allow any further air to escape and froth to disperse. Allow to stand for 24 h if this appears to be necessary. Top up with water to the hole. Entrapped air is a major source of error in this test.

By holding a finger over the hole in the cap, the pyknometer can be rolled on the bench to provide additional agitation. Ensure that no air or froth is trapped under the cap.

(6)   Weighing

Carefully dry the pyknometer on the outside, and weigh it with the water and soil to $0.5\,g\,(m_3)$.

(7) *Filling with water*

Empty the pyknometer, wash it out thoroughly, and fill with water at room temperature up to the hole in the cap. Check that the screw cap locating marks coincide when refitting the cap.

(8) *Weighing*

Dry the pyknometer on the outside and weigh to the nearest 0.5 g (m_4).

(9) *Repeat test*

Repeat stages (3)–(6) using the other specimen of soil.

(10) *Calculation*

Calculate the particle density, ρ_s, of the soil particles from the equation

$$\rho_s = \frac{m_2 - m_1}{(m_4 - m_1) - (m_3 - m_2)} \ \text{Mg/m}^3$$

where m_1 = mass of pyknometer; m_2 = mass of pyknometer and soil; m_3 = mass of pyknometer, soil and water; m_4 = mass of pyknometer full of water only.

Calculate the average of the two results obtained. If the two results differ by more than 0.05, repeat the test.

(11) *Results*

Report the particle density to the nearest 0.05 Mg/m³. The method is reported as the pyknometer jar method.

3.7 LIMITING DENSITY TESTS

3.7.1 Scope

Test to determine the limiting densities of granular soils are given in Part 4 of BS1377:1990. The procedures are based on those proposed by Kolbuszewski (1948a). These tests require only a vibrating hammer and standard soil-testing apparatus, and they give results which are reliable enough for most engineering applications.

Limiting density tests are also given in ASTM Standards (designation D4253 and 4254), where they are referred to as 'maximum and minimum index densities'. The test for maximum index density (D4253) requires the use of a vibrating table, which is not usually available in soil laboratories equipped for BS tests, but the procedure is certainly less fatiguing for the operator than the vibrating hammer method. Vibrating tables designed for making concrete test specimens in the UK do not provide the same frequency and amplitude as are required for the ASTM test on soils. However a vibrating table operating on a 60 Hz mains supply might be suitable if correctly adjusted.

Three methods are given in ASTM D4254 for determining the minimum index density. Method A is by pouring through a funnel; in Method B the soil is contained in a cylinder inside the mould, and the cylinder is suddenly withdrawn; Method C is similar to the BS test, but using a 2000 ml cylinder.

Kolbuszewski's tests were developed with clean sands — that is, sand particle sizes from 0.06 mm to 2 mm containing no silt or clay. Silty sands and silts cause difficulties in the maximum density test due to segregation and 'boiling out' of the finest particles. This results in a density lower than the true maximum density being measured; in some cases measured

Fig. 3.19   Electric vibrating hammer (Kango)

'maximum' densities less than the dry densities corresponding to optimum moisture contents in a standard compaction test have been reported. The procedure suggested here (Section 3.7.3) for silty soils may not, perhaps, give the highest density which it is possible to achieve, but it can generally be relied upon to give a repeatable result at a high density, and to avoid the above anomaly.

For the determination of the minimum density, slow pouring of dry sand in air through a funnel is often recommended (e.g. in ASTM D4254, Method 4). However, slow pouring in air does not necessarily result in the lowest possible density, as this allows each grain time to drop into voids between other grains. Minimum density is achieved when well-dispersed grains of dry sand fall with a high intensity through a small drop so that they immediately 'lock' into position in an open structure caused by the upward movement of air they displace. This is the basis of the dry shaking test (Section 3.7.5). A similar effect is aimed at in Method B of ASTM D4254 with the sudden collapse of the soil on removal of the cylinder from the mould.

3.7.2   Maximum Density — Sands (BS1377: Part 4:1990:4.2)

This test is suitable only for sands containing very little or no silt, and consisting of particles which are not susceptible to crushing. It makes use of some of the apparatus used in the vibrating-hammer method for the determination of the dry density/moisture content relationship (see Section 6.5.9). The maximum density is determined by compacting the soil, submerged under water, into a one-litre compaction mould with a vibrating hammer. The procedure is based on that recommended by Kolbuszewski.

APPARATUS

(1)   Electric vibrating hammer such as a Kango hammer of the type referred to in Section 6.5.9 (Fig. 3.19), fitted with an earth leakage circuit breaker between the hammer and the mains supply.
(2)   Steel tamper for attachment to the vibrating hammer, with a circular plate 100 mm diameter (see Fig. 3.20).

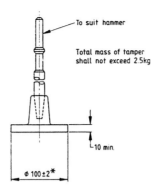

To suit hammer

Total mass of tamper
shall not exceed 2.5 kg

10 min.

φ 100 ± 2*

Fig. 3.20   Tamper for maximum density test on sands (Courtesy British Standards Institution, London)

(3)   Cylindrical metal mould (compaction mould) 105 mm internal diameter and 115.5 mm high, with detachable base-plate and extension collar (see Fig. 6.8).
(4)   Balance 10 kg capacity, reading to 1 g.
(5)   Watertight container large enough to take the compaction mould, standing on a firm base. If an ordinary galvanised bucket is used, place it on a block so that it is supported by the bottom and not just on the flange.
(6)   Two watertight containers such as plastic buckets.
(7)   Metal trays, one about 600 mm square and 80 mm deep; and one about 300 mm square and 50 mm deep.
(8)   Test sieves, 2 mm and 6.3 mm aperture, and receiver.
(9)   Steel straight-edge scraper, 300 mm long.
(10)   Small tools: spatula, scoop.
(11)   Stop clock readable to 1 second.
(12)   Oven and moisture content apparatus.
(13)   Supporting guide frame for vibrating hammer (optional) (see Fig. 6.19).
(14)   Extruder, for extracting sample from the mould (optional) (see Fig. 6.14)

PROCEDURAL STAGES

(1)   Calibrate vibrating hammer.
(2)   Prepare sample.
(3)   Assemble mould.
(4)   Place in water.
(5)   Place layer of soil in mould.
(6)   Compact by vibration under water.
(7)   Repeat stages (4) and (5) on second layer of soil.
(8)   Repeat stages (4) and (5) on third layer of soil.
(9)   Remove and allow free water to drain from sample.
(10)   Trim sample flush with mould.
(11)   Extrude sample.
(12)   Weigh sample.
(13)   Repeat test.
(14)   Calculate.
(15)   Report results.

PROCEDURE

(1) *Calibration of vibrating hammer*

Confirm that the vibrating hammer complies with the requirements of BS1377 by using the verification procedure given in Section 6.5.5.

(2) Preparation of sample

A representative sample of about 6 kg of soil passing a 6.3 mm sieve is required, to provide two separate test specimens. If necessary sieve the soil on a 6.3 mm sieve and break down retained particles to sizes between 2 mm and 6.3 mm, provided that the total proportion by mass in this size range does not exceed 10%. Mix the soil thoroughly with water on the large tray and divide into two representative portions. Place each portion into a bucket containing warm water and stir thoroughly to remove air bubbles. Cover and allow to stand (e.g. overnight) until cool.

(3) Assembly of mould

Clean and dry the mould body, base-plate and extension collar. Check and record the diameter and length of the mould body, using calipers, to 0.1 mm. Apply a thin coat of oil to the internal faces of all three components. Attach the base-plate firmly to the mould, and attach the extension collar to the mould body.

(4) Placing in water

Lower the mould assembly into the water container. Add water to about 50 mm depth, both inside and outside the mould. See that the container is on a firm solid base such as a concrete floor or plinth.

(5) Placing soil in mould

Add the soil-water mixture to nearly half-fill the mould, and spread it level. The amount of soil should be such that when compacted it will fill the mould to about one-third. Avoid loss of fines and segregation of soil particles.

(6) Compaction of first layer

Place the circular tamper on the sample, with a disc of polythene sheet between the tamper plate and soil to prevent sand particles moving upwards through the annular gap. Place a sheet of polythene or cloth around the top of the mould to protect the vibrating hammer from water which may splash up. Keeping the tamper stem vertical, compact with the vibrating hammer for at least 2 min, timed with the stop-clock, or until there is no further significant decrease in height. Throughout this time apply a steady downward pressure to give a force of about 350 N on the sample. The pressure required is greater than that necessary to prevent the hammer bouncing on the soil. It may be checked by applying the vibrating hammer (without vibration) on to a platform scale and pressing down until a reading of about 35 kg is registered. With practice one can judge the feel of the correct pressure.

(7) Second layer

Lift out the vibrating hammer carefully to avoid disturbing the soil surface. Add a second layer of soil equal to the first and repeat stage (6).

(8) Third layer

Add a third layer of soil and repeat stage (6). Add water to the container, if necessary, to ensure that the soil being compacted remains submerged. The final level of the soil should be not more than 6 mm above the top of the mould body (see Fig. 3.21).

(9) Removal of mould

Remove the mould containing the soil from the container, wipe off loose soil from the outside and allow free water to drain from the sample.

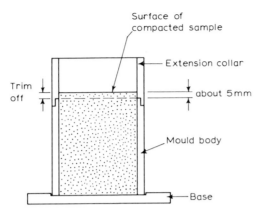

Fig. 3.21   *Soil in mould after compaction*

(10)   *Trimming sample*

Remove the extension collar carefully, and trim off the compacted soil level with the top of the mould. Refill any cavities left by removal of coarse particles using fine material well pressed in. Check with the straight-edge that the surface is level.

(11)   *Extrusion of sample*

Take off the base-plate, and remove the soil from the mould either with an extruder or by breaking it up. Place the whole sample in a weighed tray.

(12)   *Weighing*

Dry the soil in the oven at 105–110°C, weigh when cool and determine the dry mass of soil to the nearest 1 g (*m*).

(13)   *Repeat test*

Clean and dry the mould and attachments, and repeat steps (4) to (12) using the second prepared specimen. The results are acceptable if the dry masses differ by no more than 50 g; otherwise repeat the procedure using fresh batches of soil.

(14)   *Calculations*

The volume of the standard compaction mould is exactly 1000 cm^3 if the dimensions are exactly as shown in Fig. 6.8. The maximum dry density of the soil in the mould, ρ_{Dmax}, is given by the equation

$$\rho_{Dmax} = \frac{m}{1000} \ \text{Mg/m}^3$$

where *m* is the greater of the two dry masses determined in step (12).

If the dimensions are not as in Fig. 6.8 (for instance, owing to wear of the mould), then its volume (*V* cm^3) is calculated from the equation

$$V = \frac{\pi D^2 L}{4000} \ \text{cm}^3$$

where *D* = mean diameter of mould and *L* = mean length of mould (mm).

The maximum dry density, $\rho_{D,\max}$, is given by the equation

$$\rho_{D\max} = \frac{m}{V} \, \mathrm{Mg/m^3}$$

(15)   *Results*

The maximum dry density is reported to the nearest $0.01 \, \mathrm{Mg/m^3}$, and the fact that the vibrating-hammer procedure was used is recorded. The percentage material retained on the 6.3 mm sieve, and whether it was broken down or replaced, is also reported.

3.7.3   Maximum Density — Silty Soils

For the reasons given in Section 3.7.1, the above method has been found to be unsatisfactory for soils containing appreciable amounts of silt. It is suggested that a dynamic compaction test, based on the British Standard 'heavy' (4.5 kg rammer) compaction test, be used instead to establish a density approaching the 'maximum' density. As with the vibrating-hammer test (Section 3.7.2), this method is not suitable for soils containing particles susceptible to crushing.

APPARATUS

As for the 'heavy' compaction test (see Chapter 6, Section 6.5.4).

PROCEDURE

(1)   Carry out the test described in Section 6.5.4 to determine the optimum moisture content and corresponding dry density for this degree of compaction. (Compaction in five layers, 27 blows per layer, using 4.5 kg rammer with 450 mm drop.)
(2)   Take about 5 kg of the soil and adjust the moisture content to about 2% less than the optimum determined in stage (1).
(3)   Compact the soil into the mould as for the 'heavy' compaction test, but applying 80 blows instead of 27 on each of the five layers.
(4)   Measure the resulting density and moisture content.
(5)   Calculate the dry density, and report to the nearest $0.02 \, \mathrm{Mg/m^3}$ as an approximation to the maximum dry density.

The application of 80 blows per layer is suggested as a practical expedient, because with some soils there has been found to be very little increase in density when the number of blows is increased to 100 or 150 per layer. However, some soils may require a greater compactive effort to achieve the highest possible density, and this can be assessed only on the basis of experience. If appreciably greater effort is required, the moisture content may need to be more than 2% below the optimum determined from the 'heavy' compaction test. The moisture content used should be close to the 'optimum' for the degree of compaction applied, and optimum moisture content decreases with increasing compactive effort. (See Fig. 6.3.)

3.7.4   Maximum Density — Gravelly Soils (BS1377: Part 4:1990:4.3)

This test is suitable for determination of the maximum density of soils containing little or no silt, and including gravel-size particles passing a 37.5 mm sieve, provided that the particles are not susceptible to crushing. The test is similar to that described in Section 3.7.2 above, except that a larger mould (the CBR mould) is used, and a correspondingly larger diameter tamper plate is necessary.

APPARATUS

(1) As item (1) of Section 3.7.2.
(2) Steel tamper for attachment to the vibrating hammer, with a circular plate 145 mm diameter (see Fig. 6.23).
(3) Cylindrical metal mould (the CBR mould) 150 mm internal diameter and 127 mm high, with detachable base-plate and extension collar, as described in Section 6.5.4 (see Figs. 6.20 and 6.21).
(4) Balance, 20 kg capacity readable to 5 g.
(5) Watertight container large enough to take the CBR mould, standing on a firm base.
(6) and (7) (As items (6) and (7) of 3.7.2).
(8) Test sieves, 6.3 mm, 20 mm, and 37.5 mm aperture, and receiver.
(9) to (14) (As items (9) to (14) of 3.7.2).

PROCEDURAL STAGES

(The stages are the same as for Section 3.7.2.)

PROCEDURE

The procedure is similar to that described in Section 3.7.2, with the following modifications.

(1) Prepare two tests specimens each of about 8 kg. Break down any particles retained on a 37.5 mm sieve to sizes between 6.3 and 20 mm, provided that the total proportion by mass in this size range does not exceed 30%.
(2) Measure the mould to the nearest 0.5 mm.
(3) to (11) (As stages (3) to (11) of Section 3.7.2).
(12) Determine the dry mass of soil to the nearest 5 g (m).
(13) Clean and dry the mould and attachments, and repeat steps (4) to (11) above using the second prepared specimen. The results are acceptable if the dry masses differ by no more than 150 g; otherwise repeat the procedure using fresh batches of soil.
(14) If the dimensions of the CBR mould are exactly as shown in Fig. 6.20 or 6.21, its internal volume is 2305 cm^3 and the maximum dry density is given by the equation

$$\rho_{max} = \frac{m}{2305} \ \text{Mg/m}^3$$

Otherwise use the measured dimensions to calculate the volume of the mould, V (cm^3) and then

$$\rho_{max} = \frac{m}{V} \ \text{Mg/m}^3$$

(15) Include in the report the percentage of material retained on the 37.5 mm sieve, and whether it was broken down or replaced.

3.7.5 Minimum Density — Sands (BS1377: Part 4:1990:4.3 and ASTM D4254 Method C)

This test was devised by Kolbuszewski, and gives a value for the minimum density which a clean dry sand can sustain in a measuring cylinder. After shaking the soil is allowed to fall freely, and in doing so entraps air and forms a grain structure enclosing the maximum volume of voids.

The test is not suitable for sands containing more than about 10% of fine material passing a 63 μm sieve, or any particles retained on a 2 mm sieve.

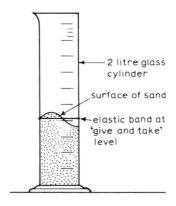

2 litre glass
cylinder

surface of sand

elastic band at
'give and take'
level

Fig. 3.22   Minimum density test on sand (dry)

APPARATUS

(1)   Glass graduated cylinder, 2000 cm³, graduated to 20 ml. (In the BS procedure a
      1000 cm³ cylinder is specified, but the author recommends the original method using
      a 2 litre cylinder.)
(2)   Rubber bung to fit the cylinder.
(3)   Balance reading to 0.1 g.
(4)   Elastic band to fit around the cylinder.

PROCEDURE

(1)   Weigh out 1000 g of oven-dried sand, and place in the cylinder. Fit the bung.
(2)   Shake the cylinder by inverting it a few times, to thoroughly loosen the sand.
(3)   Turn it upside down, pause until all the sand is at the top end, then very quickly turn
      it back again and stand it on the bench without jarring it.
(4)   If the resulting surface of the sand is level, record the volume of sand on the graduated
      scale to the nearest 10 ml. If the surface is not level, adjust the elastic band around
      the cylinder to the best give-and-take level, and record the volume at that level. Do
      not shake or jolt the cylinder while this is being done. (See Fig. 3.22.)
(5)   Repeat at least ten times.
(6)   Take the highest volume reading (V cm³) and calculate the minimum dry density from
      the equation

$$\rho_{D,min} = \frac{1000}{V} \ \text{Mg/cm}^3$$

Report the result to the nearest 0.01 Mg/m³, and the procedure as the dry shaking
method.

3.7.6   Minimum Density — Gravelly Soils (BS1377: Part 4:1990:4.5)

This test was suggested by the author for obtaining an indication of the lowest practicable
density of gravelly soil. The principle is similar to that given in Section 3.7.5, except that a
metal mould (usually a CBR mould) is used, into which the soil is dropped.

APPARATUS

(1) CBR mould (as item (3) of Section 3.7.4).
(2) Balance, 20 kg capacity readable to 5 g.
(3) Metal tray, about 600 mm square and 80 mm deep.
(4) Test sieve, 37.5 mm aperture, and receiver.
(5) Metal straight-edge scraper.
(6) Scoop.
(7) Bucket, or similar container.
(8) Weighed container (e.g. a metal tray) suitable for weighing the sample.
(9) Drying oven, 105° to 110°C.

PROCEDURES

(1) Take a representative sample of soil and remove any material retained on a 37.5 mm sieve. Prepare a representative sample, by riffling or quartering, at least 50% larger than the internal volume of the mould.
(2) Measure the internal dimensions of the mould to 0.5 mm.
(3) Thoroughly mix the oven-dried soil, to ensure an even distribution of the larger particles.
(4) Place it loosely so as to fill a bucket or similar container at least 50% larger than the mould being used.
(5) Place the mould, with base and extension attached, on a large tray on the floor.
(6) Empty the bucketful of soil rapidly into the mould from a height of about 0.5 m, so that the mould is filled in about 1 second. The surplus soil will overflow on to the tray.
(7) Carefully remove the mould extension and level the surface of the soil to the top of the mould. Avoid disturbing the soil in the mould, or jarring it. Pick off large particles individually by hand. Check the surface with the straight-edge. Any cavity left by the removal of a large particle should be filled as far as possible with one or more small particles, carefully placed.
(8) Transfer the contents of the mould to the weighed container, weigh and determine the mass of soil to the nearest 5 g (m).
(9) Re-mix the weighed soil with the excess material remaining on the tray and repeat steps (4) to (8) at least nine more times, to give at least 10 determinations.
(10) Use the lowest value of m to calculate the minimum dry density $\rho_{D,\min}$ from the equation

$$\rho_{D,\min} = \frac{m}{V} \ \text{Mg/m}^3$$

where V is the volume of the mould (cm^3).
  If the mould is the standard size, $V = 2503 \ \text{cm}^3$.
(11) Report the result to the nearest 0.01 Mg/m^3, and the procedure as rapid dry pouring. State the percentage of particles larger than 37.5 mm in the original sample, and whether they were removed, or broken down and put back.

BIBLIOGRAPHY

Abbot, A. F. (1969). *Ordinary Level Physics*, 2nd edition (SI units), Chapter 12, Heinemann, London
BS733:1965, Pyknometers, British Standards, Institution, London
Kaye, G. W. C. and Laby, T. H. (1973). *Tables of Physical and Chemical Constants*, 14th Edition. Longmans, London
Kolbuszewski, J. (1948a). 'An experimental study of the maximum and minimum porosities of sands'. *Proc. 2nd Int. Conf. Soil Mech. and Found. Eng., Rotterdam*, Vol. 1
Kolbuszewski, J. (1948b). 'General investigation of fundamental factors controlling loose packing of sands'. *Proc. 2nd Int. Conf. Soil Mech. and Found. Eng., Rotterdam*, Vol. 7
Lambe, T. W. and Whitman, R. V. (1979). *Soil Mechanics*. Wiley, New York

Loos, W. (1936). 'Comparative studies of the effectiveness of different methods for compacting cohesionless soil'. *Proc. 1st Int. Conf. Soil Mech. and Found. Eng., Cambridge, Mass., USA*, Vol. III

Pauls, J. T. and Goode, J. F. (1970). 'Suggested method of test for maximum density of noncohesive soils and aggregates'. *ASTM STP* 479

Selig, E. T. (1963). 'Effect of vibration on density of sand'. *Proc. 2nd Panamerican Conf. on Soil Mech. and Found. Eng., Brazil*, Vol. 1

Yemington, E. G. (1970). 'Suggested method of test for minimum density of noncohesive soils and aggregates'. *ASTM STP* 479

Chapter 4

Particle size

4.1 INTRODUCTION

4.1.1 Scope

A soil consists of an assemblage of discrete particles of various shapes and sizes. The object of a particle size analysis is to group these particles into separate ranges of sizes, and so determine the relative proportions, by dry mass, of each size range.

4.2.1 Test Procedures

Particle size analyses are also referred to as particle size distribution (PSD) tests, sizing tests or mechanical analysis (MA) tests. Two separate and quite different procedures are used, in order to span the very wide range of particle sizes which are encountered. These are the sieving and the sedimentation procedures. Sieving is used for gravel and sand size ('coarse') particles, which can be separated into different size ranges with a series of sieves of standard aperture openings (Section 4.6). Sieving cannot be used for the very much smaller silt and clay size ('fine') particles, so a sedimentation procedure is used instead. Measurements are made either by sampling a suspension of soil particles in water, by means of a special pipette (the pipette method, Section 4.8.2); or by determining the density of the suspension using a special hydrometer (the hydrometer method, Section 4.8.3).

For soils containing both coarse and fine particles, composite tests using both sieving and sedimentation methods have to be carried out if a full particle size analysis is required. Particle size testing can range from a simple sieving test on a 'clean' sand and gravel, to elaborate composite tests on clay–silt–sand–gravel mixtures.

Test procedures for different types of materials are similar in principle but vary in detail, and are described separately. The most difficult type of material to deal with is that referred to as 'boulder clay', which is treated as a special case.

4.1.3 Presentation of Data

Results of a particle size analysis may be presented as a table showing the percentages of particles finer than certain standard sizes. Usually, however, the results are presented graphically to show the percentages finer than any given size, plotted against the particle size on a logarithmic scale. This graphical presentation is known as the particle size distribution (PSD) curve, or the grading curve, for the material, and is described in Section 4.3.4.

There are other ways of presenting particle size distribution data graphically, such as by means of a grain size–frequency curve in which the percentages by mass between certain sizes are plotted against the logarithm of the grain size. These methods are used in other industries, such as powder technology (see, for example, Allen, 1974), but the semi-logarithmic plot to be described is the procedure generally used for soils.

4.2 DEFINITIONS

PARTICLE SIZE ANALYSIS expresses quantitatively the proportions by mass of the various
   sizes of particles present in the soil.
PARTICLE SIZE is usually given in terms of the equivalent particle diameter:

   GRAVEL: particles from 60 mm to 2 mm.
   SAND: particles from 2 mm to 0.06 mm.
   SILT: particles from 0.06 mm to 0.002 mm.
   CLAY: particles (clay minerals) smaller than 0.002 mm (2 μm).
   FINES are particles which pass a 63 μm sieve.
   CLAY FRACTION is the percentage of particles smaller than 2 μm, as determined by a
      standard sedimentation procedure.

QUARTERING   The process of obtaining a small but representative portion from a large
   sample of cohesionless material. The sample is first divided into quarters by a standard
   procedure, two of which are retained, which reduces the size of sample by half. The process
   is repeated until a sample of the required size is obtained. (See Section 1.5.5.)
RIFFLING   A similar process to quartering but using a RIFFLE BOX of appropriate size.
   However, this term is often used to indicate subdivision of material with or without a
   riffle box.
SIEVE SIZE   Refers to the aperture size of the sieve mesh — that is, the length of the side
   of the square openings through which particles pass.
SIEVE DIAMETER   Refers to the diameter of the body of the sieve (the frame) which holds
   the mesh.
EFFECTIVE SIZE (D_{10})   The particle size for which 10% of particles are finer, and 90% are
   coarser. It corresponds to $P = 10\%$ on the particle size curve (see Fig. 4.1).
UNIFORMITY COEFFICIENT (U)   The ratio of the 60% particle size to the 10% particle
   size:

$$U = \frac{D_{60}}{D_{10}}$$

   It is a measure of the slope of the line joining these two points (see Fig. 4.1).
FINENESS MODULUS   The sum of the cumulative percentages *retained* on each of a
   standard set of sieves, divided by 100. It is used as a numerical indication of grading
   characteristics, and is applied only to concrete aggregates. The relevant sieves are those
   used for testing concrete aggregates — namely 150, 300, 600 μm; 1.2, 2.4, 4.76 mm aperture
SPECIFIC SURFACE   The total surface area of particles contained in a unit mass.
FLOCCULATION   The coagulation of particles together in a suspension, giving the effect
   of larger particles.
DISPERSION   The separation of discrete particles so that they will remain separated and
   not flocculate, when in suspension.

4.3 THEORY

4.3.1 Limitations

A particle size distribution analysis is a necessary index test for soils, especially coarse soils,
in that it presents the relative proportions of different sizes of particles. From this it is
possible to tell whether the soil consists of predominantly gravel, sand, silt or clay sizes,
and to a limited extent which of these sizes ranges is likely to control the engineering
properties. Particle size curves are of greater value if supplemented by descriptive details
such as colour and particle shape, together with grain packing and fabric when observed

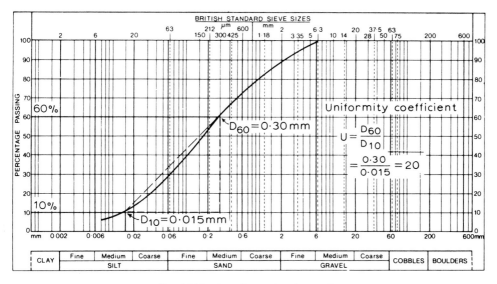

Fig. 4.1(a)   Particle size distribution chart

BS SIEVE APERTURE SIZES

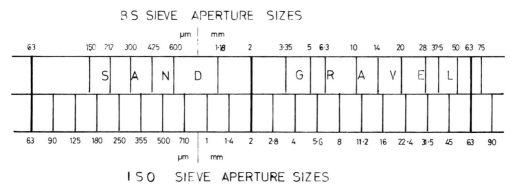

Fig. 4.1(b)   Comparison of conventional British and ISO sieve aperture sizes. Note the uniform spacing of the latter.

in the undistributed state. But engineering behaviour also depends on factors other than size of particles, such as mineral type, structure and geological history, which have a significant effect on engineering properties and cannot be assessed from particle size tests alone.

For the coarse soil fraction the proportions of the various sizes are determined from the sizes of square openings in sieves, whereas for fine soil particles they are based upon diameters of equivalent spheres obtained from a sedimentation analysis, notwithstanding that particles in the fine silt to clay size range are far from spherical. Indeed, many clay minerals consist of thin flat plate-like or elongated particles. The measurement of the true size of particles therefore becomes less accurate towards the fine end of the scale. However, the absolute magnitude of particle size is of less importance than the distribution of sizes as determined by a recognised standard procedure, and this is the sense in which particle size is applied to soils. In particular, the 'clay fraction' (the proportion of material consisting of particles smaller than 0.002 mm) is often used as an index for correlating with other engineering properties.

With some soils it is difficult to define what is meant by an 'individual particle', such as those in which the size of particles depends upon the degree of disaggregation achieved prior to testing. Examples are residual soils (see Chapter 7) and soils containing fragments

Table 4.1. CLASSIFICATION BY PARTICLE SIZE (Based on BS1377:1975)

Particle size (mm)		Designation		Test procedure
> 200		BOULDERS		Measurement of separate pieces
200 —				
		COBBLES		
60 —				
20 —	Coarse			Sieve
6 —	Medium	GRAVEL		analysis
	Fine			
2 —	Coarse			
0.6 —	Medium	SAND		
0.2 —	Fine			
0.06 —	Coarse			Sedimentation
0.02 —	Medium	SILT		analysis
0.006 —	Fine			
0.002 and less		CLAY		

of weakly cemented sandstone or similar material. With such soils it is necessary to control the extent to which particles are broken down in the preparation stage. For soil types such as peat and weathered chalk the concept of individual particles does not exist in the same sense as for sands, silts and clays.

4.3.2 Grouping by Particle Size

Soils may be divided on the basis of their dominating particle size into six arbitrary categories which are called boulders, cobbles, gravel, sand silt, clay. The size ranges of particles comprising each category, as used in Britain, were originally proposed by Glossop and Skempton (1945). These size ranges, as defined in BS5930:1981, are given in Table 4.1, which also shows the subdivision of gravel, sand and silt into coarse, medium and fine sizes. Classification by particle size in other countries may differ from British practice, and a comparison of American (ASTM) systems with the British system is shown diagrammatically in Fig. 4.2.

Usually the boulder and cobble sizes, if present, are removed and measured separately before test, although in some instances it may be necessary to include cobbles in the analysis (see Section 4.6.3). It is the gravel, sand, silt and clay sizes which are generally recognised and tested as 'soils'.

4.3.3 Influence of Particle Size

Particles encountered in most soils testing range from a maximum size of 75 mm down to about 0.001 mm, which is the normal limit of measurement in the sedimentation test. The ratio between these extreme sizes is 75 000 to 1.

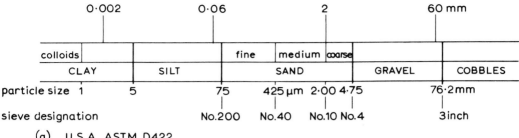

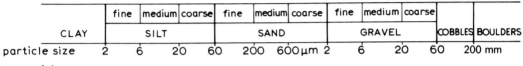

(a)   U.S.A.   ASTM D422

(b)   Great Britain   BS 1377 : 1975

Fig. 4.2   Comparison of systems for classifying particle size ranges of soils

Table 4.2.   PARTICLE SIZE, MASS AND SURFACE AREA OF EQUAL SPHERES

Equivalent soil category	Particle size (mm)	Approximate mass of particle (g)	Approximate number of particles per gram	Approximate surface area (mm²/g)	(m²/g)
Small cobble (largest 'soil' particle)	75	590	(1.7/kg)	30	
Coarse sand	1	0.0014	720	2300	
Fine sand	0.1	1.4×10^{-6}	7.2×10^5	23 000	0.023
Medium silt	0.01	1.4×10^{-9}	7.2×10^8	2.3×10^5	0.23
Clay (smallest size measured)	0.001	1.4×10^{-12}	7.2×10^{11}	2.3×10^6	2.3

As the particle size decreases, the number of particles contained in each gram of material increases, in inverse proportion to the cube of the particle size. The mass of each particle decreases in the same ratio. The total surface area of particles per unit mass (known as 'specific surface', and expressed as mm²/g or m²/g) increases in inverse proportion to the particle size. These characteristics are shown in Table 4.2 for the two extreme sizes referred to above, together with three intermediate sizes, assuming spherical particles of particle density 2.65 Mg/m³. The ratio of the extreme sizes of spherical particles in terms of mass is enormous — about 4.2×10^{14} to 1. There can be few test procedures which embrace such a wide range of values.

Because natural soil particles are rarely spherical, their specific surface is higher than those shown in Table 4.2. Thus fine sand particles may have a specific surface of about 0.03 m²/g. Individual particles of clay are flat and plate-like, which gives them specific surfaces many times greater than for spheres, depending on the type of clay mineral. Typical values for three of the commonest types are approximately as follows (Lambe and Whitman, 1979):

Kaolinite	10 m²/g
Illite	100 m²/g
Montmorillonite	1000 m²/g

The extremely high specific surface of clay particles is one of the factors responsible for the cohesive properties of clay.

4.3.4 Particle Size Distribution Curve

The type of form recommended in the British Standard for presenting particle size distribution is shown in Fig. 4.1(a). The advantage of plotting particle size data on a standard chart of this kind is that it enables engineers to recognise instantly the grading characteristics of a soil far more easily than from tabulated figures. Moreover the position of a curve on the chart indicates the fineness or coarseness of the grains; the higher and further to the left the curve lies, the finer the grains, and vice versa. The steepness, flatness and general shape indicate the distribution of grain sizes within the soil. Examples are discussed in Sections 4.4.2 and 4.4.3.

While this type of chart is used in the UK and many European countries, the practice in the USA until recently has been to use the chart the other way round, with cobbles and gravel to the left and clay to the right. Particle size curves so drawn are the mirror image of those plotted according to British practice.

In Fig. 4.1 is also shown the derivation of the 'effective size' (D_{60}) and the 'uniformity coefficient' (U) from a typical grading curve.

4.3.5 Sieve Aperture Sizes

The sieve aperture sizes shown on the particle size distribution chart in Fig. 4.1(a) are those commonly used in Britain and referred to in BS1377: Part 2:1990:9.2.2.1. An alternative set of sieves comprises the series of aperture sizes recommended by the International Standards Organisation (ISO). These are shown in Fig. 4.1(b), where the aperture sizes are plotted to the same logarithmic scale as the conventional British sizes. The ISO series provides a more logical sequence of sizes, in that each sieve aperture is $\sqrt{2}$ times the next smaller aperture (i.e. alternate sizes are related by a factor of 2). This gives a uniform spacing between each sieve size when plotted on the usual logarithmic scale, instead of the irregular and irrational spacing of the British series. BS1377:1990 permits the use of ISO sieve sizes as an alternative series. Both series include aperture sizes of 63 μm, 2 mm and 63 mm, thereby providing the same criteria for the principal elements of soil classification according to particle size.

In ASTM Test Designation D422, the following sieve aperture sizes are listed for a full set.

*3 inch	(75 mm)	*No. 4	(4.75 mm)
2 inch	(50 mm)	No. 10	(2.00 mm)
*$1\frac{1}{2}$ inch	(37.5 mm)	No. 20	(850 μm)
1 inch	(25.0 mm)	No. 40	(425 μm)
*$\frac{3}{4}$ inch	(19.0 mm)	No. 60	(250 μm)
*$\frac{3}{8}$ inch	(9.5 mm)	No. 140	(106 μm)
		No. 200	(75 μm)

A set of ASTM sieves to give a uniform spacing of points on a particle size distribution sheet, which may be used as an alternative to the above set, consists of the sieves marked with an asterisk above, plus the following.

No. 8   (2.36 mm)
No. 16   (1.18 mm)
No. 30   (600 μm)
No. 50   (300 μm)
No. 100 (150 μm)
No. 200 (75 μm)

BS and ASTM sieve aperture sizes are compared in Appendix A.4.

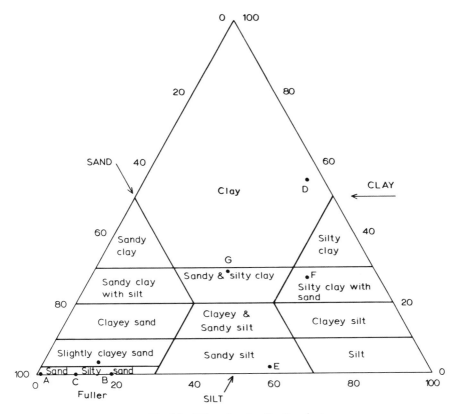

Fig. 4.3 Triangular classification chart

4.3.6 Triangular Classification Chart

A triangular classification chart is not often used but it can be convenient for comparing clay–silt–sand mixtures on the basis of the proportions of each constituent (see Fig. 4.3). Each side of the triangle is divided into 100 parts, representing the percentages of clay, silt and sand. A point within the triangle indicates the percentages of these three constituents, all of which total 100%.

The triangular chart originally introduced by the US Bureau of Reclamation used the term 'loam', which is not recognised by the British Standard classification system. The version shown in Fig. 4.3 uses the British Standard terminology.

The triangular chart shown in Fig. 4.3 is not appropriate for soils containing gravels, except when considering only the clay–silt–sand fractions. If gravelly soils are to be classified in this way, two different charts would be required, one (as shown) omitting gravel and a second chart omitting clay. Perhaps all four elements could be represented simultaneously in three dimensions in the form of a tetrahedron, but this would stretch the imagination too far.

4.4 APPLICATIONS

4.4.1 Engineering Classification

The analysis of soils by particle size provides a useful engineering classification system for soils, provided that the limitations referred to in Section 4.3.1 are taken into account.

Fig. 4.4   Samples of granular soils: (a) uniformly graded, (b) well graded, (c) poorly graded (gap graded)

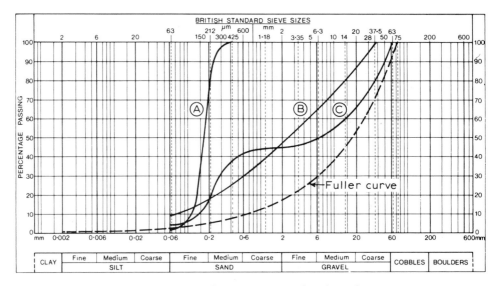

Fig. 4.5   Particle size curves — sands and gravels

For engineering purposes in Britain, soils are divided on the basis of particle size into the six categories shown in Table 4.1. Each heavy horizontal dividing line in the 'Designation' column separates materials of significantly differing engineering properties.

In practice, most natural soils do not fall entirely within one of these main size ranges, but consist of a mixture of two or more of the categories. Nevertheless grading curves provide a means by which soils can be classified and their engineering properties broadly assessed.

4.4.2   Classification of Sands and Gravels

Grading curves enable sands and gravels to be identified as being of three main types, based on the distribution of particle sizes, as follows. Typical examples are shown in Fig. 4.5.

(1)   *Uniform soils*, in which the majority of the grains are very nearly the same size (Fig. 4.4a). The grading curve is very steep, as shown by Curve A in Fig. 4.5, which represents a uniform sand. The uniformity coefficient is not much more than 1.0 (the lowest theoretically possible). A synonymous description is 'narrowly graded'.

Table 4.3 EFFECTIVE SIZES AND UNIFORMITY COEFFICIENTS

Grading curve	Effective size, D_{10} (mm)	D_{60} (mm)	Uniformity coefficient, D_{60}/D_{10}	Description
A	0.12	0.18	1.5	Uniform fine SAND
B	0.070	4.5	64	Well-graded silty SAND and GRAVEL
C	0.14	15	107	Poorly graded fine to medium SAND and GRAVEL
Fuller curve	0.66	24	36	'Ideal' grading
D	–	0.0025	–	CLAY
E	0.0051	0.060	12	Sandy SILT
F	–	0.019	–	Sandy and silty CLAY
G	<0.001 (estimated)	2.0	>2000	Gravelly sandy silty CLAY (boulder clay)

(2) *Well-graded soils*, containing a wide and even distribution of particle sizes (Fig. 4.4b). A well-graded silty sand and gravel is shown by curve B in Fig. 4.5. The smooth concave upward grading curve is typical of well-graded material.

The dashed curve in Fig. 4.5 represents a theoretical grading of a material in which the particles fit together in the densest possible state of packing (Fuller and Thompson, 1907). In this idealised material the largest particles just touch each other, while there are enough intermediate-size particles to occupy the voids between the largest without holding them apart; smaller particles occupy voids between intermediate sizes; and so on. The arrangement is illustrated diagrammatically in Fig. 3.6(c). The Fuller grading curve has the characteristic smooth shape referred to above, and is derived from the equation

$$P = 100 \sqrt{\frac{D}{D_{max}}}$$

where P is the percentage (by mass) of particles smaller than diameter D and D_{max} is the maximum particle size (75 mm in the example shown).

(3) *Poorly graded soils*, deficient in certain sizes (Fig. 4.4c). The grading curve has two distinct sections separated by a near-horizontal portion, as shown by curve C in Fig. 4.5. This is described as a gap-graded material, and in natural soils the deficiency usually occurs in the coarse sand-fine gravel range. The term 'poorly graded' can also be applied to any soil (including uniformly graded soil) which does not comply with the description 'well graded'.

Points A, B, C corresponding to the above grading curves are plotted on the triangular chart, Fig. 4.3. For B and C the gravel fraction is ignored and percentages of clay, silt and sand are enhanced to express them as percentages of material smaller than 2 mm. Effective sizes, and uniformity coefficients, are summarised in the upper part of Table 4.3.

4.4.3 Classification of Clays and silts

Soils consisting entirely of clay, or entirely of silt-size particles, are very rarely encountered. Most clays contain silt-size particles, and most material described as silt includes some clay or some sandy material, or both. A clay soil containing quite a high percentage of silt

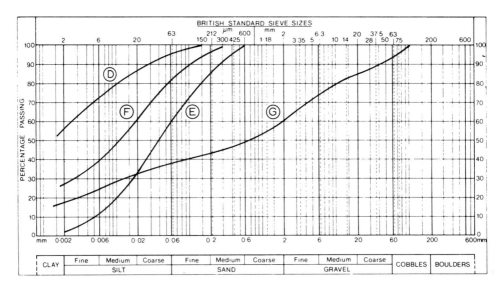

Fig. 4.6   *Particle size curves — silts and clays*

particles could have the properties of clay, and if so, would be described simply as 'clay' (see Chapter 7). This is illustrated by the large area designated 'clay' on the triangular chart, Fig. 4.3. Some typical grading curves are shown in Fig. 4.6.

Curve D is described as 'clay', although it consists of 56% clay and 44% silt size particles. However, if the clay sizes (smaller than 2 mm) consist of ground-down silt particles instead of clay minerals, this material would exhibit the properties of silt and would be described as such.

Curve E shows a well-graded soil consisting mainly of silt, with a clay fraction of less than 2%, and is described as 'sandy silt (with a trace of clay)'. Curve F has a mixture of clay, silt and sand. The effect of the clay would predominate, and the soil is described as 'silty clay with sand'. Curve G represents a well-graded soil containing particles of all sizes from cobbles down to clay. This type of soil is found, for example, in glacial till, and is often loosely called 'boulder clay', which in geological terminology is considered to be inaccurate. However, 'boulder clay' is used in this book to mean a soil, usually of glacial origin, which contains enough clay to give it cohesion and is well-graded from clay through to gravel or cobble-size particles. 'Boulder clay' in this sense does not necessarily contain boulders, but it does have the engineering characteristics of a clay. The material represented by curve G could be described as 'gravelly sandy silty clay'.

Points D, E, F representing these three grading curves are plotted in the triangular chart, Fig. 4.3. Point G relates only to the material smaller than 2 mm in the boulder clay. Effective sizes and uniformity coefficients for gradings E and G are summarised in the lower part of Table 4.3. The '10% passing' sizes for gradings D and F are outside the range of the grading chart, so effective size and uniformity coefficient for these two examples cannot be determined.

As indicated in Section 4.3.1, the grading characteristics revealed above do not furnish a complete description of cohesive soils. The physical behaviour of a clay soil is more important than its particle size distribution, and for this the Atterberg limits provide information that is more significant (see Chapter 2).

4.4.4   Engineering Practice

Some applications of particle size analysis in geotechnology and construction are as follows.

Selection of fill materials   Soils used for the construction of embankments and earth dams, in addition to other requirements, are required to be within specified limits as defined by particle size distribution curves. The various zones of an earth dam, for instance, each require different grading characteristics.

Road sub-base materials   Each layer of a road or airfield runway sub-base must comply to a particular grading specification in order to provide a mechanically stable foundation.

Drainage filters   The grading specification for a filter layer must be related in a certain way to the grading of the adjacent ground, or of the next filter layer.

Groundwater drainage   The drainage characteristics of the ground are to a large extent dependent upon the proportion of fines (silt and clay size particles) present in the soil.

Grouting and chemical injection   The most suitable grouting process to be used in soils, and the extent to which the ground can be impregnated, depend mainly on the grading characteristics of the soil.

Concreting materials   Sands and gravels for use as concrete aggregates are divided into various types on the basis of zones on a particle size distribution chart. In the exploration for sand and gravel resources, particle size analysis is the main criterion for selection of sites for potential development.

Dynamic compaction   In some situations poor ground conditions can be improved by dynamic compaction, and a particle size analysis can give an indication of the feasibility of this process,

4.5   PRACTICAL ASPECTS

4.5.1   Choice of Procedure

The procedures to be used for determining the particle size distribution in soils depend upon:

(1)   Size of the largest particles present.
(2)   Range of particle sizes.
(3)   Soil characteristics.
(4)   Stability of soil grains.

Item (1) determines the size of sample required. Item (2) determines the method to be used, such as whether or not a sedimentation test is required. The complexity of the testing is governed by (3) and depends on whether the soil is granular and 'clean' (no fines); predominantly granular with fines; or decidedly cohesive. The initial preparation of the soil for testing is dependent on (4), because special care is needed if particles are easily broken down (see Section 1.5.4).

The procedures to be described are as follows:

Simply dry sieving.
Composite dry sieving.
Wet sieving.
Special procedures for 'boulder clay' types of soil.
Sedimentation by hydrometer measurements.
Sedimentation by pipette analysis.
Composite sieving and sedimentation.

For clean sands and gravels (i.e. those containing no silt or clay) dry sieving can be used. If the material consists of sand sizes only, this can be a straightforward operation on a sample of convenient size. If the sample contains particles of medium to large gravel size, a larger sample will be required initially, but this can be reduced by quartering at some

Table 4.4   MINIMUM QUANTITIES FOR PARTICLE SIZE TESTS

Maximum size of material present in substantial proportion retained on BS sieve (mm)	Minimum mass of sample to be taken for sieving	Source
Pass 2 mm or smaller	100 g	Based on
3.35	150 g	BS1377: Part 2:1990:9.2.3
5	200 g	(Table 3)
6.3	200 g	
10	500 g	
14	1 kg	
20	2 kg	
28	6 kg	
37.5	15 kg	
50	35 kg	
63	50 kg	
75	70 kg	
100	150 kg	Author's suggestion
150	500 kg	
200	1000 kg	

stage of the sieving process. This is referred to as composite sieving. If larger particles (cobbles) are present, a very large sample will be needed, which may be quartered two or three times at suitable stages. The quantity of material required is discussed in Section 4.5.2.

Soils containing silt or clay must first be washed to remove the fine particles which pass a 63 μm sieve. This is known as wet sieving. The retained material can then be dry sieved. Since the fine material is usually run to waste with the washing water, the total quantity of fines can be determined only from the difference between the unwashed and washed masses after drying. Composite sieving is nearly always necessary unless sand is the coarsest material present.

Predominantly clayey and silty soils containing sand are first subjected to a pretreatment and dispersion process, and then washed through a 63 μm sieve. Material passing the 63 μm sieve is collected and used for a sedimentation test. The material retained is dry sieved, and the sieving curve added to the sedimentation curve to give the grading curve for the whole sample.

Cohesive soils containing gravel or larger particles require special treatment, which is described separately.

A sieving analysis should not be started without first considering which of the described procedures will be the most appropriate for the type of soil to be tested. When the procedure has been selected, sketch out the sequence of operations as a flow diagram, in the manner shown in Figs. 4.10, 4.13, 4.16, 4.18, 4.21 or 4.24. If the procedure is subsequently modified, amend the flow diagram accordingly. Calculations should be set out in a manner similar to the flow diagram, as shown in Figs. 4.11, 4.14, 4.19, 4.22 or 4.25. The sieve analysis sheet used for recording the masses retained on the sieves (Fig. 4.9) may not be suitable for this purpose except for the simplest type of calculations.

4.5.2   Quantity of Material for Test

For clays, silts and sands the size of particles are such that a sample of about 100 g is sufficient for particle size tests to be representative. However, with gravel-size particles much larger quantities are required for representative results to be obtained. If the size of the sample is too small, the inclusion or exclusion of only a few large particles can influence the whole particle size analysis considerably.

BS1377: Part 2:1990: clause 9.2.3 (Table 3) specifies the minimum quantity of material to be taken for sieve analysis. These requirements are given in Table 4.4. This relates to the

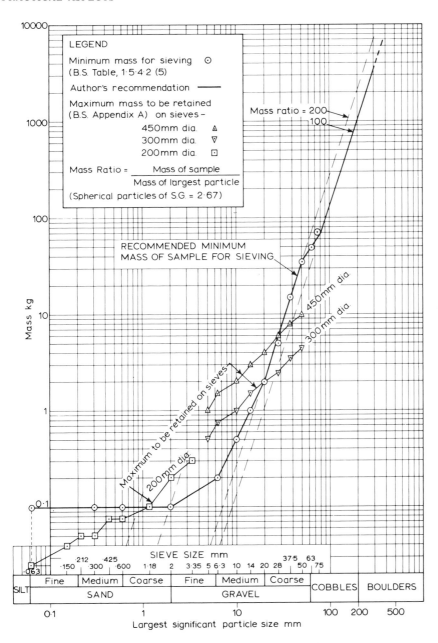

Fig. 4.7   Minimum quantity of material required for sieving

maximum size of particle present 'in substantial proportion' (more than 10% of the total sample). These minimum quantities are plotted against the size of particles, to a log scale, in Fig. 4.7. The British Standard table only goes as far as 63 mm — that is, the upper end of the gravel range.

When larger material (cobbles) is present, a useful rule is to start with a sample whose dry mass is not less than 100 times the mass of the largest particle. This relationship is indicated by the solid line on the graph in Fig. 4.7. The graph given in Fig. 10 of BS1377: Part 2:1990 follows a parallel line, dashed in Fig. 4.7, corresponding to a sample mass of

200 times the mass of the largest particle. These lines are based on the assumption that particles are spherical, with a particle density of $2.67\,Mg/m^3$.

A general rule recommended here is indicated graphically by the solid line in Fig. 4.7. This complies with the requirements of the BS below 63 mm, and above that size represents a mass ratio of about 100.

It is evident that the mass of sample required increases rapidly with increasing particle size in the cobbles range. For particles up to 200 mm across, about 1000 kg (1 tonne) of material is required. For 300 mm particles this increases to 3 tonnes. A method of dealing with very large samples of this kind of material is described in Section 4.6.3.

4.5.3   Dispersing Agents

A dispersing agent is used with a soil suspension in water in order to ensure separation, or dispersion, of discrete particles of soil, especially in the silt to clay range. A small quantity of soluble chemical is added before the sedimentation test, usually in the form of a measured quantity of a prepared stock solution.

Numerous substances have been tried for use as dispersing agents, many of them very successfully for most soils. They include

Sodium polyphosphate	Sodium hydroxide
Sodium tripolyphosphate	Sodium silicate
Sodium hexametaphosphate	Tannic acid
Sodium tetraphosphate	Starch
Sodium oxalate	Trisodium phosphate
Sodium carbonate	Tetrasodium phosphate
Sodium bicarbonate	

For most purposes it has been found that sodium hexametaphosphate (known commercially as Calgon) is one of the most suitable and convenient dispersants. The stock solution recommended by the British Standard is made up of

35 g sodium hexametaphosphate
7 g sodium carbonate
Distilled water to make 1 litre of solution

This stock solution is referred to as the *standard dispersant*, and is added to the water used for sieving or sedimentation in the proportions specified. A fresh stock solution should be prepared each month, because it is unstable, and will not keep for any length of time. The date of preparation should be marked clearly on the container, which should have a tight-fitting stopper.

The standard dispersant may not be fully effective with some tropical residual soils (defined in Chapter 7). Alternative dispersants that have been used are trisodium phosphate or tetrasodium phosphate, but a more concentrated solution (2 or 3 times stronger) of the standard dispersant may be more effective. The alternative dispersants are not usually effective with other types of soil.

Incomplete dispersion results in the formation of relatively large crumbs or 'flocs' of soil particles (aggregations of smaller particles) which in a sedimentation test fall relatively rapidly through the water, leaving a clear layer above the suspension. If this occurs, a higher concentration of dispersing agent may need to be used. Incomplete dispersion, especially in tropical residual soils, can be caused by drying the soil (even air-drying) before testing, giving erroneous results. (See Geological Society Report, 1990).

Use of a dispersant is also necessary before wet sieving of soils containing clay or silt. For this purpose the British Standard specifies covering the soil with water containing 2 g per litre of sodium hexametaphosphate, or 50 ml of 'standard dispersant' solution per litre (see Sections 4.6.6 and 4.6.7).

Table 4.5(a).   METRIC SIEVES (BS)

Construction	Aperture size: Full Set (A)	'Standard' set (B)	'Short' set C)	Suitable sieve diameters 450 mm	300 mm	200 mm
Perforated	75 mm			+		
steel place	63	+	+	+		
(square hole)	50			+		
	37.5	+		+	+	
	28			+	+	
	20	+	+	+	+	
	14				+	
	10	+			+	
	6.3	+	+		+	
	5				+	
	3.35	+			+	+
	2	+	+	(+)	+	+
	1.18	+				+
	600 μm	+	+	(+)		+
	425					+
	300	+				+
	212		+			+
	150	+				+
	63	+	+	(+)	+	+
Lid and receiver	+	+	+	+	+	+
	19 sieves	13 sieves	7 sieves			

Many other test sieves, up to 125 mm and down to 38 μm aperture size, are available. They are manufactured to BS410:1969
(+) indicates sieves useful for wet sieving of large samples

Table 4.5(b).   ISO SIEVES

Aperture size (mm)	'Alternate' set	'Short' set	Aperture size	'Alternate' set	'Short' set
90			2 mm	+	+
63	+	+	1.4 mm		
45			1 mm	+	
31.5	+		710 μm		+
22.4		+	500 μm	+	
16	+		355 μm		
11.2			250 μm	+	
8	+		180 μm		+
5.6		+	125 μm	+	
4	+		90 μm		
2.8			63 μm	+	+

4.5.4   Sample Selection and Preparation

The general procedure for the selection and preparation of a small representative sample for test from a large sample is outlined in Section 1.5. Detailed procedures for sample preparation are given in the section describing each type of test, as appropriate.

4.5.5   Selection of Sieves

The complete range of sieves specified by the British Standard is given in Table 4.5(a). It is not necessary to use every sieve for every test, but the sieves used should adequately cover the range of aperture sizes for each particular soil.

Fig. 4.8. *Test sieves*

Three selections of sieves are suggested in Table 4.5(a).

(A)   Full set — i.e. every sieve listed in the BS (19 sieves).
(B)   'Standard' set, which cover most requirements (13 sieves).
(C)   'Short' set — i.e. those sizes which subdivide soil types. These are the sizes indicated by the heavy lines on the particle size sheet (Fig. 4.1), and comprise 63 μm, 20 mm, and their submultiples (7 sieves).

For most purposes the 'standard' set is sufficient to provide a reasonable grading curve, but if the soil is uniformly graded, whether as a whole or over part of the range (giving a steeply inclined grading curve) the use of intermediate sieves will give a better-defined curve. An extra sieve can easily be used after completion of the test if the material retained on each sieve is kept separate. A full set should be used for soils that might be gap-graded.

The largest aperture sieve necessary is the one through which all the material just passes. The mass retained in this sieve is zero, and is the size corresponding to 100% passing. The next smallest aperture sieve is the first one to retain material in the test.

The range of recommended ISO sieve aperture sizes is given in Table 4.5(b). This consists of 22 sieves, but for many requirements use of alternate sieves (those which include 2 mm) would be sufficient. A 'short' set of seven sieves is also indicated in Table 4.5(b).

Test sieves are made in three standard diameters — namely 450 mm, 300 mm, 200 mm — and typical sieves are shown in Fig. 4.8. The diameters normally available are indicated in Table 4.5(a). The diameter selected should be appropriate to the quantity of material to be sieved. It is usual to riffle the material immediately before changing to sieves of smaller diameter.

4.5.6   Calibration and Checking of Sieves

CALIBRATION

Perforated metal plate test sieves with aperture sizes of 6.3 mm and upwards should be calibrated at least twice a year (depending on use) by measuring selected apertures with calibrated vernier capilers. The pattern of measurement should be in accordance with BS410, and should cover two rows of apertures along diameters at right angles and one row in the direction of a hole diagonal. The length and breadth of each hole should be measured and recorded to the nearest 0.05 mm. If the dimensions of all apertures fall within the maximum tolerance given in BS410, the sieve is satisfactory; otherwise it should be replaced.

Test sieves with aperture sizes less than 6.3 mm (including all woven wire sieves) may be calibrated by using a dry reference material which is sieved first through the working sieves, then through the reference set of sieves. A controlled sieving procedure, using a mechanical shaker, must be used. If the mass retained on any working sieve differs by more than 5% from the mass retained on the corresponding reference sieve, that working sieve should be replaced.

The reference material should be a dry quartz sand consisting of rounded or sub-rounded particles. A different portion should be used with each sieve aperture size, and approximately 50% should be retained on each sieve being checked.

Alternatively an artificial reference sample (such as glass ballotini) can be used.

ROUTINE CHECKS

Test sieves should be inspected for defects before each use. A more detailed examination should be made at regular intervals to cover signs of wear, warping, tears, splits, holes, blockages and any other defects in the mesh. The frame should also be examined for damage and to ensure proper nesting. The lid and receiver of a set of sieves should be included in the examination.

4.6 SIEVING PROCEDURES

4.6.1 Simple Dry Sieving (BS1377: Part 2:1990:9.3)

SCOPE

Dry sieving is the simplest of all methods of particle size analysis. The apparatus used, the test procedure and the method of calculation are described here in some detail because they are relevant to all other sieving methods, and are used after washing in the wet sieving method.

According to the British Standard, dry sieving may be carried out only on materials for which this procedure gives the same results as the wet-sieving procedure. This means that it is applicable only to clean granular materials, which usually implies clean sandy or gravelly soils — that is, soils containing negligible amounts of particles of silt or clay size. Normally the wet-sieving procedure (Section 4.6.4) should be followed for all soils.

If particles of medium gravel size or larger are present in significant amounts, the initial size of the sample required may be such that riffling is necessary at some stage to reduce the sample to a manageable size for fine sieving. The procedure is then referred to as 'composite sieving', which is described in Section 4.6.2.

APPARATUS

(1) Test sieves to BS410. The aperture sizes specified in BS1377:1990 are listed in Table 4.5(a). A set of sieves of each diameter requires a lid and receiver. It is advantageous to keep one set of sieves of each diameter for dry sieving and another set for wet sieving. Comments on selection of sieves are given in Section 4.5.5.
(2) Mechanical sieve shaker (optional), preferably with timing device.
(3) Balances appropriate to the quantities of material to be used.
(4) Riffle-box.
(5) Drying oven, 105–110°C.
(6) Sieve brushes, double-ended, brass or nylon bristles.
(7) Metal trays.
(8) Rubber pestle and mortar.
(9) Scoop, and miscellaneous small tools.
(10) Sieve analysis work sheet (Fig. 4.9).

SIEVE ANALYSIS OF SOIL (WET*/DRY* SIEVING)

Job: *2567*   Operator: *A. B. Smith*
Sample No.: *3/12*   Date: *6.12.78*
Site: *Elmbridge*   Depth *3·75 m*   Description: *Light brown*
                                                            fine to medium
Total mass of dry sample: *500* g   *sand*

BS test sieve size	Mass retained	Mass retained	Total Mass retained	Per cent retained	Total passing	Remarks
	g	g		%	%	
75 mm						
63 mm						
50 mm						
37.5 mm						
28 mm						
20 mm						
Passing 20 mm						

Riffled Sample Passing 20 mm

	g	g		%	%	
14 mm						
10 mm						
6.3 mm						
Passing 6.3 mm						

Riffled Sample Passing 6.3 mm *500 g*

	g			%	%	
5 mm						
3.35 mm						
2 mm						
1.18 mm	*0*			*0*	*100*	
600 micron	*20*			*4*	*96*	
425 micron						
300 micron	*170*			*34*	*62*	
212 micron						
150 micron	*235*			*47*	*15*	
63 micron	*71*			*14.2*	*0·8*	
Passing 63 micron	*3·5*			*0·7*		
Total	*499·5*			*99.9*		

* *Delete the inappropriate word*

Fig. 4.9   Sieve analysis data sheet

PROCEDURAL STAGES
(1) Select and prepare test specimen.
(2) Oven dry, cool, weigh.
(3) Select sieves,
(4) Pass through sieves.
(5) Weigh each size fraction.
(6) Calculate cumulative percentages passing each size.
(7) Plot grading curve.
(8) Report results.

The sieving process is illustrated diagrammatically in Fig. 4.10.

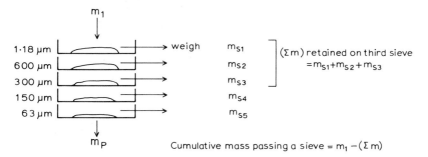

Fig. 4.10 Sequence diagram, simple dry sieving

PROCEDURE

(1) Selection

The specimen to be used for the test is obtained from the original sample by riffling, or by subdivision using the cone-and-quarter method (Section 1.5.5). The appropriate minimum quantity of material depends upon the maximum size of particles present, and is indicated in Table 4.4. See also Section 4.5.2.

(2) Preparation

The specimen is placed on a tray and is allowed to dry, preferably overnight, in an oven maintained at 105–110°C. After drying to constant weight, the whole specimen is allowed to cool, and is weighed to an accuracy within 0.1% or less of its total mass (m_1).

(3) Selection and assembly of sieves

The sieves to be used are selected to suit the size of sample and type of material, as discussed in Section 4.5.5. In this example five sieves from the 'standard' set (Table 4.5(a)) have been used.

Before each use, inspect every sieve for defects, such as tears, splits, large holes. Ensure that the sieve frames are true, and that they fit snugly one inside the other to prevent escape of particles and dust. Sieves are nested together with the largest aperture sieve at the top, and the receiving pan under the smallest aperture sieve at the bottom.

(4) Sieving

The dried soil sample is placed in the topmost sieve and is shaken for long enough for all particles smaller than each aperture size to pass through. This can be achieved most conveniently by using a mechanical sieve shaker, as described in paragraph (A) below. If a shaker is not available, sieving can be done by hand, using each sieve separately, as described in paragraph (B). General notes on sieving procedures which apply to either method are given in paragraph (C).

(A) *Use of mechanical shaker*   The whole nest of sieves with receiving pan is placed in the shaker; the dried soil is placed in the top sieve, which is then fitted with the lid; and the sieves are securely fastened down in the machine (see Figs. 1.26 and 1.27). Agitation in the shaker should be for a minimum period of 10 min. Some shakers have a built-in timing device which can be preset to switch off the motor automatically after the desired period.

(B) *Hand sieving*   The largest aperture sieve to be used is fitted with a receiver, and the dried sample is placed on the sieve. The lid should be fitted to prevent escape of dust. The sieve must be agitated by shaking so that the particles roll in an irregular motion, and until no more particles pass through the openings.

Table 4.6. MAXIMUM MASS TO BE RETAINED ON EACH SIEVE

Sieve aperture	Maximum mass		
	450 mm diameter sieves (kg)	300 mm diameter sieves (kg)	200 mm diameter sieves (g)
50 mm	10	4.5	
37.5	8	3.5	
28	6	2.5	
20	4	2.0	
14	3	1.5	
10	2	1.0	
6.3	1.5	0.75	
5	1.0	0.5	
3.35			300
2			200
1.18			100
600 μm			75
425			75
300			50
212			50
150			40
63			25

The material in the receiver is transferred to a tray and the receiver is fitted to the next sieve in the series. The process is repeated, using the material from the tray, and so on through all the sieves to be used, down to and including the 63 μm sieve.

The foregoing procedure is the ideal, but in practice several sieves can often be nested together for shaking.

(C) *Notes on procedure*   Whether shaking is carried out in a machine or by hand, it is essential to ensure at each stage that sieving is complete. On the other hand, an unduly long period of sieving must be avoided, because this would give particles additional opportunity to pass through any openings which may be slightly oversize. The British Standard specifies a minimum sieving time of 10 min, but a maximum time of 15 or 20 min should be standard practice in order to obtain consistent results.

The material retained on each sieve should be examined to make sure that only individual particles are retained. Any agglomerations of particles not naturally cemented together should be broken down with a rubber pestle in a mortar and resieved. On the larger aperture sieves individual particles may be assisted through by hand placing, but they must not be pushed through.

Sieves must not be overloaded. The mass retained on each sieve must not exceed the masses given in Table 4.6. To prevent overloading, a large test sample should be split into two or more parts before sieving. The material on an individual overloaded sieve should be split into parts each not exceeding the mass referred to in Table 4.6 and re-sieved separately.

With continual use, abrasion causes wear of the sieve material (whether metal plate or woven wire fabric), resulting in the apertures increasing in size. A periodic check should be made as part of the regular calibration procedure described in Section 4.5.6.

(5) *Weighing*

The material retained on each sieve is transferred in turn to the pan of a suitable balance, or to a weighed container. Any particles lodged in the apertures of the sieve should be carefully removed with a sieve brush, the sieve being first placed upside-down on a tray or a clean sheet of paper. These particles are added to those retained on the sieve.

Table 4.7 SIMPLE DRY SIEVING* METHODS OF CALCULATION
Initial sample mass $= m_1$

Sieve	Method (A)			Method (B)	
	Mass retained	% retained	% retained	Mass passing	% passing
(0)	0	0	100	m_1	100
(1)	m_{s1}	$\dfrac{m_{s1}}{m_1} \times 100 = R_1$	$100 - R_1 = P_1$	$m_1 - m_{s1} = a$	$\dfrac{a}{m_1} \times 100$
(2)	m_s2	$\dfrac{m_{s2}}{m_1} \times 100 = R_2$	$P_1 - R_2 = P_2$	$a - m_{s2} = b$	$\dfrac{b}{m_1} \times 100$
(3)	m_{s3}	$\dfrac{m_{s3}}{m_1} \times 100 = R_3$	$P_2 - R_3 = P_3$	$b - m_{s3} = c$	$\dfrac{c}{m_1} \times 100$
(4)	m_{s4}	$\dfrac{m_{s4}}{m_1} \times 100 = R_4$	$P_3 - R_4 = P_4$	$c - m_{s4} = d$	$\dfrac{d}{m_1} \times 100$
	etc.		etc.		etc.

Weighing of each size fraction should be to an accuracy of 0.1% or better of the total initial test sample mass. The masses retained are recorded against the sieve aperture size on the particle size test work-sheet (Fig. 4.9). The mass (m_p) passing the 63 μm sieve is also measured and recorded.

In this example the arithmetic is made simple by using an initial sample mass of 500 g, which is more than would normally be used for this type of material. The masses shown as retained on the 300 μm and 150 μm sieves exceed those given in Table 4.6, therefore these fractions should have been sieved in several portions as explained in (4) (C) above.

(6) Calculations

In order to draw up a particle size distribution curve, or to tabulate the data, it is necessary to calculate the cumulative percentage (by mass) of particles finer than each sieve aperture size — that is, passing each sieve. This can be done in two ways.

(A) *British Standard Method*   The mass retained on each sieve is expressed as a percentage of the initial sample mass (m_1). The percentage passing each sieve is then calculated by successive subtraction of each 'percentage retained' in turn. The process is summarised in the left-hand section of Table 4.7, and is illustrated by the worked example in Fig. 4.11 (top). Note that after the first sieve, each percentage retained (R_n) must be subtracted from the previous percentage passing (P_{n-1}), not from 100, because we are calculating cumulative percentages.

(B) *Alternative Method*   The mass retained on the first sieve is substracted from the initial mass (m_1) to give the mass passing the first sieve. The mass retained on each subsequent sieve is subtracted from the mass passing the previous sieve to give the mass passing each sieve. Each mass passing is then expressed as a percentage of the initial mass. The process is summarised in the right-hand half of Table 4.7, and is illustrated by the worked example in Fig. 4.11 (bottom). Note that each mass retained is subtracted from the mass passing the *previous* sieve.

The author prefers method (B) for the following reasons.

(i)   Comparison of the calculated and measured masses passing the last sieve provides a check on both the weighings and the arithmetic.

Initial mass used: 500 grams = m_1

Method A

% passing sieve 'n' = [% passing previous sieve (n−1)] − [% retained on sieve 'n']

Sieve size mm	Mass retained g	Percentage retained	Percentage passing
1.18	0	0	100%
0.600	20	$\frac{20}{500}$ x 100 = 4	100 − 4 = 96%
0.300	170	$\frac{170}{500}$ x 100 = 34	96 − 34 = 62%
0.150	235	$\frac{235}{500}$ x 100 = 47	62 − 47 = 15%
0.063	71	$\frac{71}{500}$ x 100 = 14.2	15 − 14.2 = 0.8%
Pass (m_p)	3.5	(Check:	$\frac{3.5}{500}$ x 100 = 0.7%)

Method B

mass passing sieve 'n' = [mass passing previous sieve (n−1)] − [mass retained on sieve 'n']

Sieve size mm	Mass retained g	Cumulative mass passing g	Percentage passing
1.18	0	500 − 0 = 500	100%
0.600	20	500 − 20 = 480	$\frac{480}{500}$ x 100 = 96%
0.300	170	480 − 170 = 310	$\frac{310}{500}$ x 100 = 62%
0.150	235	310 − 235 = 75	$\frac{75}{500}$ x 100 = 15%
0.063	71	75 − 71 = 4	$\frac{4}{500}$ x 100 = 0.8%
Pass (m_p)	3.5	(Discrepancy 4 − 3.5 = 0.5)	

Fig. 4.11 Calculation procedures, simple dry sieving

(ii)   Observed readings, instead of calculated percentages, are successively subtracted. There is a tendency to round off the latter, and this could lead to cumulative errors.

Both calculation procedures are shown by the examples given in Fig. 4.11. The discrepancy of 0.5 g between the calculated and weighed mass passing the 63 μm sieve is insignificant. If the difference is appreciable, the calculations, and if necessary the weighings, should be rechecked.

(7) *Plotting*

The special graph sheet referred to in Section 4.3.4 is used for plotting the particle size distribution curve. On this sheet the sieve sizes are marked by vertical dashed lines. The

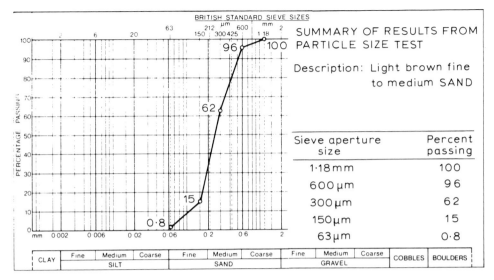

Fig. 4.12 *Results from particle size test and grading curve*

percentage smaller than any given size (i.e. the percentage passing each sieve) is plotted as the ordinate, to a linear scale, against 'sieve aperture size', and the points are connected with a smooth curve or by straight lines. The curve from the example given above is shown in Fig. 4.12, and the calculated percentages passing are shown against each plotted point. Note that the first point plotted represents 100% passing.

Several particle size curves may be drawn on one sheet, but different symbols should be used for both the plotted points and the connecting lines. Each curve should be clearly labelled, as in the examples shown in Figs. 4.5 and 4.6. Up to four particle size curves may be conveniently plotted on one sheet.

(8) Reporting results

In addition to the particle size curve and the usual sample identification data, the sheet should include the visual description of the sample. This should be the description of the sample before testing, and modified as necessary as a result of the additional information revealed by the test result. Any material removed before sieving, such as vegetation or an isolated cobble, should be reported.

Tabulated data showing the percentages passing each sieve are sometimes required instead of, or in addition to, the grading curve. An example is given in Fig. 4.12, together with the description of the sample.

The method of test is reported as dry sieving in accordance with BS1377 Part 2:1990:9.3.

4.6.2 Composite Sieving

SCOPE

The simple procedure described above should be used only with clean sands, or with the washed sand-size fraction of other soils.

When gravel is presented in appreciable quantity, it is necessary to start with a large sample and then at some stage subdivide it to give a smaller sample which is more manageable when the finer sieves are used. This process is referred to here as composite sieving. The mass of sample required initially depends upon the size of the largest particles present, as

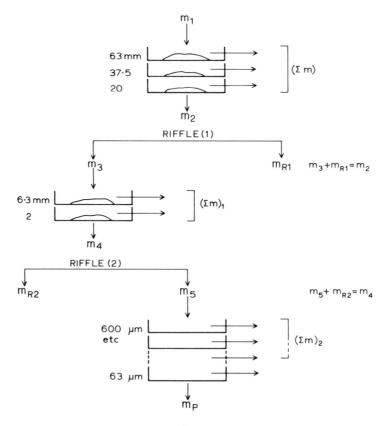

Fig. 4.13   *Sequence diagram, composite sieving*

indicated in Table 4.4 and Fig. 4.7. It may be necessary with some materials to subdivide twice, or even three times. Each subdivision must be made by proper riffling or cone-and-quartering, using the procedures described in Ssection 1.5.5 and avoiding segregation of larger particles. The term 'riffle' is used below to indicate subdivision of the sample by any appropriate method.

Riffling to reduce the sample to a smaller quantity may be done after passing the 20 mm and 6.3 mm sieves, as described in BS1377, but it is sometimes more appropriate to make the second riffle after passing the 5 mm or 2 mm sieve. The point at which riffling is done is immaterial, and should be judged to suit the sample under test. In any case, it is convenient to riffle before changing to sieves of a smaller diameter.

The process of composite sieving is shown diagrammatically in Fig. 4.13, which for the purpose of illustration omits some sieves and provides for riffling after passing the 20 mm and 2 mm sieves.

APPARATUS

The apparatus required is the same as listed in Section 4.6.1 for simple dry sieving. In addition, two large metal trays are required on which to mix and quarter the material.

PROCEDURE

The initial dry-sieving procedure, using the 'coarse gravel' size sieves down to 20 mm aperture, is similar to that for simple sieving (Section 4.6.1). Overloading of individual sieves must be avoided (see Table 4.6). Either hand shaking or the shaking machine may be used. The material passing through the 20 mm sieve is collected together and weighed. It is then riffled to a convenient size for the remaining sieves, and is weighed carefully before proceeding. Sieving through the smaller sieves using the riffled portion is carried out as for simple sieving. A second riffling will be necessary after passing, the 2 mm sieve if the mass of soil at that point is substantially greater than 150 g.

CALCULATONS

The percentage passing each sieve before riffling is calculated as for simple sieving (Section 4.6.1). After riffling the calculation is modified as explained below. The following symbols are used.

Initial mass of sample $= m_1$
Mass of soil passing the 20 mm sieve $= m_2$
Mass of riffled fraction of m_2 used for subsequent sieves $= m_3$
Mass of soil passing the 2 mm sieve $= m_5$
Mass of riffled fraction of m_5 used for subsequent sieves $= m_6$

Method (A)

The mass retained on each sieve after riffling has to be corrected, by multiplying by the factor m_2/m_3, to give the equivalent mass before riffling. The corrected masses retained are then expressed as a percentage of m_1, and the percentages passing each sieve are calculated by successive subtractions, as before. The first calculation is a subtraction from the percentage passing the 20 mm sieve.

If a second riffle is necessary (in this example after passing the 2 mm sieve), a further correction factor equal to m_5/m_6 has to be applied, so that each mass retained is multiplied by

$$\frac{m_2}{m_3} \times \frac{m_5}{m_6}$$

before calculating the percentages passing as above.

The process is summarised in Table 4.8.

Method (B)

The mass retained on the first sieve after riffling is subtracted from the riffled mass, m_3, to give the mass passing the first sieve. The mass retained on each subsequent sieve is subtracted from the mass passing the previous sieve. The mass passing each sieve is then multiplied by the factor m_2/m_3 to give the corrected mass passing, which is then expressed as a percentage of the initial mass m_1.

If a second riffle is necessary, the procedure is similar except that the mass passing each sieve is multiplied by a factor equal to

$$\frac{m_2}{m_3} \times \frac{m_5}{m_6}$$

to obtain the corrected mass passing.

The process is summarised in Table 4.9 and is illustrated by the worked example in Fig. 4.14.

Alternatively the mass retained on each sieve after riffling can be multiplied by the appropriate riffle factor first, and then subtracted from the previous mass passing, as in the simple sieving calculations.

The grading curve for this example is given in Fig. 4.15, and indicates where riffling took place. This material is sand and gravel with cobbles and virtually no silt. The presence of

Table 4.8. COMPOSITE SIEVING* CALCULATION METHOD (A)

Initial sample mass $= m_1$

Sieve	Mass retained	Corrected mass retained	% retained	% passing
(0)	0	0	0	100
(1)	m_{s1}	m_{s1}	$\dfrac{m_{s1}}{m_1} \times 100 = R_1$	$100 - R_1 = P_1$
(2)	m_{s2}	m_{s2}	$\dfrac{m_{s2}}{m_1} \times 100 = R_2$	$P_1 - R_2 = P_2$
etc.			etc.	etc.
20 mm				$P(20)$

First riffle

Mass passing sieve $(5) = m_2$

Mass after riffling $\quad = m_3$ $\qquad\qquad \dfrac{m_2}{m_3} = x$

Sieve	Mass retained	Corrected mass retained	% retained	% passing
(6)	m_{s6}	$x m_{s6}$	$\dfrac{x m_{s6}}{m_1} \times 100 = R_6$	$P(20) - R_6 = P_6$
(7)	m_{s7}	$x m_{s7}$	$\dfrac{x m_{s7}}{m_1} \times 100 = R_7$	$P_6 - R_7 = P_7$
(8)	m_{s8}	$x m_{s8}$	$\dfrac{x m_{s8}}{m_1} \times 100 = R_8$	$P_7 - R_8 = P_8$
etc.			etc.	etc.
(10)				$P(10)$

Second riffle

Mass passing sieve $(10) = m_5$

Mass after riffling $\quad m_6$ $\qquad\qquad \dfrac{m_5}{m_6} x = y$

Sieve	Mass retained	Corrected mass retained	% retained	% passing
(11)	m_{s11}	$y m_{s11}$	$\dfrac{y m_{s11}}{m_1} \times 100 = R_{11}$	$P_{10} - R_{11} = P_{11}$
(12)	m_{s12}	$y m_{s12}$	$\dfrac{y m_{s12}}{m_1} \times 100 = R_{12}$	$P_{11} - R_{12} = P_{12}$
(13)	m_{s13}	$y m_{s13}$	$\dfrac{y m_{s13}}{m_1} \times 100 = R_{13}$	$P_{12} - R_{13} = P_{13}$
etc.			etc.	etc.
63 μm	$m(63)$	$y m(63)$	$\dfrac{y m(63)}{m_1} \times 100 = R(63)$	$P_n - R(63) = P(63)$
Passing 63 μm	m_p	Check:	$\dfrac{y m_p}{m_1} \times 100$ should equal $P(63)$	

cobbles exceeding 63 mm size called for a large sample, but washing has not been included in this example.

4.6.3 Cobbles and Boulders

An extreme example of composite sieving is the particle size analysis of non-cohesive material containing large cobbles or boulders, such as river terrace deposits. For this type of material

Initial dry mass: 15 kg = 15 000 g = m_1

Sieve size	Mass retained	Cumulative mass passing	Percentage passing
75 mm	0	m_1 = 15 000	100%
63	300	15 000 − 300 = 14 700	$\dfrac{14\ 700}{15\ 000} \times 100$ = 98.0%
37.5	900	14 700 − 900 = 13 800	$\dfrac{13\ 800}{15\ 000} \times 100$ = 92.0%
20	1250	13 800 − 1250 = 12 550 = m_2	$\dfrac{12\ 550}{15\ 000} \times 100$ = 83.7%
		↓ RIFFLE (1)	
		$m_3 = 2275$ (10 275) = m_{R1}	
6.3	550	2275 − 550 = 1725	$\dfrac{1725}{2275} \times \dfrac{12\ 550}{15\ 000} \times 100$ = 63.4%
2	450	1725 − 450 = 1275 = m_5	$\dfrac{1275}{2275} \times 83.7$ = 46.9%
		↓ RIFFLE (2)	
		$m_6 = 200$ (1075) = m_{R2}	
600 μm	90	200 − 90 = 110	$\dfrac{110}{200} \times \dfrac{1275}{2275} \times \dfrac{12\ 550}{15\ 000} \times 100$ = 25.8%
212	67	110 − 67 = 43	$\dfrac{43}{200} \times 46.9$ = 10.1%
63	39	43 − 39 = 4	$\dfrac{4}{200} \times 46.9$ = 0.9%
Pass 63	$m_p = 4$		

Fig. 4.14   Calculations, composite sieving (Method B)

it may be necessary to start with a sample consisting of a tonne or several tonnes excavated from a trial pit. The quantity required can be assessed from Fig. 4.7. The procedure is illustrated in Fig. 4.16.

As the material is excavated, it is placed on a platform or tarpaulin, and pieces larger than 75 mm are set aside. If necessary, they are brushed clean of adhering fines, which must be returned to the main sample.

The large cobbles are gauged for size with a series of square wood or metal frames. Suitable aperture sizes are 100, 150, 200, 300, 400 mm, and even larger, if necessary. Each size range is weighed on site with platform scales of, say, 250 kg capacity. It is essential to mount the platform scales on a rigid level base, and to shield the scales from hot sun, wind or rain.

The main sample is weighed in batches on the platform scales. The whole weighing operation should be repeated, if practicable, to ensure that there has been no mistake in obtaining the total mass of the whole sample. The percentage of each of the cobbles and boulders sizes can be calculated immediately if the material is, for all practical purposes, dry. If not, a correction for moisture content must be made later to obtain the total dry mass.

Table 4.9. COMPOSITE SIEVING* CALCULATION METHOD (B)
Initial sample mass $= m_1$

Sieve	Mass retained	Mass passing	Corrected mass passing	% passing
(0)	0	m_1	m_1	100
(1)	m_{s1}	$m_1 - m_{s1} = (a)$	(a)	$\dfrac{(a)}{m_1} \times 100$
(2)	m_{s2}	$(a) - m_{s2} = (b)$	(b)	$\dfrac{(b)}{m_1} \times 100$
	etc.	etc.	etc.	
(5)				

First riffle

Mass passing sieve $(5) = m_2$
Mass after riffling $\quad = m_3$ $\dfrac{m_2}{m_3} = x$

(6)	m_{s6}	$m_3 - m_{s6} = (f)$	$x(f)$	$\dfrac{x(f)}{m_1} \times 100$
(7)	m_{s7}	$(f) - m_{s7} = (g)$	$x(g)$	$\dfrac{x(g)}{m_1} \times 100$
(8)	m_{s8}	$(g) - m_{s8} = (h)$	$x(h)$	$\dfrac{x(h)}{m_1} \times 100$
	etc.	etc.		etc.
(10)				

Second riffle

Mass passing sieve $(10) = m_5$
Mass after riffling $\quad m_6$ $\dfrac{m_5}{m_6} = y$

(1)	m_{s11}	$m_5 - m_{s11} = (k)$	$xy(k)$	$\dfrac{xy(k)}{m_1} \times 100$
(12)	m_{s12}	$(k) - m_{s12} = (l)$	$xy(l)$	$\dfrac{xy(l)}{m_1} \times 100$
(13)	m_{s13}	$(l) - m_{s13} = (m)$	$xy(m)$	$\dfrac{xy(m)}{m_1} \times 100$
	etc.	etc.		
		(q)		
$63\,\mu\text{m}$	$m(63)$	$(q) - m(63) = (z)$		$P(63)$
Passing $63\,\mu\text{m}$	m_p	Check: m_p should equal (z)		

The main sample, consisting of minus 75 mm material, can then be riffled, with a large riffle-box, or by cone-and-quartering. A check should be made by visual inspection that the largest particles have been divided up in representative proportions. A riffled sample of at least 50 kg is required. For convenience it should be placed in two or more lined sacks or bags for transfer to the main laboratory. Each bag should not be too heavy for one man to handle. Leakage or loss of fines during transit must be avoided; therefore unlined hessian or similar sacks are not suitable.

The rest of the particle size analysis can take place in the laboratory, the riffled sample being used. Further subdivision will be necessary at subsequent stages. A worked example is not shown, but the procedure to be followed, and the calculations, are similar to those described in Section 4.6.2, allowance being made in the calculations for every riffling stage.

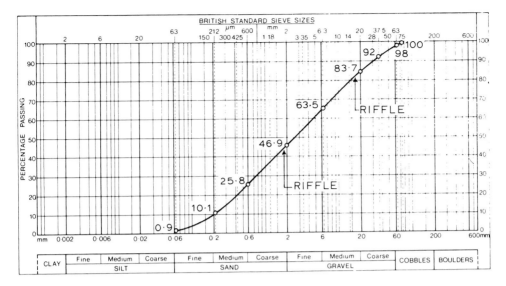

Fig. 4.15   *Grading curve, composite sieving*

If a sedimentation test is carried out on the fines fraction, a complete particle size curve ranging from cobbles or boulders down to the clay size can be constructed.

4.6.4   Wet Sieving — Fine Non-Cohesive Soils (BS1377: Part 2:1990:9.2)

SCOPE

If a soil contains silt or clay, or both, even in small quantities, it is necessary to carry out a wet sieving procedure in order to measure the proportion of fine material present. Even when dry, fine particles of silt and clay can adhere to sand-size particles and cannot be separated by dry sieving, even if prolonged. Washing is the only practicable means of ensuring complete separation of fines for a reliable assessment of their percentage. If clay is present, or if there is evidence of particles sticking together, the material should be immersed in a dispersant solution before washing, as described in Section 4.6.6 under soil category (2).

The procedure is described in detail below for non-cohesive soils containing little or no gravel.

APPARATUS

As for simple sieving (Section 4.6.1), with the addition of the following:

(9)   Evaporating dishes.
(10) Rubber tubing connected to water tap, fitted at the other end with a small spray such as the rose from a small watering can or hair spray (see Fig. 4.17).

PROCEDURAL STAGES (Non-cohesive soil)

(1)   Select and prepare test specimen.
(2)   Oven dry, cool, weigh.
(3)   Wash through 2 mm and 63 μm sieves.
(4)   Dry retained material.
(5)   Weigh.

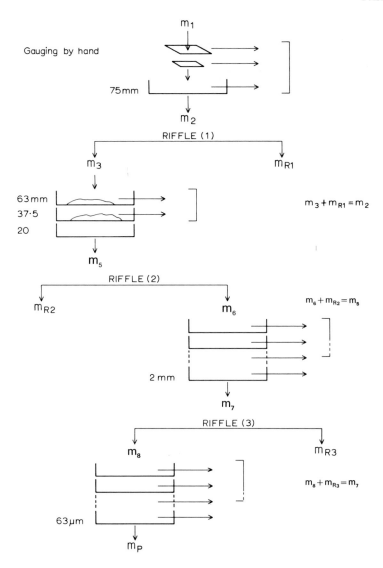

Fig. 4.16 *Sequence diagram, sieving with cobbles and boulders*

(6)   Pass through range of sieves.
(7)   Weigh each size fraction.
(8)   Calculate percentages passing each sieve and percentage of fines.
(9)   Plot grading curve.
(10) Report result.

The procedure is illustrated in Fig. 4.18.

PROCEDURE

(1) *Selection and Preparation*

As for dry sieving (Section 4.6.1).

Fig. 4.17   *Washing soil on 63 µm sieve*

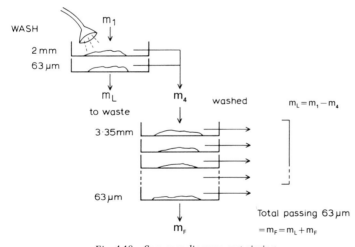

Fig. 4.18   *Sequence diagram, wet sieving*

(2)   *Oven drying, cooling, weighing*

As for dry sieving. The total initial dry mass (m_1) must be carefully measured and verified, because the fines will be washed away and their mass determined by difference. There will be no overall check by addition, as there is in dry sieving.

(3)   *Wash*

The 2 mm sieve is nested in the 63 µm sieve, but the lid and receiver are not used. An additional intermediate sieve may be included to protect the 63 µm sieve from overloading if the soil contains a high proportion of coarse or medium sand.

The soil is placed a little at a time on the 2 mm sieve, and washed over a sink with a jet or spray of clean water, as shown in Fig. 4.17. The silt and clay passing the 63 μm sieve is allowed to run to waste. When the material on the 2 mm sieve has been washed free of fines, washing on the 63 μm sieve is continued until the waste water is seen to run clear.

During this operation the sieve must not be allowed to become overloaded with soil or to overflow with water. The mass of soil retained on the 63 μm should not exceed 150 g at any one time. Table 4.6 gives the recommended maximum quantities which may be retained on each sieve. If this is likely to be exceeded, the material should be sieved in two or more portions.

The sink used for this operation should be fitted with a silt trap, as referred to in Section 1.6.6.

(4) Drying

The whole of the material retained on each sieve is allowed to drain, and is carefully transferred to trays or evaporating dishes. These are placed in an oven to dry at 105–110°C, preferably overnight.

(5) Weighing

After cooling, the whole of the dried material is put together and weighed to an accuracy of 0.1% (m_4).

(6) Sieving

The dry soil is passed through a nest of the complete range of sieves to cover the sizes of particles present, down to the 63 μm sieve. This operation may be carried out by hand, or preferably on a sieve shaker, exactly as in the dry sieving procedure (Section 4.6.1).

If the fraction passing the 2 mm sieve is large, i.e. substantially greater than 150 g, it should be accurately weighed and then subdivided to give a sample of 100–150 g, as described for composite sieving (Section 4.6.2).

(7) Weighing

The portion retained on each sieve is weighed, each to an accuracy of 0.1%.

(8) Calculations

The cumulative percentage passing each sieve is calculated in the same way as for simple sieving (Section 4.6.1, stage 6), either method (A) or method (B) being used. Note that percentages must be expressed as a percentage of the *total* initial dry mass (m_1), not the mass after washing (m_4).

The process of washing removes the clay and the finer silt particles from the material, but some particles only slightly smaller than 63 μm sieve may be retained on that sieve owing to the effect of surface tension of water. In the subsequent dry sieving these particles pass through the 63 μm sieve, and the presence of this fine material does not necessarily imply that washing was inadequate.

Therefore the total amount of fines (i.e. material passing the 63 μm sieve), m_T, in the original sample is made up of two parts: (1) that which is lost in the washing process (m_L); (2) that which passes the 63 μm sieve when dry sieved (m_p). The mass of fines lost by washing is calculated by difference:

$$m_L = m_1 - m_w$$

The mass of dry-sieve fines, m_F, is determined by weighing. Hence,

$$m_T = m_L + m_F$$

Initial dry mass: $500\,g = m_1$

Dry mass after washing on 63 μm sieve: $370\,g = m_4$

Mass lost by washing: $\overline{130\,g} = m_L$

Sieve size	Mass retained g	Percent retained	Cumulative mass passing	Percentage passing
3.35 mm	0	0	500	100%
2	20	4	$\dfrac{-20}{480}$	$\dfrac{480}{500} \times 100 = 96\%$
1.18	35	7	$\dfrac{-35}{445}$	$\dfrac{445}{500} \times 100 = 89\%$
600 μm	60	12	$\dfrac{-60}{385}$	$\dfrac{385}{500} \times 100 = 77\%$
212	145	29	$\dfrac{-145}{240}$	$\dfrac{240}{500} \times 100 = 48\%$
63	100	20	$\dfrac{-100}{140}$	$\dfrac{140}{500} \times 100 = 28\%$
Pass 63 μm	$m_F = 10$		$m_L = 130$	
			$m_F = 10$	
			$\overline{m_T = 140}$	28%

Fig. 4.19   *Calculations, simple wet sieving*

This provides a check on the sieving calculations, because m_T should be equal to the calculated cumulative mass passing the 63 μm sieve.

The percentage fines can be calculated directly, and is equal to

$$\frac{m_T}{m_1} \times 100\%$$

Calculations for a silty fine to coarse sand, which did not require riffling, using both methods (A) and (B), are given in Fig. 4.19, and the grading curve is shown in Fig. 4.20.

4.6.5 Wet Sieving — Gravelly Soils (non-cohesive) (BS1377: Part 2:1990:9.2)

OUTLINE OF PROCEDURES

If the sample contains particles of coarse gravel and cobble size, a large sample (15 kg or more) is necessary. After drying and weighing (total mass m_1), the material is sieved on a large-diameter 20 mm sieve, with a portion being taken at a time, so as not to overload the sieve (see Table 4.6). Particles retained are brushed to remove finer material which may be adhering to them, but individual particles must not be broken down. The material retained on the 20 mm sieve, after drying, if necessary, is then sieved on appropriate larger aperture sieves and the amount retained on each is weighed.

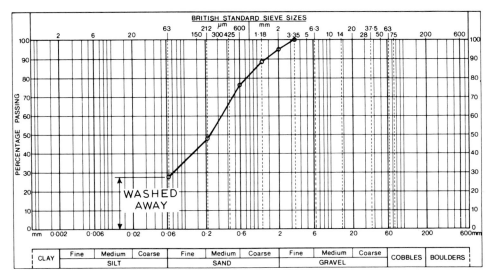

Fig. 4.20   Grading curve, simple wet sieving

The fraction passing to the 20 mm sieve, including 'brushings' from larger particles, is then weighed (m_2), and is subdivided to give a convenient mass (say 2 kg), denoted by m_3. This material is then washed on a 2 mm sieve nested in a 63 μm sieve, as described in step (3) of Section 4.6.4, followed by drying and weighing (mass m_4) as in steps (4) and (5). The procedure is illustrated in Fig. 4.2.1. The washed soil is then sieved through the next set of sieves using the dry sieving procedure, and the retained portions are weighed. If the mass passing the 6.3 mm or 2 mm sieve is about or not much more than 150 g, the soil is sieved on the remaining sieves down to the 63 μm sieve, and each portion is weighed, as described for dry sieving in Section 4.6.1. The mass of any fine material passing the 63 μm sieve is denoted by m_F.

If the mass of material passing the 6.3 mm or 2 mm sieve is appreciably greater than 150 g, riffling is necessary before using the finer sieves, as described in section 4.6.2. Masses before and after riffling are denoted by m_5 and m_6 respectively. The riffled sample is then sieved on the remaining sieves, and each portion is weighed, as in steps (6) and (7) of Section 4.6.4. The mass of any fine material passing the 63 μm sieve is now denoted by m_E.

CALCULATIONS

Even when a sample is large and is weighed initially in kilograms, calculations are easiest if all masses are expressed in grams. The percentages retained on the 20 mm and larger sieve sizes are calculated as described previously.

After riffling and washing on the 63 μm sieve, the masses retained on (method (A)) or passing (method (B)) each of the next set of sieves are corrected as before by multiplying by m_2/m_3 (not m_2/m_4), so that the cumulative mass passing each sieve can be expressed as a percentage of m_1.

In the procedure shown in Fig. 4.2.1 the total amount of fines passing the 63 μm sieve has to be calculated from two parts, comprising the mass lost by washing, m_L (equal to $m_3 - m_4$) after the first riffle, and the small amount of fine material passing the 63 μm sieve, m_E, after the second riffle.

The former must be corrected to relate to the original sample mass m_1 by multiplying by the ratio m_2/m_3. The latter must first be corrected to relate to m_3, and then to m_1, i.e. the factor is

$$\frac{m_5}{m_6} \times \frac{m_2}{m_3}$$

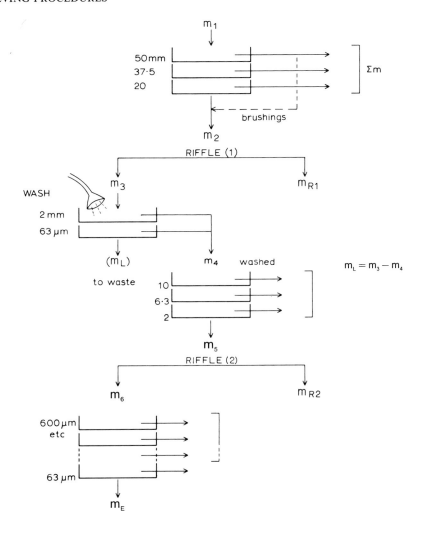

Fig. 4.21   *Sequence diagram, wet sieving of non-cohesive soil with gravel*

The total mass of fines in the original sample, m_T, is given by the equation

$$m_T = \left(m_L \times \frac{m_2}{m_3} \right) + \left(m_E \times \frac{m_5}{m_6} \times \frac{m_2}{m_3} \right)$$

where $m_L = m_3 - m_4$.

The percentage of fines in the whole sample, related to m_1, is therefore equal to

$$\left[m_L + \left(m_E \times \frac{m_5}{m_6} \right) \right] \times \frac{m_2}{m_3} \times \frac{100}{m_1} \%$$

If the second riffle is not necessary, then in the above equations $m_6 = m_5$ and m_E is replaced by m_F. The percentage of fines in the whole sample is then equal to

$$(m_L + m_F) \times \frac{m_2}{m_3} \times \frac{100}{m_1} \%$$

Initial mass = 20 kg = 20 000 g = m_1

Sieve size	Mass retained	Mass passing	Total mass passing	Percentage passing
75 mm	0	m_1 = 20 000		100%
50	1000	$\dfrac{-1000}{19\,000}$		$\dfrac{19\,000}{20\,000}$ x 100 = 95.0%
37.5	800	$\dfrac{-800}{18\,200}$		$\dfrac{18\,200}{20\,000}$ x 100 = 91.0%
20	2200	$\dfrac{-2200}{m_2 = 16\,000}$		$\dfrac{16\,000}{20\,000}$ x 100 = 80.0%

$\downarrow$

RIFFLE

$\downarrow$ $\qquad$ $\downarrow$

$m_3 = 2400 \qquad (13\,600) = m_{R1}$

$\downarrow$

—WASH ON 63 μm

$\qquad \rightarrow (m_L = 400$ to waste)

$\downarrow$

$m_4 = $ 2000 $\qquad\qquad\qquad\qquad$ 2400

10	410	$\dfrac{-410}{1590}$	$\dfrac{-410}{1990}$	$\dfrac{1990}{2400}$ x $\dfrac{16\,000}{20\,000}$ x 100 = 66.3%
6.3	400	$\dfrac{-400}{1190}$	$\dfrac{-400}{1590}$	$\dfrac{1590}{2400}$ x 80 $\quad$ = 53.0%
2	420	$\dfrac{-420}{m_5 = 770}$	$\dfrac{-420}{m_5 + m_L = 1170}$	$\dfrac{1170}{2400}$ x 80 $\quad$ = 39.0%

$\downarrow$

RIFFLE

$\downarrow \qquad \downarrow$

$m_6 = 160 \qquad (610) = m_{R2}$

Adjusted
mass retained

$\qquad\qquad\qquad\qquad\qquad\qquad\qquad\qquad$ 1170

600 μm	73	73 x $\dfrac{770}{160} = 351$	$\dfrac{-351}{819}$	$\dfrac{819}{2400}$ x 80 $\quad$ = 27.3%
212	51	51 x $\dfrac{770}{160} = 245$	$\dfrac{-245}{574}$	$\dfrac{574}{2400}$ x 80 $\quad$ = 19.1%
63	29.5	29.5 x $\dfrac{770}{160} = 142$	$\dfrac{-142}{432}$	$\dfrac{432}{2400}$ x 80 $\quad$ = 14.4%
Pass 63 μm	$m_E = $ 6.4	6.4 x $\dfrac{770}{160} = 30.8$		

Check: $(400 + 30.8) \times \dfrac{16\,000}{2400} \times \dfrac{100}{20\,000}$

$\qquad = 14.4\%$

Fig. 4.22   Calculations, wet sieving of non-cohesive soil with gravel

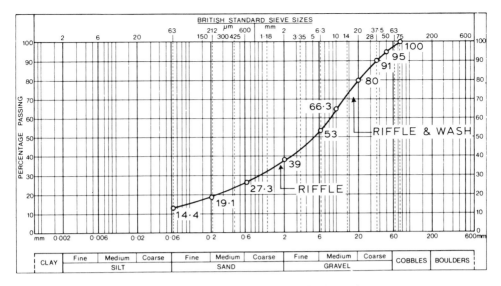

Fig. 4.23   Grading curve, silty sand and gravel

A worked example, showing calculations using method (B), is given in Fig. 4.22 and the resulting grading curve is shown in Fig. 4.23. In this example, washing was carried out after riffling. The procedure for washing before the first riffling is given in Section 4.6.7.

This example is similar to that given in Fig. 4.14, except that the presence of silt meant that washing was necessary. Washing of the whole sample without first riffling would be an unnecessarily lengthy process by the usual means. However, a large sieve-shaker can be fitted with an attachment for washing while shaking, as described by West and Dumbleton (1972). This procedure is relatively quick, and is suitable for medium gravel and sand containing small proportions of silt and clay.

4.6.6   Wet Sieving — Cohesive soils

SCOPE

The procedures for carrying out particles size tests on cohesive soils depend upon the range of particle sizes present, and for this purpose the soils can be divided into three categories:

(1)   Cohesive soils containing particles no larger than 2 mm (coarse sand size); for example, sandy and silty clays.
(2)   Cohesive soils containing particles up to 20 mm (medium gravel size); for example, gravelly sandy and silty clays.
(3)   Cohesive soils containing larger particles, such as boulder clay.

   The first two categories are dealt with separately below, and the third is described in Section 4.6.7.

COHESIVE SANDY SOILS

Cohesive soils containing particles no larger than sands present little difficulty, because a test can be carried out on a relatively small sample (100–150 g). The material is first pretreated for a sedimentation test (Section 4.8.1) and then washed on a 63 μm sieve. The retained portion is dried, then sieved as described in Section 4.6.1. The material passing the 63 μm sieve is collected for a sedimentation analysis (Section 4.8.2 or 4.8.3).

COHESIVE SOILS WITH FINE GRAVEL

Cohesive soils with up to medium-size gravel require a sample of about 2 kg. A representative sample is dried and carefully weighed, and a separate sample of the 'matrix' material (i.e. with most gravel particles removed) is set aside for the sedimentation pretreatment procedure which is described in Section 4.8.1.

The dried representative sample is spread out on a tray and covered with water containing 2 g/litre of sodium hexametaphosphate. The soil is allowed to stand for at least an hour, and is stirred frequently. This disperses the clay fraction, so that clay and silt will not adhere to larger particles.

The material is then washed a little at a time through the 2 mm sieve nested in a 63 μm sieve as described for wet sieving (Section 4.6.4), all the fines being washed to waste. When the water passing through the 63 μm sieve is virtually clear, the material retained on the sieves is dried and shaken dry through the complete range of sieves from 20 mm downwards. The material retained on each sieve is weighed to an accuracy of 0.1%. Calculations are similar to those given in Section 4.6.4. However, if riffling is carried out at some stage after washing, the calculations in Section 4.6.5 are applicable.

The grading curve from the sedimentation test, if required, is added to the sieving curve after the appropriate proportionate adjustment referred to in Section 4.8.5 has been made. The grading sheet should include a note indicating that the soil was pretreated by immersion in the dispersant solution.

4.6.7 'Boulder Clay'

SCOPE

In this context 'boulder clay' refers to soils of the type represented by curve (G) in Fig. 4.6 of Section 4.4.3. The clay content, though perhaps only about 15% of the whole, is sufficient to give the material *en masse* sufficient cohesion for it to behave as a clay. This description also applies to colliery spoil materials and the like.

A soil of boulder clay type is the most difficult on which to carry out a particle size analysis, for the following reasons:

(1) The presence of gravel or cobbles necessitates a large initial sample, probably exceeding 20 kg.
(2) Because it is cohesive, it must be pretreated with dispersant and the fines fraction washed away. This can be very time-consuming.
(3) It is of no use to dry the whole sample initially, because the clay when dried becomes almost brick-hard. It is then very difficult to break down and remove from larger particles.
(4) Because of the complexity of the test procedure, the calculations are less straightforward than those previously described.

The following procedure is the author's adaptation of the general principles outlined in BS1377: Part 2:1990:9.2. In this method the material is washed before riffling. The procedure is shown diagrammatically in Fig. 4.24.

SELECTION OF SAMPLE

The mass of the sample taken initially for testing should not be less than that indicated in Table 4.4 or Fig. 4.7 for the maximum significant particle size. If cobble sizes are present in significant quantity, a sample of 50 kg or more is required, and this is indeed a large sample for cohesive material. The quantity indicated relates to dry mass. Since this type of material is not dried at the outset, allowance must be made for the amount of water present in the soil, which can only be estimated at this stage.

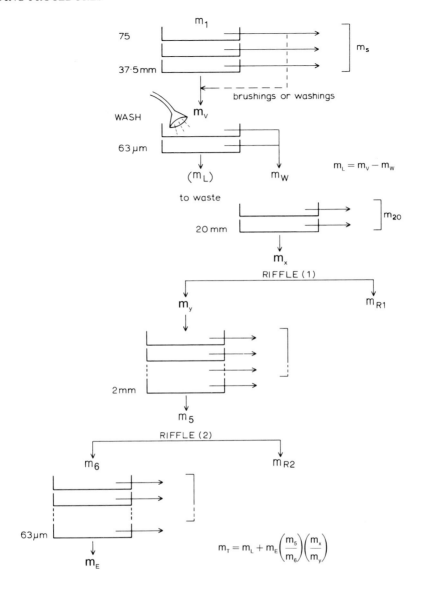

Fig. 4.24   Sequence diagram for sieving cohesive soils with gravel ('boulder clay' type)

PROCEDURE

(1) Before weighing the sample to be tested, take several small specimens (each of about 50 g) representative of the fine fraction from different points of the sample. Break or chop them up, mix together, and from this material set aside about 100 g of soil representative of the fraction finer than 2 mm for sedimentation analysis. Return the remainder to the main sample.

(2) Weigh the whole sample (W_1) in its natural undried state, and spread out on a tray or trays.

(3) Take three representative 'lumps' of material, each of about 300 g, for moisture content measurements. remove any particles larger than 20 mm and return them to the main sample.

(4)   Weigh each sample, dry overnight, weigh, and calculate the moisture contents. The average of the three ($w\%$) is the moisture content of the minus 20 mm fraction. Retain the dried material.
(5)   Add dispersant solution (water containing 2 g/litre of sodium hexametaphosphate) to the sample on the tray. Stir frequently and break down any clay lumps.
(6)   Allow to stand overnight, and add the dried material from the moisture content determination (stage 4), which should be agitated until completely broken down.
(7)   Remove all particles larger than 37.5 mm by hand, using a 37.5 mm sieve as a gauge. Brush them, if necessary, to ensure that these particles (referred to here as 'stones') have been washed free of adhering fine material, which is to remain with the main sample.
(8)   Dry the stones in the oven, and weigh when cool (m_s).
(9)   Separate the stones into size fractions on the 50 mm, 63 mm and 75 mm sieves, using, if necessary, wood frames for gauging larger sizes, such as 100 mm, 150 mm, 200 mm as required.
(10) Weigh each size fraction.
(11) Take a little at a time of the sample from stage (6), and wash on a 20 mm sieve nested on a 2 mm sieve nested on a 63 μm sieve. Use 450 mm diameter sieves if available. Allow material finer than 63 μm to run to waste. Collect about a litre of washings periodically, and continue washing until this water is seen to be clear.
(12) Oven dry the material retained on these sieves.
(13) Weigh the whole of the retained washed material (m_w).
(14) Sieve through 28 mm and 20 mm sieves (450 mm diameter).
(15) Weigh material retained on each sieve (the total denoted by m_{20}), and material passing the 20 mm sieve (m_x).
(16) Riffle m_x to obtain a suitable size sample (m_y) for the next set of sieves (300 mm diameter).
(17) Dry sieve on appropriate sieves down to 2 mm.
(18) Weigh material retained on each sieve, and material passing the 2 mm sieve (m_5).
(19) Riffle (m_5) to obtain a suitable sample (m_6) for the fine sieves (200 mm diameter).
(20) Dry sieve on selected sieves down to 63 μm size.
(21) Weigh material retained on each sieve, including any passing the 63 μm sieve (m_E).

CALCULATIONS

The notation used in the following explanation is summarised in Table 4.10. Undried masses are denoted by W; dry masses by m; moisture content by w. Symbols in parentheses represent quantities which are calculated; the others are measured. The mass of the original undried sample is denoted by W_1.
   The complete calculations are set out in Fig. 4.25.
   It is assumed that particles larger than 20 mm contain no absorbed water, or that the quantity held is negligible. The undried mass of material passing 20 mm (W_2) is therefore equal to the undried mass of the original sample with the particles larger than 20 mm removed. These consist of the stones retained on 37.5 mm, and the washed portion passing 37.5 mm and retained on 20 mm (m_{20}); therefore

$$W_2 = W_1 - (m_s + m_{20})$$

If the particles larger than 20 mm contain appreciable absorbed water in their natural state, this can be measured separately and allowed for by adjusting m_s and m_{20}.
   Since the moisture content of the minus 20 mm fraction is known ($w\%$), the equivalent dry mass (m_2) of this fraction is given by

$$m_2 = \frac{100}{100 + w} \times W_2$$

Table 4.10. NOTATION FOR BOULDER CLAY SIEVE ANALYSIS

Size fraction	Dry mass	Undried mass	Remarks
Initial total mass of sample tested	(m_1)	W_1	
'Stones' larger than 37.5 mm	m_s		
Total passing 37.5 mm	m_v		$m_v = (m_1 - m_s)$
Passing 37.5 after washing	m_w		
Washed away through 63 μm	(m_L)		$m_L = m_v - m_w$
Retained on 20 mm after passing 37.5 mm	m_{20}		$m_{20} = m_w - m_x$
Total finer than 20 mm	(m_2)	(W_2)	$m_2 = m_x + m_L$
Passing 20 mm after washing	m_x		
Riffled portion of m_x	m_y		
Washed material passing 2 mm	m_5		
Riffled portion of m_5	m_6		
Passing 63 μm	m_E		
Total finer than 63 μm	(m_T)		$m_T = m_L + m_E \left(\dfrac{m_5}{m_6} \times \dfrac{m_x}{m_y} \right)$
Initial moisture content of W_2	$w\%$		$m_2 = \dfrac{100 W_2}{100 + w}$

By adding back the dry mass of particles larger than 20 mm, the initial dry mass (m_1) of the whole sample is obtained:

$$m_1 = m_2 + m_s + m_{20}$$

The mass is used as the starting point for the calculations set out in Fig. 4.25, which shows only the dry masses.

The mass of material passing the 37.5 mm sieve (m_v), and used in the washing process, is equal to the initial mass less the mass of stones — that is,

$$m_v = m_1 - m_s$$
$$= m_2 + m_{20}$$

The calculation of percentages passing the sieves down to 20 mm is as described in Section 4.6.5. Washing on the 63 μm sieve has no effecton the masses retained on these sieves. The mass lost in the washing (m_L) is obtained by difference — that is,

$$m_L = m_v - m_w$$

The calculation procedure after riffling is similar to that given in Section 4.6.5, and corrected percentages are related to the dry mass before washing, as shown in Fig. 4.25. After the second riffle, a second multiplying factor is introduced, as explained in Section 4.6.2.

The total amount of fine material (m_T) passing the 63 μm sieve is made up of two parts, as in Section 4.6.5, comprising m_L and m_E. The latter must be increased by the ratio $(m_5/m_6) \times (m_x/m_y)$ to give the equivalent mass to add to m_L. The total mass finer than 63 μm in the whole sample is given by

$$m_T = m_L + m_E \left(\frac{m_5}{m_6} \times \frac{m_x}{m_y} \right)$$

and the percentage of fines is equal to

$$\frac{m_T}{m_1} \times 100\%$$

Initial undried mass: $24\ 970\ g = W_1$

Estimated total initial dry mass: $23\ 160\ g = m_1$

Sieve size	Mass retained	Cumulative mass	Mass passing		Percentage passing
100 mm	0	0	$m_1 = 23\ 160$	$W_1 = 24\ 970$	100%
75	714	714	$= \dfrac{-714}{22\ 446}$	$m_s + m_{20} = 3894$ $W_2 = 21\ 076$	$\dfrac{22\ 446}{23\ 160} \times 100 = 96.9\%$
50	1275	1989	$= \dfrac{-1275}{21\ 171}$	$w = 9.4\%$	$\dfrac{21\ 171}{23\ 160} \times 100 = 91.4\%$
37.5	762	$2751 = m_s$	$\dfrac{-762}{m_V = 20\ 409}$	$m_2 = \dfrac{100}{109.4} \times 21\ 076$ $= 19\ 265$	$\dfrac{20\ 409}{23\ 160} \times 100 = 88.1\%$
			WASH ON 63 μm	$m_s + m_{20} = 3894$ $\therefore\ m_1 = 23\ 159$	
			retained $\quad$ lost $m_w = 11\ 958 \quad (8451) = m_L$	Total mass passing	$\dfrac{8451}{23\ 160} \times 100 = 36.5\%$ lost
			11 958	$m_2 = 20\ 409$	
28	387	387	$\dfrac{-387}{11\ 571}$	$\dfrac{-387}{20\ 022}$	$\dfrac{20\ 022}{23\ 160} \times 100 = 86.5\%$
20	756	$1143 = m_{20}$	$\dfrac{-756}{10\ 815\ = m_X}$	$\dfrac{-756}{m_2 = 19\ 266}$	$\dfrac{19\ 266}{23\ 160} \times 100 = 83.2\%$
			RIFFLE (1)		
	3894	$= m_s + m_{20}$	$m_y = 2504 \quad (8311) = m_{R1}$		
			Corrected mass retained	$m_2 = 19\ 266$	
14	117.4	117.4	$117.4 \times \dfrac{10\ 815}{2504} = 507.1$	$\dfrac{-507.1}{18\ 758.9}$	$\dfrac{18\ 759}{23\ 160} \times 100 = 81.0\%$
6.3	331.2	448.6	$331.2 \times \dfrac{10\ 815}{2504} = 1430.5$	$\dfrac{-1430.5}{17\ 328.4}$	$\dfrac{17\ 328}{23\ 160} \times 100 = 74.8\%$
2	477.2	925.8	$477.2 \times \dfrac{10\ 815}{2504} = 2061.1$	$\dfrac{-2061.1}{15\ 267.3}$	$\dfrac{15\ 267}{23\ 160} \times 100 = 65.9\%$
		$m_y = 2504$ $\dfrac{925.8}{1578.2}$	$m_5 = 1578.2$		
			RIFFLE (2)		
			$m_6 = 105.3 \quad (1472.9) = m_{R2}$		
				15 267.3	
600 μm	38.2	38.2	$38.2 \times \dfrac{1578.2}{105.3} \times \dfrac{10\ 815}{2504} = 2472.5$	$\dfrac{-2472.5}{12\ 794.8}$	$\dfrac{12\ 795}{23\ 160} \times 100 = 55.2\%$
212	41.3	79.5	$79.5 \times \dfrac{1578.2}{105.3} \times \dfrac{10\ 815}{2504} = 2673.1$	$\dfrac{-2673.1}{10\ 121.7}$	$\dfrac{10\ 122}{23\ 160} \times 100 = 43.7\%$
63	22.5	102.0	$102.0 \times \dfrac{1578.2}{105.3} \times \dfrac{10\ 815}{2504} = 1456.3$	$\dfrac{-1456.3}{8665.4}$	$\dfrac{8665}{23\ 160} \times 100 = 37.4\%$
Passing 63	$3.3 = m_E$	$105.3 = m_6$	$3.3 \times \dfrac{1578.2}{105.3} \times \dfrac{10\ 815}{2504} = 214$		
			$m_L = 8451$ Check: $m_T = 8665$		

Fig. 4.25   Calculations for 'boulder clay' type soil

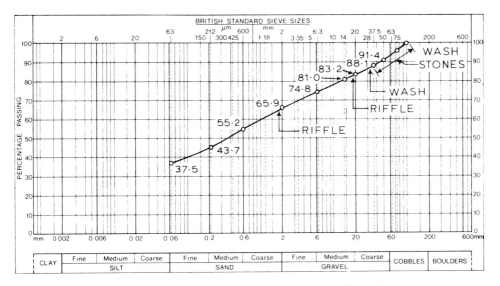

Fig. 4.26 Grading curve from sieving test on 'boulder clay'

The above equations might seem more difficult than the step-by-step arithmetical procedure shown in Fig. 4.25, which is easier to follow when calculating by hand. However, the equations may be useful when a computer program for these and similar sieving calculations is being written.

The grading curve from the sieving test on this material is shown in Fig. 4.26. The complete grading curve, including a sedimentation analysis, is given in Fig. 4.49, and shows a well-graded mixture of all particle sizes from cobbles down to and including clay.

4.7 SEDIMENTATION THEORY

4.7.1 Introduction

The theory of sedimentation is based on the fact that large particles in suspension in a liquid settle more quickly than small particles, assuming that all particles have similar densities and shapes. The velocity which a falling particle eventually reaches is known as its terminal velocity. If the particles are approximately spherical, the relationship between terminal velocity, V, and particle diameter, D, is given by *Stokes' Law*, named after Sir George Stokes (1891). This states that the terminal velocity is proportional to the square of the particle diameter, or

$$V \propto D^2$$

Although clay particles are far from spherical (see Section 4.3.1), the application of Stokes' Law based on diameters of equivalent spheres provides a basis for comparison of the particle size distribution in fine soils which is sufficiently realistic for most prctical purposes.

4.7.2 General Principles

In a sedimentation test a suspension of a known mass of fine soil particles of various sizes is made up in a known volume of water. The particles are allowed to settle under gravity,

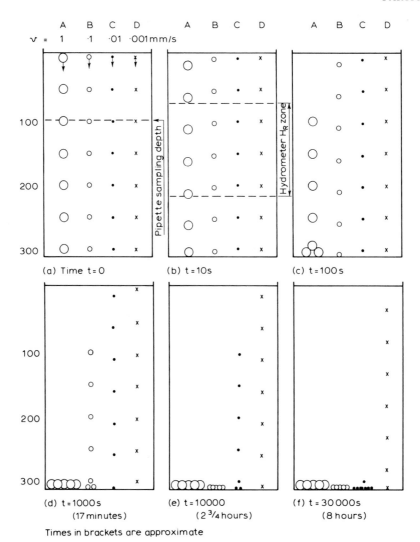

Fig. 4.27   *Representation of sedimentation process*

Table 4.11.   TERMINAL VELOCITIES OF PARTICLES IN SUSPENSION

Particle	Terminal velocity (mm/s)	Approximate diameter (μm)
Coarse silt	1	35
Medium silt	0.1	12
Fine silt	0.01	3.5
Clay	0.001	1.2

and this is the process known as sedimentation. From certain measurements made at known intervals of time, the distribution of particle sizes can be assessed.

A model which illustrates the sedimentation process is represented diagrammatically in Fig. 4.27. Only four different sizes of particles are represented, their terminal velocities and approximate equivalent diameters being shown in Table 4.11.

A real soil contains particles of many different sizes, but the principles of sedimentation can be understood by considering what happens to each of these four sizes after various intervals of time as they are allowed to settle out of a suspension in water in a container just over 300 mm deep.

Immediately after the suspension has been shaken up (time $t = 0$), all particles are uniformly distributed throughout the depth of suspension, as in Fig. 4.27(a). If we assume that each particle reaches its terminal velocity within a very short time, after 10 s the coarse silt particles have each fallen 10 mm, the fine silt particles 1 mm and the particles finer than that hardly at all (Fig. 4.27b). After 100 s the situation is as in Fig. 4.27(c) and the coarse silt particles starting at the 200 mm mark have settled to the bottom. After 1000 s (about 17 min) all the coarse silt particles have reached the bottom (Fig. 4.27d). All the solid particles now remaining in suspension are smaller than 35 μm, so a sample taken from anywhere in the suspension would contain only those particles smaller than this size. At the same instant a sample taken from just above the 100 mm mark would only contain particles smaller than 12 μm (medium silt).

After 10 000 s (about $2\frac{3}{4}$ h) from the start all medium silt particles have reached the bottom and all fine silt particles are below the 100 mm mark (Fig. 4.27e). The upper 100 mm, therefore, contains only clay particles in suspension. After 30 000 s (about 8 h) silt particles of all sizes have reached the bottom, leaving only the clay sizes in suspension (Fig. 4.27f). The smallest size we are considering, 1.2 μm, will require 300 000 s (about $3\frac{1}{2}$ days) to reach the bottom, but smaller particles will take much longer. A suspension containing an appreciable amount of clay remains cloudy indefinitely.

By applying Stokes' Law to the sedimentation model described above, we can calculate the maximum diameter of particles remaining above a particular depth after a certain interval of time from the start. The mass of solid particles present can be determined either by sampling from a specified depth (as in the pipette test), or by measuring the density of the suspension with a hydrometer (the hydrometer test).

The sampling depth of 100 mm used in the standard pipette test is indicated in Fig. 4.27(a). The region marked 'H_R zone' in Fig. 4.27(b) indicates the range of effective depths at which the density of suspension is measured during the period of a typical hydrometer test.

4.7.3 Stokes' Law

According to Stokes' Law, the terminal velocity v of a spherical particle falling freely in a fluid is given by

$$v = \frac{D^2 g(\rho_s - \rho_L)}{18\eta} \qquad (4.1)$$

where D = diameter of particle; ρ_s = mass density of the solid particle; ρ_L = mass density of the fluid (liquid); η = dynamic viscosity of the fluid; g = acceleration due to gravity.

The application of Stokes' Law to the process of sedimentation is based on the following simplifying assumptions.

(1) The condition of viscous flow in a still liquid is maintained.
(2) There is no turbulence — that is, the concentration of particles is such that they do not interfere with one another.
(3) The temperature of the liquid remains constant.
(4) Particles are small spheres.
(5) Their terminal velocity is small.
(6) All particles have the same density.
(7) A uniform distribution of particles of all sizes is formed within the liquid.

Table 4.12. SYMBOLS, UNITS AND CONVERSION FACTORS FOR SEDIMENTATION FORMULA

Quantity	Symbol	Coherent unit	Practical unit	Multiplying factor
Viscosity of water (see Note (1))	η	Ns/m^2	mNs/m^2 ($=$mPa s)	$\frac{1}{1000}$
Height of fall	H	m	mm	$\frac{1}{1000}$
Acceleration due to gravity	g	m/s^2	m/s^2	1
Time	T	s	min	60
Densities (see Note (2))	ρ_s, ρ_w	kg/m^3	Mg/m^3	1000
Particle diameter	D	m	mm	$\frac{1}{1000}$

(1) *Viscosity.* The practical SI unit of dynamic viscosity is mPa s (millipascal second) or mN s m^{-2} (millinewton second per square metre), which is equal to the traditional metric unit cP (centipoise).
(2) *Densities.* The density of water (ρ_w) may be taken here as 1.000 Mg/m^3.

Equation (4.1) can be re-written

$$D = \left\{ \frac{18\,\eta}{g} \cdot \frac{v}{(\rho_s - \rho_L)} \right\}^{1/2} \tag{4.2}$$

If the particle falls a distance H in time T, its mean velocity $= H/T$. If the liquid is water, we may write

$$\rho_L = \rho_w$$

Equation (4.2) then becomes

$$D = \left\{ \frac{18\,\eta}{g} \cdot \frac{H}{T(\rho_s - \rho_w)} \right\}^{1/2} \tag{4.3}$$

Using the practical units given in Table 4.12, and writing t (min) for T, Eq. (4.3) can be written as

$$\frac{D}{1000} = \left\{ \frac{18 \times \eta}{1000\,g} \cdot \frac{H}{1000 \times t \times 60(\rho_s - \rho_w) \times 1000} \right\}^{1/2} \text{ mm}$$

or

$$D = \left\{ \frac{3 \times 10^{-4} \times \eta}{g} \cdot \frac{H}{t(\rho_s - \rho_w)} \right\}^{1/2} \tag{4.4}$$

Putting $g = 9.807$ m/s^2 and $\rho_w = 1.000$ Mg/m^3

$$D = \left\{ \frac{30.59 \times 10^{-6} \times \eta H}{t(\rho_s - 1)} \right\}^{1/2} \text{ mm}$$

or

$$D = 0.005\,531 \left\{ \frac{\eta H}{t(\rho_s - 1)} \right\}^{1/2} \text{ mm} \tag{4.5}$$

Equation (4.5) provides the basis for the derivation of sampling times for the pipette test (Table 4.16), and for calculations in the hydrometer method. Values of η and ρ_w for water temperatures from 0–40°C are given in Table 4.13. Intermediate values may be obtained by interpolation, either arithmetically or graphically.

Table 4.13. VISCOSITY AND DENSITY OF WATER

Temperature (°C)	Dynamic viscosity, η (mPa s)	Density, ρ_w (Mg/m³)
0	1.7865	0.999 84
5	1.5138	0.999 95
10	1.3037	0.999 70
15	1.1369	0.999 09
20	1.0019	0.998 20
25	0.8909	0.997 04
30	0.7982	0.995 65
40	0.6540	0.992 22

Based on Kaye and Laby (1973) (see also Fig. 3.1).

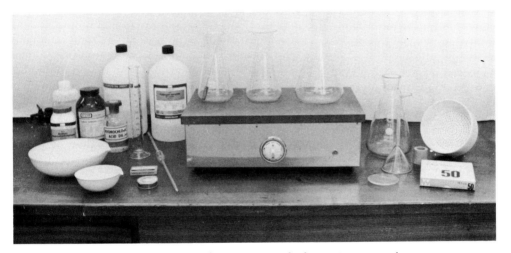

Fig. 4.28   *Apparatus for pretreatment of sedimentation test sample*

4.8 SEDIMENTATION PROCEDURES

4.8.1 Pretreatment

Before proceeding with either the pipette or hydrometer sedimentation test, any organic matter present must be removed by chemical treatment. The soil must also be treated with dispersant and thoroughly agitated to ensure that discrete particles are separated. This process is known as pretreatment. The principle is the same whether it is for a pipette test or a hydrometer test, but there are a few differences in detail which are mainly due to the different sizes of sample required for the two tests.

The procedure described below applies to both the pipette and hydrometer sedimentation tests. Where the two types of test require different procedures, these are given under the appropriate heading.

APPARATUS

Most of the items listed below are shown in Fig. 4.28.

(1) Mechanical shaker (end-over-end or vibratory) capable of keeping up to 75 g of soil in suspension in 150 ml of water. See Section 1.2.5 (10).

(2) Wide-mouth conical beaker, 1000 ml or 650 ml.

(3)   Measuring cylinder, 100 ml.
(4)   Wash bottle and distilled water.
(5)   Glass stirring rod with rubber 'policeman'.
(6*)   Centrifuge (if available) capable of holding bottles of 250 ml capacity.
(7*)   Polypropylene centrifuge bottles, 250 ml.
(8*)   Buchner funnel, 150 mm, and filtration flask of 1 litre capacity with rubber bung to take the funnel.
(9*)   Filter pump or vacuum supply with vacuum tubing.
(10*) Whatman No. 50 filter paper.
(11*) Evaporating dish, about 150 mm diameter.
(12)   Sieves 200 mm diameter; 63 μm, 212 μm, 600 μm, 2 mm apertures.
(13)   Riffle-box, and small tools for sample mixing and dividing.
(14)   Oven, 105–110°C, and desiccator.
(15)   Electric hot-plate.

Required for pipette test only

(16)   Balance, reading to 0.001 g.
(17)   Pipette, or measuring cylinder, 25 ml.
(18)   Glass beaker, about 500 ml.
(19)   Glass filter funnel, 100 mm.

Required for hydrometer test only

(20)   Balance, reading to 0.01 g.
(21)   Pipettes, 100 ml and 50 ml, or measuring cylinders.
(22)   Four porcelain evaporating dishes about 150 mm diameter.
(23)   Measuring cylinder 500 ml.

REAGENTS

(1)   Hydrogen peroxide (20 volume solution).
(2)   'Standard' dispersant solution — that is, 33 g of sodium hexametaphosphate and 7 g of sodium carbonate in distilled water to make 1 litre (see Section 4.5.3).
(3)   Hydrochloric acid ($\underline{c}$(HCl) = 1 mol/litre).

PROCEDURAL STAGES

(1)   Select and prepare test specimen.
(2)   Treat for organic matter.
(3)   Treat for calcareous matter.
(4)** Filter and dry.
       or
(5)** Centrifuge.
(6)   Disperse.
(7)   Sieve.

TEST PROCEDURE

(1) *Selection of sample*

The test specimen is obtained from the air-dried original sample by riffling, or by cone-and-quartering, using the fraction which passes a 2 mm sieve. The appropriate size of specimen required is suggested in Table 4.14, but the mass used depends upon the type of soil. Too

---

* Items (8)–(11) are alternatives to items (6) and (7) if a centrifuge is not available.
** See footnotes on pages 208 and 209.

Table 4.14. QUANTITY OF MATERIAL FOR
SEDIMENTATION TESTS

Material	Pipette test	Hydrometer test
Sample after initial riffling	60 g	200 g
Test specimen:		
Sandy soils	30 g	100 g
Clayey soils	12 g	50 g

much soil can prolong a test unnecessarily; too little can give unreliable results. Experience is the best guide; if in doubt, trial tests should be made.

Determine the initial dry mass of the specimen (m_0) to the nearest 0.001 g for the pipette test, and to the nearest 0.01 g for the hydrometer test. All subsequent weighings should be made to the same degree of resolution. The soil should be oven dried at 105°C only when it is known from precedent that oven drying does not affect the particle size properties, in which case the dry test specimen itself can be weighed.

In some soils, especially some organic soils and tropical residual soils, the particle size properties can be affected by drying, even if air-dried at room temperature. (See Geological Society Report, 1990). It is then necessary to weigh the undried soil and to calculate the dry mass, m_0, from the moisture content determined on a duplicate specimen. In some instances it might be acceptable to air-dry the soil first, and then calculate the dry mass from the air-dry moisture content. After pre-treatment, drying at 60°C is specified in the BS as a compromise, but here again drying should be avoided with these special soils and the dry mass calculated from a moisture content measurement. This will require a rather larger specimen at the outset.

(2)   Pretreatment for organic matter

Pretreatment is not necessary for inorganic soils. Where the effect of pretreatment on the results is uncertain, parallel tests should be carried out with and without pretreatment on two similar specimens.

If organic matter is present, the test specimen should be pretreated as described below.
If pretreatment is not required, omit this stage.

Pipette method   Place the soil in a 650 ml conical beaker and add 50 ml of distilled water. Boil the suspension gently until the total volume is reduced to about 40 ml. Use a similar beaker alongside containing 40 ml of water as a comparison in order to judge the right amount. Allow to cool, and then add 75 ml of hydrogen peroxide.

Hydrometer method   Place the soil in a 1000 ml conical beaker, or a 500 ml beaker if it is known that a large amount of hydrogen peroxide will not be needed. Add 150 ml of hydrogen peroxide and stir gently for a few minutes with a glass rod.

Warning — Soils containing manganese oxides or sulphides may react violently with hydrogen peroxide, especially when heated.

Cover with a cover-glass and allow to stand overnight. A tight-fitting cork must not be used, otherwise the pressure of evolved gas may burst the flask.

Next morning heat the flask and contents gently, either on a low-heat hot-plate or on a low gas flame. Agitate frequently by stirring or by shaking with a rotary motion, but avoid frothing over. If necessary, add more hydrogen peroxide in increments of about 100 ml until the oxidation process is complete. Very organic soils may require several additions of hydrogen peroxide, and the oxidation process may take 2 or 3 days.

As soon as frothing has subsided, reduce the volume of liquid to about 50 ml by boiling, which decomposes any excess hydrogen peroxide. The flask must not be boiled dry. Allow the mixture to cool.

Table 4.15. AMOUNT OF ACID FOR PRETREATMENT
(hydrochloric acid, c (HCl) = 1 mol/litre)

	Pipette test	Hydrometer test
Initial quantity	10 ml	100 ml
Subsequent increments	10 ml	25 ml

If appropriate, hydrochloric acid treatment is carried out next (stage 3). If acid treatment is not used, the next stage is to extract the soil by filtration (stage 4) or by means of a centrifuge (stage 5).

(3)   *Pretreatment for calcareous matter*

This process is not called for in BS1377: Part 2:1990, but its use might sometimes be appropriate. A check should always be made for reaction with hydrochloric acid by dropping a few spots of HCl on to a small portion of the sample. If there is no effervescence, acid treatment is not required. A visible reaction indicates the presence of calcareous compounds which could act as a cementing agent, preventing separation of individual grains. If the effect of these compounds on the final result is to be assessed, parallel tests should be carried out on duplicate specimens with and without acid treatment.

When the soil and water mixture has cooled, add the required volume of hydrochloric acid (see Table 4.15). Stir with a glass rod for a few minutes and allow to stand for 1 h.

Check the pH of the suspension with a universal or wide-range pH paper, or with litmus paper. If the reaction has ceased, the suspension should show an acid reaction (pH less than 5, or blue litmus turns red). If not, add further increments of hydrochloric acid, as indicated in Table 4.15, followed by stirring and standing as above, until an acid reaction remains.

(4)   *Filtration and Drying

Set up a Buchner funnel fitted with a Whatman No. 50 filter paper on a vacuum filtration flask. It may be convenient to arrange simultaneous operation of, say, six filration flasks by connecting them to a manifold outlet on the vacuum service line. The vacuum outlet should be fitted with a water trap to prevent water being inadvertently drawn into the main vacuum line and to the vacuum pump.

Transfer the contents of the conical beaker to the Buchner funnel, rinsing the beaker with a little distilled water to ensure that no soil is lost.

Wash thoroughly with distilled water. If acid treatment has been carried out, the washing must continue until all traces of acid are removed. When a pH paper indicator or blue litmus paper no longer gives an acid reaction, washing is complete.

Transfer the soil residue from the filter paper to a previously weighed borosilicate glass evaporating dish, using a fine jet of distilled water from a wash-bottle. Ensure that no soil is lost.

Dry in an oven at 60° to 65°C, cool, and weigh to the nearest 0.001 g or 0.01 g. Calculate the dry mass of pre-treated soil. This mass, denoted by m_p, is used in subsequent calculations, and should be carefully checked.

If the initial untreated dry mass, m_0, was measured, the pretreatment loss is equal to $(m_0 - m_p)$, and the percentage loss which would be reported is equal to $[(m_0 - m_p)/m_0] \times 100\%$.

---

*This procedure may be used in place of stage (5) if a centrifuge is not available, and must be followed if acid treatment (stage 3) has been done. The apparatus is shown in Fig. 4.29.

Fig. 4.29 Vacuum filtration apparatus

(5) *Centrifuging*

A centrifuge, if available, provides the most rapid and convenient means of recovering the soil after hydrogen peroxide pretreatment.

Weigh the dry 250 ml centrifuge bottle with stopper to the nearest 0.001 g or 0.01 g. Transfer the soil suspension from the conical beaker into the centrifuge bottle, using a fine jet of distilled water from a wash-bottle, taking care not to lose any soil partcles. Adjust the volume of water in the bottle to about 200 ml, and fit the stopper. Place in the centrifuge. Usually four to six bottles can be fitted simultaneously, but opposite tubes must be balanced using bottle containing plain water if necessary. Run the centrifuge for 15 min at about 2000 rev/min. Remove the bottle and decant the clear supernatant liquid (i.e. the layer of liquid containing no suspended solids). Place the bottle and contents in the oven set at 60° to 65°C with the stopper removed and allow to dry overnight.

If a centrifuge which takes 250 ml bottles is not available, the soil suspension may be divided, if necessary, into two or more smaller bottles to suit a smaller centrifuge.

Next morning replace the stopper, cool in a desiccator, and weigh accurately. Calculate the dry mass of pretreated soil (m_p).

Also calculate the pretreatment loss as in stage (4) above, if required.

(6) *Dispersion*

If the centrifuge (stage 5) has been used, dispersion by mechanical shaking can be carried out in the centrifuge bottle. If the soil has been filtered (stage 4), transfer it from the evaporating dish to a suitable container, such as a conical flask with stopper, as indicated below.

Pipette method   Add 100 ml of distilled water to the soil in the centrifuge bottle or evaporating dish. If in the evaporating dish, transfer the suspension to the flask without any loss of soil particles. Shake vigorously until all the soil is in suspension. Add 25 ml of the standard dispersing solution (Section 4.5.3), using a pipette.

Hydrometer method   Add 100 ml of the standard dispersing solution (Section 4.5.3), using a pipette, to the soil in the centrifuge bottle or evaporating dish. If in the evaporating dish,

---

* This procedure should not be used if acid treatment (stage 3) has been done, because it is then necessary to wash on a filter paper to remove all traces of acidity. The soil is retained on the filter paper as described in stage (4) above.

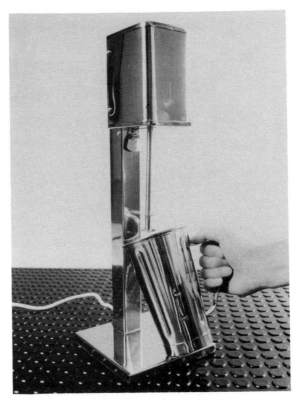

Fig. 4.30   High-speed stirrer

transfer the suspension to the flask without any loss of soil particles. Shake vigorously until all the soil is in suspension.

Fit the centrifuge bottle or flask, tightly stoppered, on to the mechanical shaker. Shake for at least 4 h, or overnight if convenient. However, if excess shaking would cause breakdown of soil particles (e.g. with shales), the period of shaking should be less than 4 h. The shaking period should be based on experience and knowledge of the soil.

If a mechanical shaker of this type is not available, the following alternative procedures could be used for dispersion.

(a)   The ASTM Standard specifies dispersion in the cup of a mechanical stirrer (Horlicks mixer), fitted with a wire baffle (Fig. 4.30). The soil suspension is stirred at high speed for 15 min.

(b)   Alternatively, a Vibro-Stirrer has been found to be equally effective. This type of stirrer is fitted with a blade which oscillates through an angle of 10° at 5000 Hz (Fig. 4.31). One advantage of this device is that it can be used in the beaker or conical flask used for pretreatment.

(c)   A hand method which could be used if a motorised unit is not available is as follows. The soil suspension is placed in a mortar and rubbed vigorously with a rubber pestle. After settling for 2 min, the liquid suspension is decanted through a 63 μm sieve into the receiver to which it is fitted. More distilled water is added to the mortar and the pestling and decanting processes are repeated until the decanted liquid is clear. Care must be taken to prevent breakdown of individual particles.

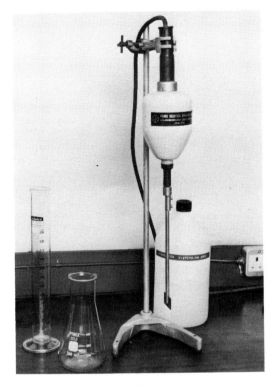

Fig. 4.31 *Vibro-stirrer*

(7) *Wet Sieving*

Transfer the soil and suspension to a 63 μm sieve, nested on a receiver, without loss of soil.

Wash the soil with a jet of distilled water from a wash-bottle, until all fine material is washed through the sieve. The amount of water should not exceed 150 ml for the pipette test or 500 ml for the hydrometer test. The material collected in the receiver is transferred to a cylinder for the appropriate sedimentation test as described in Section 4.8.2 or 4.8.3.

Transfer the material retained on the 63 μm sieve to an evaporating dish, dry and weigh (mass m_s g).

Sieve through larger aperture sieves as appropriate, and weigh the amount retained on each sieve, as for simple dry sieving (Section 4.6.1). These masses are used subsequently for the construction of the grading curve in the sand size range.

Any material passing the 63 μm sieve when dry sieved should be added to the sedimentation cylinder.

4.8.2 Pipette Analysis (BS1377:1990: Part 2:9.4)

SCOPE

In this test a special sampling pipette is used to obtain a sample of soil, pretreated as in Section 4.8.1, from a suspension in water at three fixed time intervals. The mass of soil in each sample is determined, from which the percentages of coarse silt, medium silt, fine silt and clay can be calculated. Additional samples at selected intervals, up to a maximum of six, may be taken if a better spread of readings is required, but the relevant particle sizes are then determined by calculation.

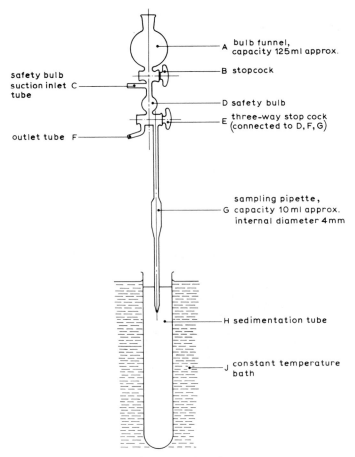

Fig. 4.32   Sampling pipette for sedimentation test

The British Standard refers to this procedure as the standard or primary method for fine particle analysis. However, the apparatus required is expensive and delicate, and is not suitable for use in a site laboratory running control tests during construction.

APPARATUS

The following apparatus is required in addition to that used for pretreatment.

(1)   Sampling pipette (Fig. 4.32) mounted on a stand with a suitable lowering and raising device (Fig. 4.33). The pipette, having a capacity of about 10 ml, must be calibrated before use, as described under stage (7) below. It is sometimes referred to as an Andreasen pipette.

(2)   Two glass cylinders, 50 mm diameter, approximately 350 mm long, graduated at 500 ml, with rubber bungs to fit (the sedimentation cylinders).

(3)   Constant temperature bath maintained at 25°C to ± 0.5°C, deep enough for immersing the sedimentation cylinders to the 500 ml mark.

(4)   Stop-clock reading to 1 s.

(5)   Glass rod about 12 mm diameter and about 400 mm long.

(6)   Nine numbered glass weighing bottles, about 25 mm diameter and 50 mm high, fitted with numbered ground glass stoppers. The mass of each should be determined to the nearest 0.001 g after drying at 105° C and cooling in a desiccator.

(7)   Thermometer covering the range 0 to 50°C, reading to 0.5°C.

Fig. 4.33   Apparatus for pipette sampling

PROCEDURAL STAGES

(1) Prepare suspension.
(2) Take samples with pipette.
(3) Dry and weigh solid matter.
(4) Take and measure calibration sample.
(5) Calculate.
(6) Plot and present results.
(7) Calibration of pipette (need not be repeated when the internal volume is known).

TEST PROCEDURE

(1) *Preparation of suspension*

Transfer the suspension of pretreated soil passing to 63 μm sieve (obtained as described in Section 4.8.1) with aid of a glass funnel, from the receiver into a 500 ml sedimentation cylinder. No soil must be lost in this operation. Make up the water level in the cylinder to the 500 ml graduation mark by adding distilled water.

Place the sedimentation cylinder in the constant-temperature bath, set at 25°C. Into a second sedimentation cylinder place 25 ml of the dispersant solution and make up to the 500 ml graduation mark with distilled water. Place this tube in the same constant-temperature bath.

Insert a rubber bung into each cylinder to obtain a watertight fit, but avoid using excessive force which may split the glass and cause serious injury to the hand. Shake the cylinders and allow them to stand in the constant-temperature bath until they have reached the bath temperature; about 1 h is usually sufficient.

Several sedimentation cylinders (up to six or eight) may be tested at the same time. When all cylinders are in the constant-temperature bath, the water level of the bath should just reach their 500 ml graduation marks.

Remove the cylinder containing the dispersant solution, shake thoroughly and replace it in the bath. Remove each sedimentation cylinder in turn and, if necessary, stir the sediment with the glass rod so that all the soil goes into suspension. Shake the cylinder vigorously by applying about 120 end-over-end cycles during a period of 2 minutes, and immediately replace upright in the constant temperature bath.

At the instant when the cylinder containing soil suspension is replaced in the bath, start the stop-clock. The instant of placing the tube in the bath is zero time $(t = 0)$ for that sample.

Each cylinder can be timed separately, a stop-clock being used for each, but only one clock is necessary if they are started at regular intervals of, say, 5 min.

(2)   Sampling with Pipette

Pipette samples are normally taken at three specified time intervals from zero time for each sample. These times depend upon the density of the particles in the suspension, and are given in Table 4.16. The third sampling operation is at about 7 h from zero, so it is convenient to shake up and start a batch of tests first thing in the morning.

Move the sampling pipette on its supporting frame over the sedimentation cylinder. Lower the pipette until the tip just touches the surface of the water in the cylinder, and record the reading R_0 (mm) on the graduated scale. Repeat for each sedimentation tube of the batch, recording each level separately.

About 15 s before a pipette sample is due (Table 4.16), lower the pipette G steadily with tap E closed (see Fig. 4.32) into the suspension until the tip is 100 mm below the surface.

Table 4.16.   PIPETTE SAMPLING TIMES (25°C)

Particle density of silt and clay fraction	Times after shaking of starting sampling operation					
	1st sample		2nd sample		3rd sample	
	min	s	min	s	h	min
2.50	4	30	50	30	7	35
2.55	4	20	49	0	7	21
2.60	4	10	47	30	7	7
2.65	4	5	46	0	6	54
2.70	4	0	44	30	6	42
2.75	3	50	43	30	6	30
2.80	3	40	42	0	6	20
2.85	3	35	41	0	6	10
2.90	3	30	40	0	6	0
2.95	3	25	39	0	5	50
3.00	3	20	38	0	5	41
3.05	3	15	37	0	5	33
3.10	3	10	36	0	5	25
3.15	3	5	35	0	5	18
3.20	3	0	34	30	5	10
Equivalent particle diameter mm	0.02		0.006		0.002	

This will be when the scale reading is equal to $(R_0 + 100)$ mm (or $R_0 - 100$ if the scale reads upwards). The lowering operation should take about 10 s, and must cause no turbulence in the suspension.

Open tap E at the exact sampling time, and draw a sample (volume V_p ml) up into the pipette, until the pipette and the bore in tap E are filled with suspension. A small amount may also be drawn into bulb D. The easiest way of drawing the sample up is by sucking through a length of rubber tubing attached to the inlet C. This operation should take about 10 s. Withdraw the pipette steadily from the suspension, taking about 10 s.

If any suspension has been drawn up into bulb D, wash it away into a beaker down outlet tube F by opening tap E to connect D and F. Allow distilled water to flow from bulb funnel A into D and out through F until no suspension remains.

Place a weighed glass weighing bottle under the end of the pipette and open tap E so that the contents run into the bottle. Wash any suspension on the inside of the pipette into the weighing bottle by allowing distilled water to flow from A via B, D and E into the pipette G. The last few drops of suspension can be removed from the end of the pipette by gently blowing through the tube connected to C.

(3) Determination of dry solid matter

Place the weighing bottle and contents in an oven at 105°C until the sample is evaporated to dryness.

Cool in a desiccator, and weigh carefully to the nearest 0.001 g. Accurate weighing is essential, because the quantity of soil recovered is very small. Check the mass of the weighing bottle again after cleaning out and oven drying.

From the mass of the empty bottle, determine the mass of soil in the sample by difference. This is denoted by m_1, m_2, m_3 for the first, second and third sampling operation, respectively. These masses are used for the calculations described in stage (5).

(4) Calibration sampling

At any convenient time between the stipulated sampling times, take a pipette sample from the sedimentation tube containing dispersant only, exactly as described for the soil suspensions. The time of this sampling need not be recorded.

Transfer the pipette sample to a glass weighing bottle, dry, cool and weigh accurately, exactly as for the soil suspensions. Hence determine the mass of solid material, m_r, in the dispersant sample.

(5) Calculations

The percentages finer than each sieve size in the gravel (if present) and sand size ranges are calculated as described in the appropriate section under sieving.

Data obtained from the pipette test enable the percentages of coarse silt, medium silt, fine silt and clay present in the pretreated sample (m) to be calculated, as follows. A worked example is given in Fig. 4.34.

The mass of solid material in the whole 500 ml of suspension at any sampling time can be calculated by proportion from the measured mass in the pipette volume, V_p. If the mass in 500 ml is denoted by W_1, W_2, W_3 (corresponding to sample masses m_1, m_2, m_3), then $W_1 = m_1 \times (500/V_p)$ g, and similarly for W_2 and W_3.

Similarly the mass of solid material in 500 ml of the dispersant solution, W_r, is given by the equation

$$W_r = \frac{m_r}{V_p} \times 500 \text{ g}$$

The percentage by mass (K) of particles smaller than the particle diameter (see below)

Mass of sample after pretreatment m_p = 25.36 g   Particle density = 2.65 Mg/m³   Volume of pipette

Mass retained on 63 μm sieve m_s = 3.88 g   Temperature 25°C   9.656 ml
(all finer than 212 μm sieve)

Pipette sample no. and time	Pipette mass calculation	grams	Mass in 500 ml in suspension	Percentage calculation	Particle diameter D mm
(1)	Weighing bottle no. (25)		$0.2956 \times \dfrac{500}{9.656} = 15.307$	W_1 15.307	
	Bottle + sample	5.5980		W_r 1.336	
4 min. 5s	Bottle	5.3024		$(W_1 - W_r)$ $\overline{13.971 \times 100}$ = 55.1%	0.02
	Mass of sample m_1	0.2956	$m_1 \times \dfrac{500}{V_p} = W_1$	25.36	
(2)	Weighing bottle no. (44)		$0.1742 \times \dfrac{500}{9.656} = 9.020$	W_2 9.020	
	Bottle + sample	5.6344		W_r 1.336	
46 min.	Bottle	5.4602		$(W_2 - W_r)$ $\overline{7.684 \times 100}$ = 30.3%	0.006
	Mass of sample m_2	0.1742	$= W_2$	25.36	
(3)	Weighing bottle no. (76)		$0.1104 \times \dfrac{500}{9.656} = 5.717$	W_3 5.717	
	Bottle + sample	5.9262		W_r 1.336	
6 hours	Bottle	5.8158		$(W_3 - W_r)$ $\overline{4.381 \times 100}$ = 17.3%	0.002
54 min.	Mass of sample m_3	0.1104	$= W_3$	25.36	
(4)	Weighing bottle no. (18)		$0.0258 \times \dfrac{500}{9.656} = 1.336$		
	Bottle + sample	5.4418			
About	Bottle	5.4160		m_s $\overline{3.88 \times 100}$ = 15.3%	(63 μm)
30 min.	Mass of sample m_r	0.0258	$= W_r$	25.36	
				100 − 15.3 = 84.7%	0.063

Fig. 4.34   *Pipette sedimentation test results and calculations*

corresponding to the first sampling operation is calculated from the equation

$$K = \frac{W_1 - W_r}{m} \times 100\%$$

where $m(g)$ is equal to the initial sample dry mass, m_0, if pretreatment was omitted; or is equal to the dry mass after pretreatment, m_p, if the soil was pretreated. (Either includes the mass of any material retained on the 63 μm sieve.)

The percentages smaller than the diameter corresponding to the subsequent samplings are calculated in a similar manner.

The equivalent particle diameter, D(mm), corresponding to each listed time of sampling is given in the bottom line of Table 4.16. If sampling times other than those given in Table 4.16 are used, the equivalent particle diameter can be calculated from equation (4.5), which can be written

$$D = 0.005531 \sqrt{\left\{ \frac{\eta}{(\rho_s - 1)} \cdot \frac{H}{t} \right\}}$$

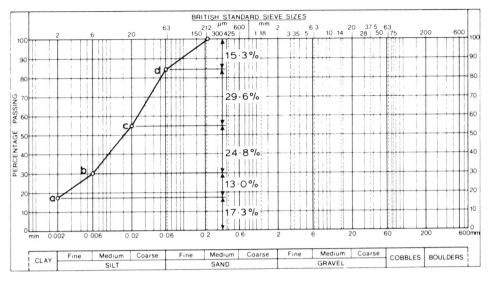

Fig. 4.35   Particle size curve from pipette analysis

or

$$D = K_1 \sqrt{\frac{H}{t}}$$

where

$$K_1 = \sqrt{\left(\frac{\eta}{\rho_s - 1} \right)}$$

Values of K_1 are the same as those used for the hydrometer test, and are tabulated for various temperatures and particle densities in Table 4.17. If the sampling depth H is exactly 100 mm, the above equation becomes

$$D = \frac{10 K_1}{\sqrt{t}}$$

In these equation, D and H are in millimetres and t is measured in minutes.

For the gravel and sand fractions (total mass m_s), calculate the percentage finer than each sieve aperture size from the mass retained on each sieve as for simple dry sieving (Section 4.6.1). Express the masses passing as a percentage of the initial sample dry mass, m_0, or pretreated dry mass, m_p, as appropriate.

(6) Plotting and presentation

The percentages finer than each size, from gravel down to fine silt, can be tabulated, expressing the values to the nearest 1%.

Alternatively, and more usually, the percentages obtained from sedimentation and sieving are plotted on a particle size distribution chart to give a continuous grading curve for the whole material. Results from the worked example are plotted in this way in Fig. 4.35.

(7) Calibration of pipette

The internal volume (V_p ml) of the sampling pipette is determined as follows.

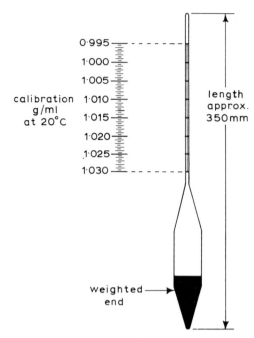

calibration
g/ml
at 20°C

0·995
1·000
1·005
1·010
1·015
1·020
1·025
1·030

length
approx.
350mm

weighted
end

Fig. 4.36   Soil hydrometer

Clean the pipette thoroughly and dry it. Immerse the nozzle in distilled water, with tap B closed, and open tap E (Fig. 4.32). Using a rubber tube attached to outlet C, suck water into the pipette until it rises above E. Close tap E and remove the pipette from the water. Pour off surplus water in the cavity above E through outlet F into a small beaker. Discharge the water in the pipette and tap E into a glass weighing bottle of known mass, and determine its mass. The internal volume V_p ml of the pipette and tap is equal to the mass of water in grams, calculated to the nearest 0.01 g.

Make three determinations of the volume, take the average and express the volume V_p to the nearest 0.05 ml.

4.8.3   Hydrometer Analysis (BS1377: Part 2:1990:9.5 and ASTM D422)

SCOPE

In this method a density hydrometer of special design is used to measure the density of a soil, pretreated as in Section 4.8.1, in a suspension in water at various intervals of time. From these measurements the distribution of particle sizes in the silt range (60–2 μm) can be assessed. The test is not usually performed if less than 10% of the material passes the 63 μm sieve.

This method can give results which are sufficiently accurate for most engineering purposes. The techniques are less exacting than those required for the pipette method. The hydrometer method has the additional advantage that it can be performed without much difficulty in a small field laboratory. If a main central laboratory also uses this procedure, the results obtained by both are directly comparable.

APPARATUS

The following apparatus is required in addition to that used for pretreatment.
(1)   Soil hydrometer of the type illustrated in Fig. 4.36. A detailed specification is given in BS1377: Part 2:1990:9.5.2.1. An essential requirement is that the scale reading is graduated to indicate density, in g/cm³ or g/ml, at scale intervals of 0.0005 g/cm³.

When used in pure water at 20°C the hydrometer should indicate the density of water at that temperature, i.e. 0.9982 g/cm^3. Each hydrometer is identified by a unique number. Before use each hydrometer must be calibrated, and a calibration curve relating 'effective depth' to 'hydrometer reading' must be drawn (see Section 4.8.4).

In ASTM D422, two types of hydrometer are specified, reference 151 H or 152 H in ASTM Specification E 100. Individual calibration is not required for these hydrometers.

(2) Two 1000 ml glass measuring cylinders about 360 mm high, marked at 1000 ml, with stoppers or rubber bungs (the sedimentation cylinders).

(3) Constant-temperature bath capable of being maintained at 25°C±0.5°C, deep enough for immersing the sedimentation cylinders to the 1000 ml mark.

(4) Stop-clock reading to 1 s.

(5) Glass rod about 12 mm diameter and about 400 mm long.

(6) Thermometer covering the range 0 to 50°C, reading to 0.5°C.

PROCEDURAL STAGES

(1) Prepare suspension.
(2) Take hydrometer readings.
(3) Correct hydrometer readings.
(4) Calculate.
(5) Present results.

TEST PROCEDURE

(1) *Preparation of suspension*

Transfer the suspension of pretreated soil passing the 63 μm sieve, obtained as described in Section 4.8.1, from the receiver into a 1000 ml sedimentation cylinder without losing any soil. Make up the suspension to the 1000 ml mark exactly with distilled water.

Place the sedimentation cylinder in the constant-temperature bath, set at 25°C. Place a second cylinder containing 100 ml of dispersant solution, made up to the 1000 ml mark with distilled water, in the constant-temperature bath. This solution will be of the same composition as the liquid in the sedimentation cylinder. If more than the usual amount of dispersant is used for the test, the concentration in the second cylinder should be the same.

Allow the cylinders to stand in the bath until they have reached the bath temperature; about 1 h is usually sufficient. Several sedimentation cylinders (up to six or eight) may be tested at the same time. When all cylinders are in the constant-temperature bath, the water level in the bach should just reach the 1000 ml graduation marks.

Insert a rubber bung into each cylinder to obtain a watertight fit, but avoid using excessive force which may split the glass and cause serious injury to the hand. Remove the cylinder containing the dispersant solution, shake thoroughly and replace it in the bath. Remove each sedimentation cylinder in turn and, if necessary, stir the sediment with the glass rod so that all the soil goes into suspension. Shake the cylinder vigorously by applying about 60 end-over-end cycles during a period of 2 minutes, and immediately replace upright in the constant temperature bath. As soon as it is in the upright position, start the stop-watch or timer clock (zero time, $t = 0$ for that sample).

(2) *Hydrometer readings*

Remove the rubber bungs from the cylinders, insert the hydrometer steadily and allow it to float freely (Fig. 4.37). It must not be allowed to bob up and down, or to rotate, when let go. However, a quick rotational twist with the fingers on the end of the stem will dislodge any air bubbles which may adhere to the side.

Take readings of the hydrometer, in the manner described in stage (3), at the top of meniscus level at the following times from zero: $\frac{1}{2}$, 1, 2, 4 min.

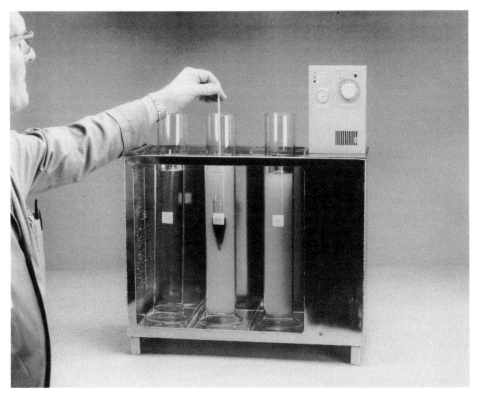

Fig. 4.37   Immersion of hydrometer in sedimentation cylinder

Remove the hydrometer slowly, rinse it in distilled water at water-bath temperature and place it in the cylinder containing dispersant solution standing in the water-bath. Record the hydrometer reading at the top of the meniscus (denoted by R'_0).

Insert the hydrometer in the soil suspension for further readings at the following times from zero, and remove and place it back in the cylinder containing dispersant solution after each reading: 8, 15, 30 min; 1, 2, 4, 8 h; overnight (about 16 h); thereafter (if necessary) twice daily. It is not essential to keep rigidly to these times, provided that the actual time of each reading is recorded on the hydrometer test sheet.

Insertion and withdrawal of the hydrometer into the suspension must be done carefully. Each operation should take about 10 s, and when released the hydrometer should be in its steady floating position.

Disturbance of the suspension, either by the hydrometer or by vibration, must be avoided. If a heater/stirrer unit is fitted to the constant-temperature bath, this must be mounted so that no vibration is transmitted to the sedimentation cylinder.

If six or eight samples are being treated together as a batch, it is best to start them at intervals of 5 min from the start of the first. Between each reading the hydrometer should be dipped into the cylinder containing dispersant solution.

The temperature of the suspension should be checked at intervals, but if a reliable constant-temperature bath is used, there should normally be no significant change in temperature throughout the test. A constant temperature of 25°C is preferable to one of 20°C, because it eliminates the necessity of cooling in all but the hottest climates, where a higher working temperature would be appropriate.

(3) *Correction of hydrometer readings*

(a) *Hydrometer readings*   Each density reading taken on the hydrometer must first be

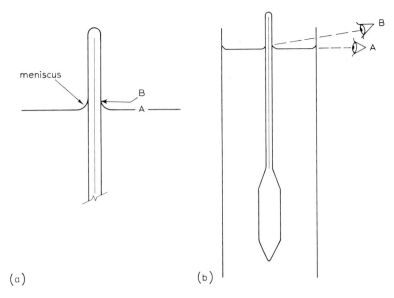

Fig. 4.38   Reading a hydrometer

expressed as a hydrometer reading, R'_h, corresponding to the level of the upper rim of the meniscus. This is done by subtracting 1 from the density and moving the decimal point three places to the right (i.e. multiply by 1000). For example, a density of 1.0325 would be a hydrometer reading of $R'_h = 32.5$.

(b) *Meniscus correction*   A hydrometer is calibrated to read correctly at the surface of the liquid in which it is immersed (level A in Fig. 4.38a). Since soil suspensions are not transparent enough to permit a reading to be taken at this level, the scale has to be read at the upper rim of the meniscus. This is shown at B in Fig. 4.38(a). It is therefore essential that the meniscus be fully developed, which means that the hydrometer stem must be perfectly clean.

The meniscus correction (C_m) has to be added to R'_h in order to obtain the true reading R_h, because density readings on the stem increase downwards. The correction C_m is a constant for a given hydrometer, and is determined as follows.

The hydrometer is inserted in a 1000 ml cylinder about three-quarters full of distilled water. The plane of the surface of the liquid is seen as an ellipse from just below the surface. The eye is raised until the surface is seen as a straight line, and the scale marking at which this plane intersects the hydrometer stem is noted (reading A in Fig. 4.38b). By looking from just above the plane of the liquid surface, the scale marking at the level of the upper limit of the meniscus is noted (reading B). The difference between the two scale readings, multiplied by 1000, is the meniscus correction:

$$C_m = (B - A) \times 1000$$

For example:

$$\text{If reading } A = 0.9985$$
$$\text{reading } B = 0.9990$$

$$(B - A) = 0.0005$$

$$C_m = +0.5$$

This is a typical value for C_m, but it must be determined for every hydrometer. The true hydrometer reading R_h is given by

$$R_h = R'_h + C_m$$

(c) *Datum reading*   No further corrections are required by BS1377:1990 because the hydrometer reading in the dispersant solution, R'_0, provides the datum to which all other readings are related, provided that it is at the same temperature as the soil suspension. The hydrometer density reading in the dispersant solution may be less than 1.000, but the same rule as described above must be followed even if the value of R'_0 is found to be negative.

For example, if the hydrometer reading in the dispersant solution is 0.9985,

$$R'_0 = (0.9985 - 1.000) \times 1000$$

$$= (-0.0015) \times 100$$

i.e., $R'_0 = -1.5$

As explained in stage (4), the modified reading R_d is used only for calculation of percentages of particles smaller than a given size. The value of $R_h = R'_h + C_m$ (i.e. meniscus correction only) is used at all temperatures for computing the particle diameter D, whether by calculation, by using tables or from the nomographic chart, because here the hydrometer is acting as a measuring rod to determine the effective depth at which the density reading applies.

(4)   *Calculations*

(a)   *True hydrometer reading*   Calculate the true hydrometer reading, R_h, for each observed reading, R'_h, by adding the meniscus correction, C_m, as described above. (Column 4 of Fig. 4.39).

(b)   *Effective depth*   From the hydrometer calibration curve (obtained as described in Section 4.8.4), obtain the effective depth, H_r (mm) corresponding to each value of R_h (column 5 of Fig. 4.39).

(c)   *Equivalent particle diameter.*   Calculate the equivalent particle diameter, D (mm), for each reading from equation (4.5) derived in Section 4.7.3:

$$D = 0.005531 \sqrt{\frac{\eta H_r}{(\rho_s - 1)t}}$$

where η is the dynamic viscosity of water (mPas) at the test temperature (see Table 4.13); ρ_s is the particle density (Mg/m^3); t is the elapsed time (min). Values of D are entered in Column 7 of Fig. 4.39.

To avoid repetitive calculations, the above equation can be written

$$D = K_1 \sqrt{\frac{H_R}{t}} \text{ mm}$$

as explained in Section 4.8.2, Stage (5). Values of K_1 for a range of working temperatures and particle densities are given in Table 4.17. This Table is based on Table 3 given in ASTM D422 (with some corrections), but the values have been calculated to be compatible with SI units of measurement.

(d)   *Modified hydrometer reading*   Calculate the modified hydrometer reading, R_d (i.e. the reading related to the reading in the dispersant solution as datum) for each observed reading

Hydrometer no.	52284	
Test temperature	25°C	
Meniscus correction	C_m = + 0.5	
Reading in dispersant solution 0.9997 R'_0 = −0.3		

Particle density	G_s = 2.65
Viscosity of water	η = 0.8909 mPas
Initial dry mass of soil	59.37
Dry mass after pretreatment	m = 58.88 g
Pretreatment loss	= 0.49 g
$\frac{0.49}{59.37} \times 100$	= 0.83 %

(1) Date Time	(2) Elapsed time t minutes	(3) Hydrometer reading R_h'	(4) True reading R_h	(5) Effective depth H_R mm	(6) Modified reading R_d	(7) Particle diameter D μm	(8) Percentage finer than D K %
1.6.78 0935	0						
	0.5	30	30.5	88.9	30.3	$4.064 \times \sqrt{\dfrac{88.9}{0.5}} = 54.2$	2.728 × 30.3 = 82.7%
	1	29.5	30	91.0	29.8	$4.064 \times \sqrt{\dfrac{91.0}{1}} = 38.8$	2.728 × 29.8 = 81.3
	2	28.5	29	95.1	28.8	$4.064 \times \sqrt{\dfrac{95.1}{2}} = 28.0$	2.728 × 28.8 = 78.6
	4	27	27.5	101.2	27.3	$4.064 \times \sqrt{\dfrac{101.2}{4}} = 20.4$	2.728 × 27.3 = 74.5
0943	8	24.5	25	111.5	24.8	$4.064 \times \sqrt{\dfrac{111.5}{8}} = 15.2$	2.728 × 24.8 = 67.7
0950	15	22.5	23	119.7	22.8	$4.064 \times \sqrt{\dfrac{119.7}{15}} = 11.5$	2.728 × 22.8 = 62.2
1005	30	20	20.5	130.0	20.3	$4.064 \times \sqrt{\dfrac{130.0}{30}} = 8.46$	2.728 × 20.3 = 55.4
1035	60	17.5	18	140.2	17.8	$4.064 \times \sqrt{\dfrac{140.2}{60}} = 6.21$	2.728 × 17.8 = 48.6
1135	120	14.5	15	152.5	14.8	$4.064 \times \sqrt{\dfrac{152.5}{120}} = 4.58$	2.728 × 14.8 = 40.4
1335	240	11	11.5	166.9	11.3	$4.064 \times \sqrt{\dfrac{166.9}{240}} = 3.39$	2.728 × 11.3 = 30.8
1705	450	8	8.5	179.2	8.3	$4.064 \times \sqrt{\dfrac{179.2}{450}} = 2.56$	2.728 × 8.3 = 22.6
2.6.78 0915	1420	5.5	6	189.4	5.8	$4.064 \times \sqrt{\dfrac{189.4}{1420}} = 1.48$	2.728 × 5.8 = 15.8

Calculations: Col.(4) $R_h = R_h' + C_m = R_h' + 0.5$ Col.(6) $R_d = R_h' - R_0'$

Col.(5) $H_R = 214 - 4.1 R_h$ (or read off calibration curve)

Col.(7) $D = 0.005531 \sqrt{\dfrac{\eta H_R}{(-1)t}} = 0.005531 \sqrt{\dfrac{0.8909 H_R}{1.65t}}$ mm $= 4.064 \sqrt{\dfrac{H_R}{t}}$ μm

Col.(8) $K = \dfrac{\rho_s}{m(-1)} \times R \times 100 = \dfrac{2.65}{58.88 \times 1.65} \times R \times 100 = 2.728 \times R\%$

Fig. 4.39   Hydrometer sedimentation test data, results and calculations

from the equation

$$R_d = R_h' - R_0'$$

Note that if R_0' is negative, R_d will be greater than R_h'. Enter the values in column 6 of Fig. 4.39.

(e) *Percentage finer than D*   Calculate the percentage by mass of particles smaller than the equivalent particle diameter D (represented by K) from the equation

$$K = \frac{100\rho_s R_d}{m(\rho_s - 1)}$$

Table 4.17. VALUES OF CONSTANTS K_1, K_2 FOR HYDROMETER TEST CALCULATIONS

Temperature °C	Particle density, ρ_s (Mg/m³)								
	2.45	2.50	2.55	2.60	2.65	2.70	2.75	2.80	2.85
	Values of K_1								
16	0.04832	0.04759	0.04683	0.04607	0.04538	0.04471	0.04408	0.04345	0.04288
17	0.04775	0.04699	0.04623	0.04551	0.04481	0.04415	0.04351	0.04288	0.04231
18	0.04718	0.04639	0.04563	0.04494	0.04424	0.04358	0.04298	0.04234	0.04177
19	0.04661	0.04582	0.04506	0.04437	0.04370	0.04304	0.04244	0.04184	0.04127
20	0.04604	0.04525	0.04452	0.04383	0.04317	0.04250	0.04190	0.04133	0.04076
21	0.04547	0.04471	0.04399	0.04329	0.04263	0.04200	0.04139	0.04082	0.04026
22	0.04494	0.04418	0.04345	0.04279	0.04212	0.04149	0.04092	0.04035	0.03978
23	0.04440	0.04367	0.04294	0.04228	0.04165	0.04101	0.04045	0.03988	0.03931
24	0.04389	0.04317	0.04244	0.04177	0.04114	0.04054	0.03997	0.03940	0.03886
25	0.04339	0.04266	0.04196	0.04130	0.04067	0.04007	0.03950	0.03896	0.03842
26	0.04291	0.04218	0.04149	0.04082	0.04022	0.03962	0.03905	0.03852	0.03798
27	0.04244	0.04171	0.04101	0.04038	0.03978	0.03918	0.03861	0.03807	0.03757
28	0.04196	0.04124	0.04057	0.03997	0.03934	0.03874	0.03820	0.03766	0.03716
29	0.04149	0.04079	0.04013	0.03950	0.03890	0.03833	0.03779	0.03725	0.03675
30	0.04105	0.04035	0.03972	0.03909	0.03848	0.03792	0.03738	0.03684	0.03633
	Values of K_2								
	169.0	166.7	164.5	162.5	160.6	158.8	157.1	155.6	154.1

(Based on Table 3 of ASTM D422)

where ρ_s is the particle density (Mg/m³); m is equal to the initial sample dry mass, m_0, if pretreatment was omitted; or is equal to the dry mass after pretreatment, m_p, if the soil was pretreated. (Either includes the mass of any material retained on the 63 μm sieve.)

Enter the values in column 8 of Fig. 4.39.

To avoid repetitive calculations, the above equation can be written

$$K = K_2 \left(\frac{R_d}{m} \right) \%$$

where

$$K_2 = \frac{100\rho_s}{\rho_s - 1}$$

Values of K_2, which depend only on the particle density, ρ_s, are given at the foot of Table 4.17.

The value of K calculated for each hydrometer reading is plotted against the corresponding particle size, drawn to a logarithmic scale, exactly as for a grading curve determined by sieving. The same graph sheet is used, the sizes of particles being extended downards to about 1 μm. Usually the test is terminated at about 2 μm, which is the lower limit of the silt size range. The intersection of the particle size curve with the 2 μm ordinate gives the percentage which is referred to as the 'clay fraction'.

The above method of calculating percentages applies only when the whole of the sample passes a 2 mm sieve, and is used for pretreatment. If the hydrometer test specimen is obtained by quartering down from a larger sample which has been put through larger aperture sieves, the calculated percentages must be adjusted as described in Section 4.8.5.

(5) Presentation of results

Calculated percentages finer than each determined size are plotted against the corresponding particle diameter on the same sheet as that used for a sieving analysis. A smooth curve is drawn through the plotted points. If a sieving analysis has also been carried out, a single continuous curve is drawn to give the particle size distribution of the whole sample (see Section 4.8.5).

Details of pretreatment, the size of sample used and the particle density used in the calculations are added to the particle size distribution sheet, together with a visual description of the soil.

Initial total dry mass after pretreatment 58.88 g

Dry mass retained on 63 μm sieve   5.92 g

Sieve size μm	Mass retained g	Cumulative mass passing	Percent passing
600	0	58.88	100%
425	1.46	$\dfrac{-1.46}{57.42}$	$\dfrac{57.42}{58.88} \times 100 = 97.3\%$
212	1.74	$\dfrac{-1.74}{55.68}$	$\dfrac{55.68}{58.88} \times 100 = 94.7\%$
150	0.55	$\dfrac{-0.55}{55.13}$	$\dfrac{55.13}{58.88} \times 100 = 93.5\%$
63	1.94	$\dfrac{-1.94}{53.19}$	$\dfrac{53.19}{58.88} \times 100 = 90.3\%$
Pass	$\dfrac{0.23}{5.92}$	$\dfrac{-0.23}{52.96}$	

Fig. 4.40   Sieving data and calculations relating to hydrometer test

(6)   *Typical results and calculations*

A typical set of results obtained from a hydrometer sedimentation test is given in Fig. 4.39, and the sieving data for the dried pretreated portion retained on the 63 μm sieve are given in Fig. 4.40.

The method of calculation of particle diameters, and of percentages finer than each particle size, from the sedimentation test is shown in Fig. 4.39.

The percentages corresponding to each particle size are plotted on the particle size sheet in Fig. 4.41, for both the sieving test and the sedimentation test. The masses retained on the sieves are calculated as percentages of mass m_p, the total dry mass *after* pretreatment. In this example the first point from the sedimentation test does not lie on the smooth curve connecting the sieving curve with the remaining points. This is not uncommon, and the reason for the apparent discrepancy is discussed in Section 4.8.5.

4.8.4   Calibration of Hydrometer

No two hydrometers are exactly alike, and every hydrometer has a unique identification number. This number must be recorded on the calibration data sheets, and on every test sheet for which it is used.

The hydrometer must be calibrated in the cylinder in which it is to be used. This is because the cross-sectional area, A, of the cylinder comes into the calibration calculations. In practice, the sectional area varies but little between similar measuring cylinders of one batch; nevertheless each cylinder used should be checked. The sides of the cylinder should be parallel, so that the sectional area is constant throughout its length.

To determine the sectional area A, measure the distance, L, in millimetres between two well-spaced graduations (such as 100 and 1000 ml) on the cylinder. The volume included between these two marks is 900 ml, so the sectional area, A, is given by

$$A = \frac{900}{L} \times 1000 \, mm^2$$

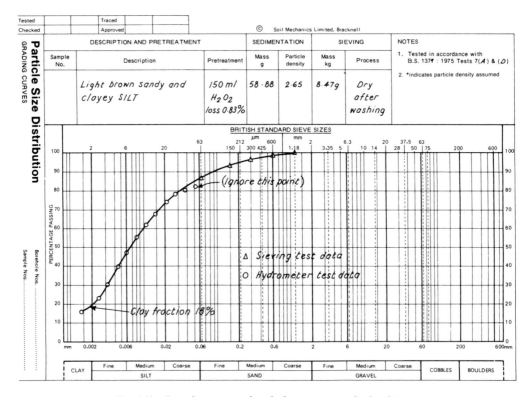

Fig. 4.41   *Particle size curve from hydrometer test and related sieving*

If the cylinder does not have a graduation mark at 100 ml, determine the level by adding a measured 100 ml (or 100 g) of water. Mark a piece of adhesive tape at the bottom of the meniscus when the cylinder is standing on a level surface.

On the hydrometer itself the distance from the neck of the bulb to the lowest calibration mark is measured, to the nearest millimetre, with a steel ruler. This is denoted by N in Fig. 4.42. The distances l_1, l_2, etc., from this calibration mark to each of the other major marks are measured as shown, to the nearest millimetre, and tabulated. The distance H corresponding to each graduation mark is given by $(N + l_1)$, $(N + l_2)$, etc. For the lowest mark, $H = N$.

The distance from the neck to the bottom of the bulb is measured, in millimetres, and is denoted by h. This measurement can be made either by laying the hydrometer flat on a sheet of paper and projecting down on to the paper using a set-square, or by holding the hydrometer vertically and projecting across with a square to a metre-stick held vertically by a burette stand.

The volume of the hydrometer bulb, V_h, can be measured by weighing the hydrometer to the nearest 0.1 g and equating the mass in grams to its volume in ml. Alternatively, the rise in level of water in a 1000 ml cylinder, initially filled to the 800 ml mark, can be measured. In both methods there is a small error due to the inclusion of the stem or part of it, but this can be neglected for practical purposes.

If the hydrometer bulb is of symmetrical shape, no further measurements are necessary, but if it is not symmetrical, the position of the centre of volume of the bulb must be determined. This can be done with sufficient accuracy by projecting the shape of the bulb on to a sheet of paper and estimating the position of the centre of gravity of the outline. The distance of the centre of gravity of the bulb from the neck is denoted by h_g (see Fig. 4.42), and for a symmetrical bulb $h_g = \frac{1}{2}h$.

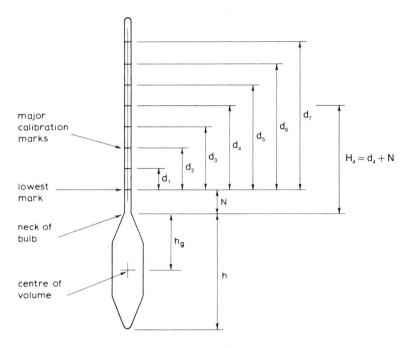

For a symmetrical bulb, $h_g = \dfrac{h}{2}$

Fig. 4.42 Measurements for calibration of hydrometer

The effective depth H_R (mm), corresponding to each major graduation mark R_h, is calculated from the equation

$$H_R = H + (h - h_g) - \frac{1000 V_h}{2A}$$

If the hydrometer bulb is symmetrical, this equation becomes

$$H_R = H + \tfrac{1}{2}\left(h - \frac{1000 V_h}{A}\right)$$

In practice, calculation of the area A, and the use of zeros which cancel out, can be avoided by re-writing this equation (using L as defined above) in the form

$$H_r = H + \tfrac{1}{2}\left(h - \frac{V_h}{900} L\right)$$

Values of H_R are plotted against R_h on ordinary graph paper, and a smooth curve is drawn through the points as shown in Fig. 4.43. The curve usually approximates to a straight line over the range used. This relationship takes into account the effective depth of the suspension at the level being considered at a given time, and allows for the rise of liquid in the cylinder due to displacement by the hydrometer.

By measuring the slope of the calibration line, and reading off its intercept on the H_R axis, the equation of the calibration line can be written in the form

$$H_r = j_1 - j_2 R_h$$

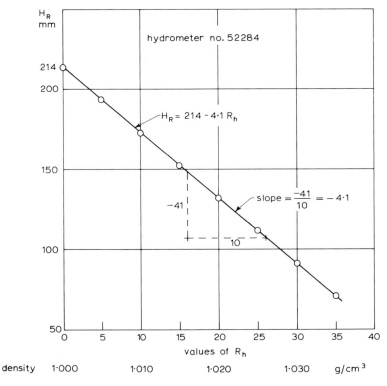

Fig. 4.43 *Hydrometer calibration curve*

where j_1 is the intercept on the H_r axis corresponding to $R_h = 0$; j_2 is the slope of the line (which is mathematically negative, hence the minus sign).

A graph such as that shown in Fig. 4.43, which is derived from the data and calculations shown as an example in Fig. 4.44, must be derived for each hydrometer. The calibration equation for this example is

$$H_r = 214 - 4.1 R_h$$

The calibration data, calculations and graph given here are for illustrative purposes only.

Since the graphical representation of the hydrometer is approximately linear, it can easily be incorporated into a computer programme. This enables the values of the equivalent particle diameter, D, and the percentages finer than D, to be calculated directly from the observed readings of R_h and time t, after entering values of C_m, η, ρ_s and m which are constants for a particular test.

4.8.5 Combined Sieving and Sedimentation

If the sample under test consists of particles ranging in size from sand size or larger down to silt or clay size, the results of sieving and sedimentation tests must be combined to give a single continuous curve. The method of calculation depends on the type of soil, and the following categories cover most requirements:

(1) Soils containing particles up to but not exceeding 2 mm.
(2) Non-cohesive soils containing particles larger than 2 mm.
(3) Cohesive soils containing gravel or larger sizes, including 'boulder clay'.

Date of Calibration 5.4.90   Hydrometer no. 52284
Calibrated by R.M.S.   Cylinder no. 17

Cylinder

Distance from 100 ml to 1000 ml graduation marks L = 318 mm

area of cross-section $A = \dfrac{900}{318} \times 1000 = 2830 \text{ mm}^2$

Hydrometer

Mass = 81.2 g

Volume $V_h = 81.2 \text{ cm}^3$

Height of bulb h = 158 mm

$\dfrac{V_h L}{900} = \dfrac{81.2 \times 318}{900} = 28.7 \text{ mm}$

$158 - 28.7 = 129.3$

$\therefore \dfrac{1}{2}\left(h - \dfrac{V_h L}{900}\right) = \dfrac{129.3}{2} = 64.6$

Neck to lowest graduation (1.030)
  N = 26 mm

say 65 mm

Measurements

Scale mark (g/cm³)	Reading R_h	Distance from lowest mark d (mm)	$d + N$ $= H$ (mm)	$H + \dfrac{1}{2}\left(h - \dfrac{V_h L}{900}\right)$ $= H_R$ (mm)
1.030	30	0	26	91
1.025	25	21	47	112
1.020	20	41	67	132
1.015	15	61	87	152
1.010	10	82	108	173
1.005	5	103	129	194
1.000	0	123	149	214
0.995	− 5	144	170	235

Fig. 4.44   Calibration of hydrometer

Each calculation procedure is described below. They apply equally whether the sedimentation curve was obtained by the pipette analysis or the hydrometer procedure.

SOILS CONTAINING PARTICLES UP TO 2 mm

With this type of soil a representative portion of the whole sample is taken for the pretreatment procedure (Section 4.8.1) followed by sedimentation (Section 4.8.2 or 4.8.3).

  The material retained on the 63 μm sieve after washing is dried, resieved and weighed as described in Section 4.6.4. Percentages passing each sieve are calculated on the basis of the initial dry mass of soil (m_0) or the dry mass of soil (m_p) remaining *after* pretreatment. The fine material washed through the 63 μm sieve is used for the sedimentation test, but each size fraction is calculated as a percentage of the mass (m) of soil used (m_0 or m_p), *not* the mass transferred to the sedimentation cylinder. An example of the calculation procedure is given in Figs. 4.39 and 4.40.

  The calculated percentages can be plotted directly on to the particle dize sheet *without further correction*, against the relevant particle size. A smooth curve is drawn through the plotted points, as in Fig. 4.41. The first two or three points calculated from hydrometer readings sometimes do not lie on a smooth curve connecting the sieving and the sedimentation portions of the graph. This is partly because in the early stage of the sedimentation test the assumptions made in the theory based on Stoke's Law (Section 4.7.3) may not be strictly valid. In addition, some of the coarse silt particles are retained on the 63 μm sieve when wet sieved, owing to the effects of surface tension. These particles reappear when dry sieved and unless they are added to the sedimentation cylinder, there is a small deficiency in this size range. The initial readings should be ignored if they do not lie on a smooth curve continued from the sieving curve.

(2) NON-COHESIVE SOILS CONTAINING PARTICLES LARGER THAN 2 mm

These soils require a sample of more than 100 g, the minimum quantity depending upon the maximum sieze of the particles (see Table 4.4). It is usually necessary to riffle the portion passing the 2 mm sieve to obtain a sample of a size suitable for pretreatment and sedimentation. As far as the sieving test is concerned, calculations of percentages are similar to those described for composite sieving (Section 4.6.2). However, allowance should be made for the fact that part of the sample has been lost in the pretreatment process, as follows.

Let m_1 = dry mass of original sample; m_2 = dry mass passing 2 mm sieve; m_3 = mass of riffled portion of m_2 used for pretreatment, fine sieving and sedimentation; m_p = dry mass of test sample after pretreatment. The loss due to pretreatment, expressed as a percentage of the riffled mass, m_3, is equal to

$$\frac{m_3 - m_p}{m_3} \times 100\%$$

If this loss is small, say 1% or less, sieving percentages can be based on the mass m_1 for particle sizes greater than 2 mm, and on the pretreated mass m_p for particle sizes from 2 mm to 63 μm. The latter are corrected to percentages of the original mass by multiplying by the factor m_2/m_1, as explained in Section 4.6.2. Any errors introduced by this simplification will be insignificant.

If the pretreatment loss is not small, a further correction should be applied. It can be assumed that the percentage of material which would be removed by pretreatment of the unriffled sample, m_2, is the same as that removed from the riffled portion, m_3. The mass removable by pretreatment would therefore be

$$\frac{m_3 - m_p}{m_3} \times m_2$$

If it can be assumed that particles larger than 2 mm consist of minerals, such as quartz, which are unaffected by pretreatment, this loss would be the same if the whole of the original sample, m_1, where pretreated. The mass of the whole sample remaining after pretreatment (m_0) would be given by

$$m_0 = m_1 - \left(\frac{m_3 - m_p}{m_3}\right) \times m_2$$

The mass passing the 2 mm sieve would be reduced by the same amount to a corrected value of m_4, where

$$m_4 = m_2 - \left(\frac{m_3 - m_p}{m_3}\right) \times m_2$$

The mass m_0 should be used in place of m_1 for calculating sieving percentages of particles larger than 2 mm. Percentages retained on sieves below 2 mm should be corrected by the factor m_4/m_0 instead of m_2/m_1.

Calculations illustrating this procedure are given in the upper part of Fig. 4.45.

The sedimentation test is carried out on the material washed through the 63 μm sieve, as

Sieve size	Mass retained	Mass passing		Percentage passing
		Measured	Corrected	
20 mm	0	$m_1 = 2517$	$m_0 = 2482$	100
6.3	484	$\dfrac{-484}{2033}$	$\dfrac{-484}{1998}$	$\dfrac{1998}{2482} \times 100 \qquad = 80.5$
2	403	$m_2 = \dfrac{-403}{1630}$	$m_4 = \dfrac{-403}{1595}$	$\dfrac{1595}{2482} \times 100 \qquad = 64.3$

RIFFLED

$m_3 = 91.26$ (1538.74)

			Corrected	Percentage passing
		Pretreated	Pretreatment loss	Equivalent loss from original sample
		$m = 89.32$ (loss 1.94)	$= \dfrac{1.94}{91.26} \times 100 = 2.13\%$	$= \dfrac{2.13}{100} \times 1630 \qquad = \ \ 34.7\ g$

WASH on 63 µm

Retained 51.64   (Sedimentation test 37.68)

$\therefore\ m_0 = 2517 - 34.7 = \underline{2482.3\ g}$

and $m_4 = 1630 - 34.7 = \underline{1595.3\ g}$

Sieve size	Mass retained	Mass passing	Percentage passing
		$m = 89.32$	
600 µm	21.52	$\dfrac{-21.52}{67.80}$	$\dfrac{67.80}{89.32} \times \dfrac{1595}{2482} \times 100 \quad = \quad 48.8$
212	14.68	$\dfrac{-14.68}{53.12}$	$\dfrac{53.12}{89.32} \times 64.3 \qquad\qquad = \quad 38.2$
63	15.31	$\dfrac{-15.31}{37.81}$	$\dfrac{37.81}{89.32} \times 64.3 \qquad\qquad = \quad 27.2$
		37.68	
Passing 63	$\dfrac{0.13}{51.64}$	add to sedimentation test $\dfrac{0.13}{37.81}$	

Sedimentation tests

(a)   Hydrometer method

Fully corrected hydrometer reading for a given particle diameter, derived as in Fig. 4.39 $= R$

$$P = \frac{G_s}{89.32\,(G_s - 1)} \times R \times 64.3$$

Example:

$R_h = 13.5$ at $t = 60$ minutes

$R = 13.5 - 0.7 = 12.8$

$H_R = 214 - (4.1 \times 13.5) = 158.65$

$D = 0.004064 \sqrt{\dfrac{158.65}{60}} = 0.00661$ mm

$$P = \frac{2.67}{89.32 \times 1.67} \times 12.8 \times 64.3$$
$$= 14.7\%$$

(b)   Pipette method

Percentage finer than each size, calculated as in Fig. 4.34 $= p$

$P = p \times 64.3$

NOTE: The mass used in the sedimentation calculations is the mass after pretreatment (m = 89.32 g), *not* the mass (37.81 g) transferred to the sedimentation cylinder.

Fig. 4.45   Calculations for combined sieving and sedimentation tests

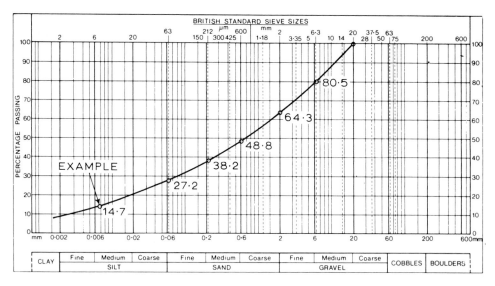

Fig. 4.46   *Grading curve from combined sieving and sedimentation test data*

described in Section 4.8.2 or 4.8.2. The percentage of each size fraction is first calculated on the basis of the mass of soil after pretreatment (m_p), not the mass passing the 63 μm sieve. This is then corrected to the percentage of the whole sample by multiplying by the factor m_2/m_1, or m_4/m_0 if the pretreatment loss was significant. The procedures for both hydrometer and sieving tests are indicated in the lower portion of Fig. 4.45.

The complete grading curve is shown in Fig. 4.46.

(3) 'BOULDER CLAY'

The procedure for carrying out a sieve analysis on materials of boulder clay type is described in Section 4.6.7. This enables the grading curve to be plotted down to 63 μm, as shown in Fig. 4.26 and represented by curve (S), portion b–c in Fig. 4.47.

The subsample of material smaller than 2 mm (Section 4.6.7; Selection of sample) is treated as described in method (1) above to obtain a combined sieving and sedimentation curve from 2 mm down to the clay size range. This is shown as curve (F) in Fig. 4.47. It may be assumed that the distribution of particles finer than 63 μm in the fine soil fraction (F) is representative of the original sample, but the distribution from 63 μm up to 2 mm may not be properly representative of the whole.

Curves (F) and (S) are combined by adding the sedimentation curve to the main sieving curve, joining them at 63 μm. If necessary, curve (F) should first be smoothed by rejecting the points derived from the early hydrometer readings, as described in method (1). The sedimentation curve is scaled down by the ratio of percentages passing 63 μm read off from the two curves, as follows.

Let P_1 = percentage passing 63 μm as measured by wet sieving on the total sample; P_2 = percentage smaller than 63 μm on the 'fines' fraction used for sedimentation; y = percentage smaller than any given particle size on the hydrometer test curve; x = percentage smaller than the same particle size on the complete corrected grading curve. Then x (%) is calculated from the equation

$$x(\%) = \frac{P_1}{P_2} \times y(\%)$$

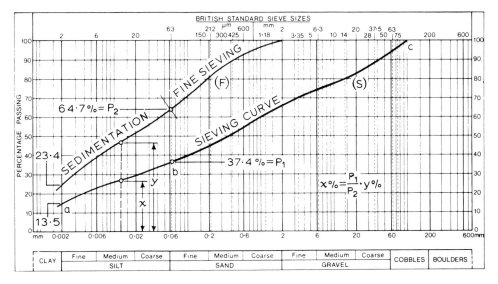

Fig. 4.47 Combination of separate sieving and sedimentation curves ('boulder clay' type of soil)

As an example, in Fig. 4.47

$$P_1 = 37.4\%$$

$$P_2 = 64.7\%$$

At the particle diameter shown by the dashed line, $y = 46.5\%$.
The percentage ($x\%$) for this diameter on the complete curve is given by

$$x = 64.5 \times \frac{37.4}{64.7} = 26.9\%$$

Similarly the clay fraction, expressed as a percentage of the whole sample, is equal to

$$23.4 \times \frac{37.4}{64.7} = 13.5\%$$

The corrected sedimentation curve is that denoted by the portion a–b in Fig. 4.47, and a–b–c is the grading curve for the whole sample.

BIBLIOGRAPHY

Allen, T. (1974). *Particle Size Measurement*. Chapman & Hall, London
BS410:1969, 'Test sieves'. British Standards Institution, London
BS5930:1981, 'Site investigations'. British Standards Institution, London
Casagrande, A. (1931). *The Hydrometer Method for Mechanical Analysis of Soils and other Granular Materials*. Cambridge, Mass.
Casagrande, A. (1947). 'Classification and identification of soils. Discussion of grain-size classifications and of methods for representing the results of mechanical analysis'. *Proc. Am. Soc. Civ. Eng.*, June 1947
Fuller, W. B. and Thompson, S. E. (1907). 'The laws of proportioning concrete'. *Trans. Am. Soc. Civ. Enf.*, Vol. 59
Glossop, R. and Skempton, A. W. (1945). 'Particle size in silts and sands'. *Proc. Inst. Civ. Eng. Paper No. 5492*

Kaye, G. W. C. and Laby, T. H. (1973). *Tables of Physical and Chemical Constants*, 14th edition. Longmans, London

Lambe, T. W. (1951). *Soil Testing for Engineers*. Wiley, New York

Neville, A. M. (1975). *Properties of Concrete*, 2nd edition (metric), Chapter 3. Pitman, London

Rothfuchs, G. (1935). 'Particle size distribution of concrete aggregates to obtain maximum density'. *Zement*, Vol. 24, No. 1

Stokes, Sir George G. (1891). *Mathematical and Physical Paper III*. Cambridge University Press.

Transport and Road Research Laboratory. Road Note 29: 'A guide to the structural design for pavements for new roads'. HMSO, London

West, G. and Dumbleton, M. J. (1972). 'Wet sieving for the particle size distribution of soils'. TRRL Report No. LR 437. Transport and Road Research Laboratory, Crowthorne, Berks.

Chapter 5

Chemical tests

5.1 INTRODUCTION

5.1.1 Scope

The detailed chemical composition of soil is of little interest for civil engineering purposes, but the presence of certain constituents can be very significant. These include organic matter, sulphates, carbonates and chlorides. The pH reaction (acidity or alkalinity) of the groundwater can also be of importance.

Chemical testing in a soil laboratory is usually limited to routine tests for the determination of the following:

(1) Acidity or alkalinity (pH value).
(2) Sulphate content.
(3) Organic content.
(4) Carbonate content.
(5) Chloride content.
(6) Total dissolved solids (in water).

The relevant tests are given in BS1377: Part 3:1990, and are described in this chapter, together with some additional procedures.

Chemical tests for the presence of other substances would normally be carried out by a specialist chemical testing laboratory, as could the tests referred to above if adequate facilities are not available in the soil laboratory. In any case these chemical tests should be performed by a chemist or by a technician who has been specially trained in chemical testing procedures. Any unusual behaviour or observation should be referred to a qualified chemist for further analysis.

5.1.2 Types of Test

The tests described in this chapter are listed in Table 5.1, which includes notes on their use and their limitations.

5.1.3 Relevance and Accuracy of Chemical Tests

In the quantitative chemical analysis of soils by far the biggest source of possible error is in the selection of the test sample. Usually a very small sample of dried soil is required at the outset, and it is essential that this sample be truly representative of the original sample. The proper procedure of mixing, riffling and quartering, described in Section 1.5.5, must be rigidly adhered to. Short cuts in this procedure lead to inconsistent results.

Results of chemical tests on soils should be regarded as an indication of the order of magnitude of constituents for classification purposes rather than as precise percentages. The British Standard test procedures provide accurate enough results for most soils, but with

Table 5.1 CHEMICAL TESTS FOR SOILS AND GROUNDWATER

Section reference	Type of test	Procedure	Reference ('BS' implies BS1377:1990)	Uses	Limitations and comments
5.5	pH value	Indicator papers	Supplier's instructions	Simple and quick. Useful for determining approximate pH range for a more sensitive test.	Gives approximate values only.
		Colorimetric (Kuhn's method)	Manufacturer's instructions	Quick field test for soils. Apparatus available as a kit.	Requires colour comparison with chart printed in British standard. Values given to nearest 0.5.
		Lovibond Comparator	Manufacturer's instructions	Colour comparison with standard coloured discs gives pH to nearest 0.2. Range of indicators available.	
		Electrometric	BS Test Part 3:9	BS 'standard' method. Accurate to 0.1 pH unit or better.	Requires a special electrical apparatus, although low-priced portable battery models are available. Electrodes age slowly, and should be checked periodically with buffer solutions.
5.6	Sulphate content	Total sulphates in soils	BS Part 3:5.2, 5.5 BRE Digest 250	Accurate if performed with care and with proper chemical testing facilities. Gives the total amount of sulphates present, including calcium sulphate, which is insoluble in water.	If the measured sulphate content is greater than 0.5%, the water-soluble sulphates should also be measured.
		Water-soluble sulphates in soils (gravimetric)	BS Part 3:5.3, 5.5	Accuracy as above. Gives the amount of water-soluble sulphates only, which are those most likely to attack concrete.	
		Water-soluble sulphates in soils (ion exchange)	BS Part 3:5.3, 5.6	Quick, easy	Cannot be used if chloride, nitrate or phosphate ions are present. Requires a special ion-exchange resin which needs reactivating frequently.
		Sulphates in groundwater (ion exchange)	BS Part 3:5.4, 5.6	As above.	As above.
		Sulphates in groundwater (gravimetric)	BS Part 3:5.4, 5.5	As for water-soluble sulphates in soils.	

		Method	Standard	Description	Remarks
5.7	Organic content	Peroxide oxidation		Eliminates organic matter before sedimentation particle size tests.	Has limited action on undecomposed plant remains (e.g. roots and fibres).
		Dichromate oxidation	BS Part 3:3	Accurate, if proper chemical testing facilities used. Suitable for all soils. Presence of carbon and carbonates does not affect results. Fairly rapid, suitable for small batches.	Presence of chlorides affects results but a correction can be applied if chlorides are measured separately. Their effect can be overcome by adding mercuric sulphate.
5.8	Carbonate content	Rapid titration	BS Part 3:6.3	For carbonate content exceeding 10%	Accuracy no better than 1% carbonates
		Gravimetric	BS Part 3:6.4	Requires precision weighing and chemical testing facilities.	Method as used for hardened concrete.
		Calcimeter (Collin's modification of Schleiber's apparatus)	BS1881: Part 124	Compact, simple, fairly quick. Measures the volume of carbon dioxide evolved.	An approximate method, but accurate enough for most engineering purposes. Atmospheric pressure must be known.
5.9	Chloride content	Reaction with silver nitrate (Volhard's method): Water soluble Acid soluble	BS Part 3:7.2 BS Part 3:7.3	Titration process requiring proper chemical testing facilities. Designed for concrete aggregates.	Several standardised reagents are required
		Mohr's titration method	Bowley (1979)	Simpler than Volhard's method. Designed for concrete aggregates.	Both methods require an analytical balance.
5.10.2	Total dissolved solids	Evaporation	BS Part 3:8	Simple procedure.	Requires very accurate weighing.
5.10.3	Loss on ignition	Ignition	BS Part 3:4	Destroys all organic matter. Suitable for sandy soils containing little or no clay or chalk.	High temperature breaks down certain minerals in clay, and carbonates, and removes water of crystallisation.
5.10.4	Concentration of certain salts	Indicator papers	Manufacturer's instructions	Very simple, quick, inexpensive.	Gives approximate indication only; not for accurate work. Presence of salts other than those being tested might affect readings.

some soils (especially tropical soils) there is the possibility that the presence of other constituents could have an undetermined effect on the chemical process for the measurement of a particular substance.

However, this does not mean that care and accuracy are not important. All chemical testing procedures require extreme care, high accuracy and clean working conditions. The atmosphere must be dust-free, and this means that operations such as grinding soil, sieving and mixing of dry soil should not be carried out in the chemical testing area. Smoking should not be permitted.

Processes such as acid extraction require the use of a fume-cabinet and air-extractor fan to take away obnoxious gases and fumes. If a separate laboratory for chemical testing is impracticable, a portion of bench space under an air-extraction hood should be allocated solely to chemical tests so as to minimise the effect of fumes, and to keep glassware and other delicate apparatus away from soils testing equipment.

Duplicate determinations should always be carried out for each test measurement.

5.1.4 Practical Aspects

SAFETY

Handling and use of chemicals requires extreme care, and the technician must be aware of the possible hazards involved. Some of the precautions which should be observed are outlined in Section 1.6.8, which should be studied before proceeding with chemical tests.

BATCH TESTING

The apparatus listed for each test is that required for a single analysis. In a commercial laboratory many tests, such as those for the determination of sulphate and chloride content, have to be carried out in batches, typically six at a time. In a specialist chemical laboratory 20 or more tests may be carried out in a batch. Where several tests are run concurrently, additional numbers of items such as beakers, funnels and crucibles are required, but there is no need to duplicate the major items, such as ovens and burettes.

CONCENTRATION OF SOLUTIONS

The traditional methods of expressing amounts-of-substance concentrations of solutions in terms of molarity (M) and normality (N) are now obsolescent and are not recognised by the International System of Units (SI) nor by British Standards. The SI derived unit for the amount-of-substance concentration of volumetric solutions is the mole per litre (symbol mol/l). The mole (SI basic unit for amount of substance) is defined in Appendix A.1.3, and for most practical purposes is the same as the molecular mass in grams. A chemical formula must always be associated with the units mole and mol/l, in order to identify the molecular species considered; the name is not sufficient. The formula (in brackets) is preceded by the symbol $\underline{c}$, meaning 'concentration of'.

A solution with a concentration of 1 mol/litre of a substance contains the molecular weight, in grams, of the substance in distilled water to make up one litre of solution. The masses of constituents required to give 1 mol/l solutions of substances used for the tests described in this chapter are listed in Table 5.3.

As an example, a solution containing 40 g of sodium hydroxide per litre would be written

$$\underline{c}\,(NaOH) = 1\,mol/l$$

Solutions of different concentrations are designated by using the appropriate factor; for example, a solution containing 10 g/litre of NaOH is written

$$\underline{c}\,(NaOH) = 0.25\,mol/l$$

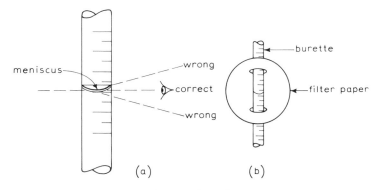

Fig. 5.1   Reading a burette

The unit mol/l is numerically the same as the obsolescent molarity (M); or the (obsolete) normality (N) divided by the relevant valency.

When a litre of standard solution is made up, the constituent should first be dissolved in about three-quarters of that volume of distilled water. This is to allow for the resulting increase in volume. When the constituent is dissolved, further distilled water is added to make 1 litre of solution.

All chemicals used should be of analytical reagent (AR) grade. In a few instances, where the highest accuracy is necessary, a calibration check on a 'standard' solution is required as part of the test procedure.

USE OF BURETTE

Volumetric analysis usually requires the use of a burette for the measurement of the volume of a solution used, which is more accurate than the use of a measuring cylinder. For reliable results attention must be paid to the following points:

(1)   The burette must be clean.
(2)   The tap must not leak.
(3)   The tap should be lubricated with only a trace of a lubricant such as Vaseline (*not* silicone grease).
(4)   The jet should be fine enough for the rate of discharge not to exceed about 20 ml/min.
(5)   The burette must be properly clamped to a burette stand so that it is exactly vertical.

Burettes are usually calibrated at intervals of 0.1 ml, and numbered (downwards) every 1 ml. When the liquid level in a burette is being read, observe the bottom of the curve of the dark part of the meniscus, as indicated in Fig. 5.1(a). The eye must be at the same level as the meniscus, to avoid parallax errors.

A sheet of white filter paper slipped around the burette as in Fig. 5.1(b) makes the level of liquid easier to read. Generally, readings to the nearest half division (0.05 ml) are accurate enough for soils testing purposes.

Fill the burette from the open end, using a small funnel. Make sure that any air trapped in the tap or jet is removed by running a little liquid through, before taking the first reading.

PIPETTES

For the delivery of a predetermined volume of liquid in multiples of 5 or 10 ml up to 50 ml, a pipette is convenient, quick and sufficiently accurate for soils testing purposes. A bulb pipette is most often used, but a graduated narrow-tube pipette is required for accurately measuring volumes of less than 1 or 2 ml. They should be used with a suction syringe; acids and other chemical solutions should *never* be drawn into a pipette by mouth suction.

FILTERING UNDER VACUUM

Flasks used for vacuum filtration must be purpose-made of thick glass to withstand the external atmospheric pressure (see Section 1.6.10). When a Buchner funnel is used, the holes are covered with a filter paper which must be of the correct diameter. The stem of the funnel should extend below the level of the side-arm to which the vacuum line is connected.

A conical funnel fitted with a cone of paper is not recommended for filtration by suction.

POLICEMAN

A 'policeman' is a wiper consisting of a rubber sleeve that is slipped on to the end of a stirring rod, covering the end of the rod. A piece of rubber tubing may not be adequate. The policeman is used for wiping the inside of a beaker when a precipitate or solid particles is being transferred to another vessel. It should be made of gum rubber, and should be discarded as soon as it shows signs of going hard.

5.1.5   Initial Preparation of Soil

The size of the soil sample normally required for chemical tests is very small, and it is essential that the specimen used for the test is properly representative of the soil as a whole. (See Section 5.1.3).

The initial preparation procedure described below is common to all chemical tests, except that there are a few variations (e.g. drying temperature) which are indicated as appropriate. The crushing procedure is applicable only to certain tests.

APPARATUS

Balance, readable to 1 g.
Balance, readable to 1 mg or 0.1 mg (as required for the relevant test).
Drying oven, capable of being controlled to maintain the temperature specified.
Desiccator and desiccant.
Test sieve, 2 mm aperture.
Test sieves, as specified: 425 μm aperture
                                            63 μm aperture.
Riffle boxes of suitable sizes (including the smallest size shown in Fig. 1.25, with slot width openings of about 7 mm).
Pestle and mortar, or suitable mechanical crusher.
Rubber-headed pestle.
Large metal tray suitable for mixing soil.
Small weighed corrosion-resistant metal trays suitable for containing soil for oven-drying.
Weighed corrosion-resistant containers for oven-drying.

INITIAL PREPARATION

(1)   From the initial soil sample, obtain a representative sub-sample of a mass not less than the following.

Fine-grained soil	100 g
Medium-grained soil	500 g
Coarse-grained soil	3 g

The three categories of soil are as defined in Section 1.1.7. The sub-sample should be prepared as described in Section 1.5.

(2)   Dry this sample (usually in an oven) at the specified temperature, and cool in a desiccator. In this and subsequent drying operations, dry to constant mass, i.e. after weighing return to the oven for a further period of 4 hours, then cool and weigh again.

The difference in successive weighings should not exceed 0.1% of the specimen mass. Appropriate drying temperatures for the various tests are as follows.

pH value	air dry (room temperature)
Sulphate content	70°–80°C
Organic content	50° ± 2.5°C
Carbonate content	105°–110°C
Chloride content	105°–110°C
Loss on ignition	50° ± 2.5°C

(3) Weigh the dried sample to within 0.1% of its mass, and record the mass (m_1) in grams.

(4) Sieve the sample on a 2 mm sieve, guarded if necessary by a sieve or sieves of larger aperture. Crush any retained particles, other than stones, to pass the 2 mm sieve.

(5) Reject the stones, but remove any fine material adhering to them and add it to the fine fraction. Weigh the material passing the 2 mm sieve to 0.1% (m_2).

(6) Take care to ensure that no fines are lost during the above, and subsequent, operations.

(7) Sub-divide the material passing the 2 mm sieve by successive riffling, using an appropriate riffle box, to produce an initial sample of the mass specified for the test. Normally this should be sufficient to provide at least two test specimens.

CRUSHING PROCEDURE

The following steps are used only when specified for the test.

(8) Sieve the riffled sample on the 425 μm sieve, and crush all retained particles to pass the sieve. Mix the crushed material thoroughly with the material already passing the sieve.

Throughout this and subsequent operations, mix the material thoroughly before dividing, and avoid segregation during riffling.

(9) Sub-divide this sample by riffling to obtain test samples each of the required mass.

(10) Place each portion in a glass weighing bottle which has been weighed to the nearest 0.001 g, and dry in the oven at the specified temperature to constant mass (i.e. differences in successive weighings at intervals of 4 hours should not exceed 0.1% of the specimen mass).

(11) Allow to cool to room temperature in the desiccator.

(12) Weigh to the nearest 0.001 g, and calculate the mass of dry soil by difference.

5.2 DEFINITIONS AND DATA

5.2.1 Definitions

ION A charged atom, molecule or radical whose migration effects the transport of electricity through an electrolyte.

ELECTROLYTIC DISSOCIATION The (reversible) breaking down of a substance into ions on dissolution in a suitable liquid.

pH VALUE The logarithm to base 10 of the reciprocal of the concentration of hydrogen ions in an aqueous solution. It provides a measure of the acidity or alkalinity of the solution, on a scale reading from 0 to 14, on which 7 represents neutrality.

INDICATOR A substance which is capable of giving a clear visual indication of the completion of a chemical reaction in a solution being titrated, usually by means of a change in colour.

TITRATION The addition of a solution from a graduated burette to a known volume of a second solution, until the chemical reaction between the two is completed. If the strength of one of the solutions is known, that of the other can be calculated from the volume of liquid added.

MOLAR SOLUTION   A solution containing the gram-molecular weight (molecular weight in grams) of a substance in distilled water to make up one litre of solution.

MOLE   The amount of substance, as represented by the molecular mass in grams.

5.2.2   Atomic Masses

Atomic masses of elements referred to in this chapter, and their symbols, are given to three significant figures in Table 5.2.

5.2.3   Solutions

The 'amount of substance' in molar solutions of substances used in the tests described in this chapter are listed in Table 5.3.

The approximate composition of sea-water, expressed as percentages by mass of dissolved salts, is given in Table 5.4.

Table 5.2.   ATOMIC MASSES (to three significant figures)

Element	Symbol	Atomic mass
Aluminium	Al	27.0
Barium	Ba	137
Bromine	Br	79.9
Calcium	Ca	40.1
Carbon	C	12.0
Chlorine	Cl	35.5
Chromium	Cr	52.0
Copper	Cu	63.5
Hydrogen	H	1.01
Iron	Fe	55.8
Lead	Pb	207
Magnesium	Mg	24.3
Mercury	Hg	201
Nitrogen	N	14.0
Oxygen	O	16.0
Phosphorus	P	31.0
Potassium	K	39.1
Silicon	Si	28.1
Silver	Ag	108
Sodium	Na	23.0
Sulphur	S	32.1

Table 5.3.   MOLAR SOLUTION

Solution	Constituent	Grams of constituent in 1 litre to give a 1 mol/litre solution
Hydrochloric acid	HCl	36.5
Sulphuric acid	H_2SO_4	98
Nitric acid	HNO_3	63
Sodium hydroxide	NaOH	40
Silver nitrate	$AgNO_3$	170
Potassium dichromate	$K_2Cr_2O_7$	294

Table 5.4 MAIN CONSTITUENTS OF SEA-WATER

Dissolved salt		Typical percentage by mass
Sodium chloride	NaCl	2.71
Magnesium chloride	$MgCl_2$	0.38
Magnesium sulphate	$MgSO_4$	0.16
Calcium sulphate	$CaSO_4$	0.13
Potassium sulphate	K_2SO_4	0.09
Other	(mainly $CaCO_3$, $MgBr_2$)	0.02
	Total dissolved salts	3.49

5.3 THEORY

5.3.1 Acidity, Alkalinity and pH

THE pH SCALE

All liquids containing water contain at least two kinds of free ions (atoms or groups of atoms) carrying electric charges. These are the hydrogen ions, which are positively charged, and the hydroxyl ions, which are negatively charged. This results from the electrolytic dissociation of some of the water molecules represented by the reversible action

$$H_2O \rightleftharpoons H^+ + OH^-$$

When the numbers of these two kinds of ions are equal, the liquid is said to be neutral.

One litre of pure freshly distilled water contains one ten-millionth of a gram (10^{-7} g) of hydrogen ions (H^+) and the same number of hydroxyl ions (OH^-). The addition of acid to the water increases the concentration of the H^+ ions and decreases the concentration of OH^- ions. The water then gives an acid reaction and the active *acidity* increases in proportion to the increase in the concentration of hydrogen ions.

The addition of alkali has the opposite effect, and the active alkalinity increases in proportion to the decrease in hydrogen ion concentration.

At a given temperature the product of the concentration of the H^+ and OH^- ions is constant, so if one is known the other can be calculated. It is usual to refer only to the hydrogen ion concentration, which is expressed in grams of active (ionised) hydrogen per litre of liquid. Since these values are minutely small, the concentration is expressed more conveniently on a logarithmic scale, known as the pH scale. The 'p' stands for the mathematical power, and the 'H' for hydrogen ions. The pH value is the logarithm to base 10 of the reciprocal of the hydrogen ion concentration in grams per litre. This means that the pH value is the index, or power of ten, of the hydrogen ion concentration with the negative sign changed to positive. The value can vary with temperature.

Pure distilled water has an H^+ concentration of 10^{-7} g/litre, and its pH value is 7, which is neutral (neither acid nor alkaline). A solution having a pH of less than 7 is acid, and having a pH greater than 7 is alkaline. Since this is a logarithmic scale, a decrease of one unit of the pH scale represents an increase of H^+ ions by a factor of 10; a decrease of two units by a factor of 100; and so on.

The acidity referred to above is the 'active' acidity, which may be described as the intensity of the acidity, and the pH provides a measure of this intensity. The 'total' acidity, or amount of acid present, is a different property, which can be measured quantitatively by titration.

The pH values of some solutions are shown diagrammatically on the pH scale in Fig. 5.2. Freshly distilled or de-ionised water usually shows a slight acid reaction (pH of 6.6–7.0).

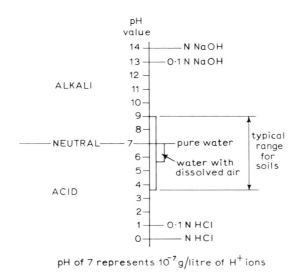

pH of 7 represents 10^{-7} g/litre of H^+ ions

Fig. 5.2   Scale of pH values

On exposure to air this can fall to 6.0 or less, owing to rapid absorption of carbon dioxide, which produces acidity.

It is very difficult to measure accurately the pH of pure water.

INDICATORS

Certain dyes, known as pH indicators, change colour in a definite manner, according to the acidity or alkalinity of the solution in which they are mixed. This feature is used in the colorimetric methods of determining pH. The best-known indicator is litmus, which is red in an acid solution and blue in an alkaline solution. However, it is not sensitive enough for measurement of pH, because it may require a pH as low as 4.6 to indicate acid, and up to 8.4 to indicate alkali. Indicators used for pH measurement show a complete colour change over a small range of pH values. By using a universal indicator and comparing the colour with the BDH colour chart, the pH can be estimated to the nearest 0.5.

A more accurate assessment can be made with the Lovibond comparator, the procedure for which is described in Section 5.5.4. The colour change of an indicator is identified by comparison with a number of permanent glass colour standards mounted in a rotatable plastics disc. The number marked on the matching standard is the pH value of the solution under test.

There are many different indicators available covering different ranges of pH. Each indicator must be used with the appropriate disc of standard colours. Some indicators have a double range of pH values, for which two different discs are required. Some of the indicators most useful for soils and groundwaters are listed in Table 5.5 together with the pH ranges over which they are applicable. The universal or wide-range indicators are intended to give an approximate pH value only, so that the appropriate narrow-range indicator may be selected for a more accurate assessment.

An indicator is sensitive to colour change only within the limits of its pH range. Beyond these limits there is no further colour change, and another indicator is necessary. There is some overlap between indicators covering adjacent ranges, so that a reading at the limit of one indicator can be confirmed by repeating the test with the next.

INDICATOR PAPERS

Indicator papers are available in booklet form, and consist of strips of absorbent paper

Table 5.5. INDICATORS FOR SOIL TESTING AND pH MEASUREMENT

Purpose	Indicator	pH range	Lovibond disc reference
General indicators	litmus — blue	<5	
		>8 — red	
	bromophenol blue	2.8–4.6	
	methyl orange	2.8–4.6	
	methyl red	4.4–6.0	
	thymol blue	8.0–9.6	
	phenolphthalein	8.4–10.0	
Indicator papers for pH	full-range	1–14	
	narrow ranges	1–4	
		4–6	
		6–8	
		8–10	
		10–12	
		12–14	
Indicators for use with Lovibond comparator:	full range*	1–13	2/IZE
Wide-range	Universal*	4–11	2/1P
	B.D.H. Soil*	4–8	2/1N
	B.D.H. 9011	7–11	2/1M
	B.D.H. 1014	10–14	2/BB
Narrow-range	bromophenol blue	2.8–4.4	2/1B
	bromocresol green	3.6–5.2	2/1C
	bromocresol purple	5.2–6.8	2/1G
	bromothymol blue	6.0–7.6	2/1H
	phenol red	6.8–8.4	2/1J
Double-range	cresol red	1.2–2.8	2/1Y
		7.2–8.8	2/1K
	m-cresol purple	1.0–2.7	2/1W
		7.6–9.2	2/1Z
	thymol blue	1.2–2.8	2/1A
		8.0–9.5	2/1L

*Suitable for many purposes not requiring high accuracy.

impregnated with an indicator. Both universal and narrow-range indicator papers are available. The most useful ranges are given in Table 5.5.

ELECTRICAL CONDUCTIVITY

The electrical conductivity of the hydrogen ions (H^+) in a very dilute solution is almost double that of the hydroxyl ions (OH^-). Electrical conductivity of a solution can therefore be related to its pH, and although the converse is far more complex, this property is made use of in the electrometric method of determining pH, described in Section 5.5.2.

5.3.2 Sulphates

The water-soluble sulphates usually found in soils are sodium sulphate (Na_2SO_4) and magnesium sulphate ($MgSO_4$). Calcium sulphate ($CaSO_4$) is commonly found as gypsum, and is only slightly soluble in water but is readily soluble in dilute hydrochloric acid. Treatment with acid is therefore necessary if the total amount of sulphates is required.

The approximate maximum solubilities in water of the three salts mentioned, expressed in terms of SO_3 per litre at about 20°C, are as follows:

sodium sulphate (Glauber's salt)   240 g/litre
magnesium sulphate (Epsom salt)   180 g/litre
calcium sulphate (gypsum, or selenite)   1.2 g/litre

In the gravimetric methods described in Section 5.6.5 the dissolved sulphates are precipitated as insoluble barium sulphate as a result of a reaction with barium chloride in slightly acid conditions. The chemical reaction with magnesium and calcium sulphate may be represented by the equation

$$\left.\begin{matrix} Mg \\ Ca \end{matrix}\right\} SO_4 + BaCl_2 \rightarrow BaSO_4{\downarrow} + \left.\begin{matrix} Mg \\ Ca \end{matrix}\right\} Cl_2$$

(soluble)   (soluble)   (insoluble, precipitated)   (soluble)

The reaction is similar with sodium sulphate, but the equation is

$$Na_2SO_4 + BaCl_2 \rightarrow BaSO_4{\downarrow} + 2NaCl$$

The precipitate of barium sulphate is filtered out, dried and weighed. From the atomic masses the equivalent percentage of SO_3 in the original sample is calculated, as follows.
Molecular mass of barium sulphate ($BaSO_4$):

Element	Atomic mass (approx.)	No. of atoms	
Ba	137	×1=	137
S	32	×1=	32
O	16	×4=	64
		Molecular mass	233

Mass of SO_3:

Element	Atomic mass (approx.)	No. of atoms
S	32	×1=32
O	16	×3=48
		80

Thus the mass of SO_3 will be $80/233 = 0.343$ times the mass of barium sulphate precipitated. If the mass of precipitate is m_4 and the mass of soil used in m_3, the percentage of SO_3 in the soil used is given by

$$\frac{m_4 \times 0.343}{m_3} \times 100\% = 34.3 \times \frac{m_4}{m_3}\%$$

If the mass m_3 was taken not from the original sample, but from the fraction passing a 2 mm sieve, the calculated percentage must be multiplied by m_2/m_1 to convert it to a percentage of the original sample, where m_1 = mass of selected sample before sieving; m_2 = mass of sample passing the 2 mm sieve.

It is customary to express sulphates in terms of SO_3. The BRE Digest 250 (1981) and other references which give practical advice on concrete in sulphate-bearing soils all base their recommendations on SO_3 content.

If sesquioxides are present the addition of ammonia causes them to be precipitated out, so that they can be removed before the reaction with barium chloride.

During the final filtration the presence of remaining soluble chlorides is indicated by turbidity when a drop of the washings is tested with silver nitrate solution:

$$CaCl_2 + 2AgNO_3 \rightarrow 2AgCl\downarrow + Ca(NO_3)_2$$

$$BaCl_2 + 2AgNO_3 \rightarrow 2AgCl\downarrow + Ba(NO_3)_2$$
$$\text{(white precipitate}$$
$$\text{if chloride}$$
$$\text{is present)}$$

Washings must continue until no turbidity is indicated; otherwise the chlorides will be included in the final weighing.

5.3.3 Organic Matter

Organic matter contains carbon, which may occur in complex chain compounds with hydrogen, oxygen, nitrogen and other elements. In a test these compounds are broken down in various ways, depending on the process used.

IGNITION TEST

In the ignition test the carbon burns to combine with oxygen forming carbon dioxide, which is driven off:

$$C + O_2 \rightarrow CO_2\uparrow$$
$$\text{(in complex organic compounds)}$$

The other constituents of the organic compounds break down, and most are also lost as gases.

PEROXIDE TEST

In the peroxide test hydrogen peroxide (H_2O_2) releases nascent oxygen which vigorously oxides most of the organic matter present:

$$H_2O_2 \rightarrow H_2O + O\uparrow$$

In each process the mass of organic matter present is assumed to be equal to the mass lost, which is expressed as a percentage of the dry mass of the soil.

DICHROMATE OXIDATION

In the dichromate oxidation method it is assumed that the organic matter in soil contains 58% of carbon by mass, and that approximately 77% of that carbon is oxidised by the action of potassium dichromate. These factors are taken into account in the equation for the determination of organic matter content:

$$\text{percentage organic matter} = \frac{0.67 \times V}{m_3}$$

where V is the volume of potassium dichromate used to oxidise the organic matter in the soil, of initial mass m_3. It is measured by titration with ferrous sulphate.

V is calculated from the equation

$$V = 10.5 \times \left(1 - \frac{y}{x}\right)$$

where y = total volume if ferrous sulphate used in the test; x = volume used in standardisation test.

If the mass of soil used, m_3, is taken from the mass m_3 passing a 10 mm sieve from an initial mass m_1, the calculated percentage must be multiplied by m_2/m_1 to give the organic matter content as a percentage of the whole original sample.

5.3.4 Carbonates

When hydrochloric acid is allowed to react with a carbonate, such as calcium carbonate (which is the main form of carbonates in soils), the chloride is formed and carbon dioxide is evolved:

$$CaCO_3 + 2HCl \rightarrow CaCl_2 + H_2O + CO_2 \uparrow$$

The mass of carbon dioxide evolved is related to the mass of carbonates in the test specimen, and the test result is usually expressed in terms of percentage CO_2 by mass. The mass of carbon dioxide evolved can be determined either directly or indirectly, as follows.

(a) By mass — the CO_2 is absorbed in a granular absorbent and the mass is obtained directly by difference (Section 5.8.3). No other calculations are necessary.
(b) By volume — the CO_2 is collected and its volume is measured, enabling the mass to be calculated if its temperature and pressure is known. (Sections 5.8.4 and 5.8.5).
(c) By titration — the mass of CO_2 is calculated from the volume of hydrochloric acid used in the reaction, after determining the volume of excess acid by titrating against sodium hydroxide. (Section 5.8.2).

TITRATION METHOD

In the rapid titration procedure (Section 5.8.2) the mass of carbon dioxide is calculated from the amount of acid used for the reaction represented by the equation given above. A known volume of hydrochloric acid solution of a known concentration is added to the soil to ensure that the reaction reaches finality, and then the volume of excess acid remaining is determined by titration against sodium hydroxide solution of a known concentration. The mass of acid used in the reaction is obtained by difference.

The concentrations of the solutions are as follows.

$$\underline{c}\,(HCl) = H \text{ mol/l}$$

$$\underline{c}\,(NaOH) = B \text{ mol/l}$$

Molecular mass of carbon dioxide:

C	12
O (16×2)	32
CO_2	44

In the test, 100 ml of the HCl solution is added to the soil specimen. Part of the acid (V ml of the original solution) reacts with all the carbonates to produce CO_2 and water. The volume of acid solution, equivalent to H mol/l, remaining unused is therefore $(100 - V)$ ml in a total volume of liquid which remains the same (100 ml). From this, 25 ml is taken for titration against sodium hydroxide, so that the titration portion contains $0.25(100 - V)$ ml of the H mol/l solution of HCl. This is neutralised by a measured volume (V_2 ml) of NaOH solution (B mol/l). When the reaction is just completed this volume must be equal to $H/B \times$ (volume of acid solution), i.e.

$$V_2 = \frac{H}{B} \times 0.25(100 - V)$$

Hence

$$V = 100 - \frac{4BV_2}{H} \, \text{ml}$$

From the reaction equation given above it can be seen that 2 molecules of HCl produce 1 molecule of CO_2. This means that 2 litres of the HCl solution are required to produce $44H$ grams of CO_2. Therefore a volume V ml of HCl will produce

$$\frac{44H}{2 \times 1000} \times V \, \text{grams of } CO_2$$

Hence the mass of CO_2 produced is equal to

$$\frac{22H}{1000}\left(100 - \frac{4BV_2}{H}\right) \text{grams}$$

Expressed as a percentage of the mass in grams of the soil specimen used for the test, the proportion of CO_2 is equal to

$$\frac{22}{1000m}(100H - 4BV_2) \times 100\% = \frac{8.8(25H - BV_2)}{m}\%$$

This equation is used for calculating the test result.

VOLUMETRIC METHOD

The measurement of volume in this method (Sections 5.8.4 and 5.8.5) is very sensitive to small changes in temperature and pressure. One of the practical difficulties of this type of test is the maintenance of a known steady temperature and pressure. The apparatus described in Section 5.8.4 (Fig. 5.13) was developed to fulfil these conditions.

The mass of 100 ml of carbon dioxide collected in this apparatus depends on the volume of hydrochloric acid used for the test, as well as the temperature and pressure. At the standard conditions (20°C and 760 mmHg), if 20 ml of acid is used, the mass of 100 ml of carbon dioxide is 200 mg. If the atmospheric pressure is higher, or the temperature lower, the mass of gas will be greater, and vice versa. The mass of 100 ml of carbon dioxide (W_2) can be obtained from a table.

$$W_1 = \text{mass of solid sample tested (g)}$$
$$W_2 = \text{mass of 100 ml of } CO_2 \text{ (mg)}$$
$$V_g = \text{volume of } CO_2 \text{ evolved (ml)}$$

$$\text{mass of } CO_2 \text{ evolved} = \frac{W_2}{100} \times V_g \text{ mg}$$

$$= \frac{W_2 \times V_g}{100\,000} \text{ g}$$

Expressed as a percentage of W_1,

$$CO_2 \text{ content} = \frac{W_2 \times V_g}{100\,000 \times W_1} \times 100\%$$

or

$$\text{carbonate content (as } CO_2) = \frac{W_2 \times V_g}{1000\,W_1}$$

In a simplified procedure (Section 5.8.5) the volume of acid used is first determined from the prevailing pressure and temperature conditions, so that no subsequent correction is necessary.

5.3.5 Chlorides

The methods for the determination of chloride content described in Section 5.9 depend upon the exchange reaction which takes place between silver nitrate and the chloride salt in solution. For sodium chloride this is represented by the equation

$$NaCl + AgNO_3 \rightarrow NaNO_3 + AgCl\downarrow$$
(soluble) (soluble) (soluble) (precipitate)

WATER-SOLUBLE CHLORIDES

In the BS1377 test for water-soluble chlorides (Volhard's method; Section 5.9.3) an excess of silver nitrate is used to precipitate the chloride, and the quantity of the unreacted portion of silver nitrate is determined from a titration with potassium or thiocyanate. Silver thiocyanate is precipitated, until all the silver has been used up. In the acidified solution the next few drops of ammonium thiocyanate react with the ferric alum indicator to produce ferric thiocyanate, which gives the permanent brown colour and indicates that the end point has been reached.

In this test, c $(Ag\,NO_3) = 0.100\,\text{mol/l}$
   c $(KSCN \text{ or } NH_4SCN) = C\,\text{mol/l}$

Atomic masses: Ag 108, Cl 35.5

A measured volume (V_2 ml) of $AgNO_3$ is added to the water extract, of which an unknown volume V ml reacts with the chlorides. The excess silver nitrate then reacts with a volume V_3 ml of the thiocyanate (C mol/l) solution to just reach finality.

1 litre of the thiocyanate solution reacts with $108C$ g of Ag. Therefore V_3 ml of thiocyanate solution reacts with

$$\frac{108C}{1000} \times V_3 \text{ g of Ag}$$

This is contained in a volume of $AgNO_3$ (0.1 mol/l) solution equal to

$$\frac{1000}{10.8} \times \frac{108C}{1000} \times V_3 = 10CV_3 \text{ ml}$$

Therefore, the volume of $AgNO_3$ solution, V ml, which reacted with the chloride is equal to the difference, i.e.

$$V = V_2 - 10CV_3$$

1 litre of silver nitrate (0.1 mol/l) solution contains 10.8 g of Ag which combines with 3.55 g of Cl.
Therefore, V ml of $AgNO_3$ reacts with

$$\frac{3.55}{1000} \times V \text{ g of Cl}$$

i.e.

$$\text{mass of Cl} = \frac{3.55}{1000}(V_2 - 10CV_3)$$

In the 2:1 water:soil extract, 100 ml of solution contains the extract from 50 g of soil. The proportion of chloride ions, Cl, expressed as a percentage of 50 g of soil by mass, is equal to

$$\frac{3.55}{1000}(V_2 - 10CV_3) \times \frac{100}{50}\% = 0.0071\,(V_2 - 10CV_3)$$

In the above calculations the atomic masses have been rounded to three significant figures which is why the multiplying factor is 0.0071 instead of 0.00709 as given in BS1377.

ACID SOLUBLE CHLORIDES

In the BS1377 test for acid-soluble chlorides (Section 5.9.5) a smaller specimen is used for obtaining the acid extract but the chemical principles are similar to those outlined above for Volhard's method.
The mass of the test specimen is 5 g, so the percentage by mass of chloride ions is equal to

$$\frac{3.55}{1000}(V_2 - 10CV_3) \times \frac{100}{5} = 0.071\,(V_2 - 10CV_3)\%$$

MOHR'S METHOD

In Mohr's method (Section 5.9.4) the silver nitrate is added to the chloride in a neutral solution which also contains potassium chromate. The silver has a far greater affinity for the chloride than for the chromate, so the above reaction takes place until all the chloride is combined as silver chloride. The next few drops of silver nitrate react with the potassium chromate, producing a red colour which remains even after the titration flask is swirled, indicating that the end point has been reached:

$$2AgNO_3 + K_2CrO_4 \rightarrow 2KNO_3 + Ag_2CrO_4$$
$$\text{(red)}$$

Provided that the same intensity of colour is observed in the test solution as in the 'blank', the difference between the volumes of silver nitrate used in the two flasks is the volume required to react with the chloride.

The principle of the calculations is similar to that for Volhard's method, but the mass of soil used and the concentration of silver nitrate are different.

In this test 100 g of soil is mixed with 200 ml of water, from which 25 ml is taken for test. Therefore the extract in this portion represents 12.5 g of soil.

$$\text{Silver nitrate solution: } \underline{c}\ (AgNO_3) = 0.02\ \text{mol/l}$$

The volume of this solution which reacts with the chlorides, V ml, is known by difference.

The proportion of chloride ions, expressed as a percentage of 12.5 g of soil by mass, is equal to

$$\frac{35.5 \times 0.02\ V}{1000} \times \frac{100}{12.5}\ \% = 0.00568\ V\ \%$$

5.4   APPLICATIONS

5.4.1   pH Value

Excessive acidity or alkalinity of the groundwater in soils can have detrimental effects on concrete buried in the ground. Even a moderate degree of acidity can cause corrosion of metals. Measurement of the pH value of the groundwater reveals these potential dangers so that remedial measures can be taken.

In the stablisation of soils for roads, some resinous materials are unsuitable for alkaline soils, yet may be satisfactory with neutral or slightly acid soils.

As well as being used for the above purposes, the pH value is usually determined whenever the sulphate content is measured.

5.4.2   Sulphate Content

Groundwater containing dissolved sulphates can attack concrete, and other materials containing cement (e.g. cement-stabilised soil), placed in the ground or on the surface. A reaction takes place between the sulphates and the aluminate compounds in cement, causing crystallisation of complex compounds. The expansion which accompanies crystallisation induces internal stresses in the concrete, which results in mechanical disintegration.

Measurement of the sulphate content enables the ground conditions to be classified according to potential sulphate attack. Appropriate precautionary measures, such as the use of sulphate-resisting cement or of a richer, denser concrete mix, can be taken during construction.

Sulphates in soil can also cause disintegration of precast members, such as slabs and concrete pipes, and can lead to corrosion of metal pipes placed in contact with the soil.

The soluble sulphates (sodium and magnesium) are much more aggressive to concrete than calcium sulphate, which is relatively insoluble in water. Therefore if the predominating sulphate present in the soil is calcium sulphate, the test based on the extract of total sulphates (Section 5.6.2) is likely to give a pessimistic indication of the danger due to sulphates. If the total sulphate content exceeds 0.5% the sulphate content of a 1:1 soil–water extract Section 5.6.3) should be determined. Because of its low solubility (Section 5.3.2) calcium sulphate will show a sulphate content in the aqueous extract of not more than 1.2 g/litre (0.12%). A sulphate content in excess of this figure in the soil–water extract indicates the presence of other and more harmful salts.

Although the solubility of calcium sulphate is low, appreciable quantities can be dissolved away in the long term if the groundwater can be continually replenished.

Table 5.6. CLASSIFICATION OF SOIL AND GROUNDWATER ACCORDING TO SULPHATE CONTENT

Class	Concentration of sulphates expressed as SO_3			Type of cement recommended for buried concrete (summary)
	In soil		In groundwater SO_3 (g/litre)	OPC = Ordinary Portland Cement RHPC = Rapid Hardening Portland Cement PBFC = Portland Blast Furnace Cement SRPC = Sulphate Resisting Portland Cement pfa = pulverised fuel ash
	Total SO_3 (%)	SO_3 in 2:1 aqueous extract (g/litre*)		
1	<0.2	<1.0	<0.3	OPC or RHPC — or with slag or pfa PBFC
2	0.2–0.5	1.0–1.9	0.3–1.2	OPC or RHPC — or with slag or pfa PBFC SRPC
3	0.5–1.0	1.9–3.1	1.2–2.5	OPC or RHPC with slag or pfa SRPC
4	1.0–2.0	3.1–5.6	2.5–5.0	SRPC
5	>2	>5.6	>5.0	SRPC with protective coatings

Reproduced from BRE Digest 250, 'Concrete in sulphate bearing soils and ground waters', by permission of the Controller HMSO, Crown Copyright.

Further information on sulphates in soils is given in Building Research Establishment Digest 250 (1981). The classification of soil and groundwater according to sulphate content is reproduced in Table 5.6, together with an indication of the type of cement which may be required for buried concrete. Reference should be made to BRE Digest 250 for details of the requirements for concrete mixes.

5.4.3   Organic Matter content

Organic matter in soil is derived from a wide variety of animal and plant remains, so there can be a great variety of organic compounds. They all can have undesirable effects on the engineering behaviour of soils. These can be summarised as follows:

(1)   Bearing capacity is reduced.
(2)   Compressibility is increased.
(3)   Swelling and shrinkage potential due to changes in moisture content is increased.
(4)   The presence of gas in the voids can lead to large immediate settlements, and can affect the derivation of consolidation coefficients from laboratory tests.
(5)   The gas can also give misleading shear strength values derived from total stress tests.
(6)   The presence of organic matter (e.g. in peat) is usually associated with acidity (low pH), and sometimes with the presence of sulphates. Detrimental effects on foundations could result if precautions are not taken.
(7)   Organic matter is detrimental in soils used for stabilisation for roads.

A measure of the organic content of soils is necessary in order to make allowance for these effects.

5.4.4   Carbonate Content

Knowledge of the carbonate content of soils is useful for the following reasons:

(1)   Carbonate content can be used as an index to assess the quality of chalk as a foundation

material. A high carbonate content means a low clay mineral content, and usually indicates a relatively high strength.

(2)   In cemented soils and soft sedimentary rocks the carbonate content can indicate the degree of cementing.

(3)   In roads construction chalky subgrades are susceptible to frost action.

(4)   The carbonate content of chalk or limestone is an indication of its suitability for the manufacture of cement.

5.4.5   Chloride Content

The chloride content is most often used as an indication of whether or not the groundwater is sea-water, or whether the soil has been affected by sea-water. In some coastal situations, notably in the Middle East, the concentration of sodium chloride in the groundwater can be very much higher than that in sea-water. High concentrations can also be present in soils and permeable rocks not now directly in contact with sea-water.

Aqueous solutions of chlorides cause corrosion of iron and steel, including steel reinforcement in concrete. If the presence and concentration of chlorides is known, suitable preventative measures can be taken in the design and construction of buried or underwater reinforced concrete structures.

5.4.6   Corrosion of Metals

Adverse ground conditions can cause corrosion of metals buried in the ground as well as initiating attack on concrete. Steel and cast-iron pipes, and steel sheet piles and tie-bars, are perhaps the most common examples of buried metalwork. In addition, steel bars in reinforced concrete members may become exposed if the surrounding concrete is attacked.

Many factors contribute to the corrosion of iron and steel in the ground, but simple chemical tests can often indicate whether corrosion is likely to develop. An acid environment (low pH) is always potentially aggressive, but corrosion of iron and steel can take place in neutral or alkaline conditions if sulphate-reducing bacteria are also present. These bacteria flourish under anaerobic conditions (i.e. where oxygen is absent), and their presence is indicated by the presence of sulphides as well as sulphate conditions (i.e. in the presence of oxygen). The presence of chlorides can accelerate the corrosion process, even in alkaline (high pH) conditions.

Other important procedures for the assessment of soil corrosivity are tests for the measurement of electrical resistivity and of redox potential (referred to as electro-chemical tests). These procedures are beyond the scope of most soil testing laboratories and are not given here, but are included in BS1377: Part 3:1990, Clauses 10 and 11.

Tests for the presence of sulphides, aggressive bacteria, and other corrosive agents apart from those covered in this chapter, require the facilities of a specialist chemical testing laboratory.

5.5   TESTS FOR pH

5.5.1   Indicator Papers

To determine the pH of water, simply dip a strip of the indicator paper into the water and lay it on a white tile or similar white non-absorbent surface. After 30 s compare its colour with the colour chart on the packet or dispenser. The number on the colour which matches the test strip most closely is the pH value of the water.

If the approximate pH of the water is not known to begin with, use a 'universal' or wide-range paper first. Then select a narrow-range paper appropriate to the approximate value indicated. It should be possible to assess the pH to 0.5 unit with a narrow-range paper.

If the water is turbid, it is better to place a drop of water on one side of the paper and observe the same spot on the reverse side for comparison with the colour chart.

To test the pH of a soil, place a quantity of the soil in a test-tube, add distilled water and shake vigorously until all the soil is in suspension. Dip the test paper in the water (if clear), or place a spot of the water on the paper, and observe the pH value as explained above. The quantity of soil used should be such that the ratio of water to soil by volume is about 5 for clay, 3 for silt and 2 for sand.

5.5.2 Electrometric Method (BS1377: Part 3:1990:9)

The electrometric determination of the pH value of a soil suspension in water, or of groundwater, can be the most accurate method, and gives a direct reading to 0.05 pH unit, or with some instruments to 0.02 pH unit. The equipment is relatively expensive, and the electrodes must be maintained in perfect condition for the readings to be reliable.

This type of instrument is intended for use in a main laboratory where pH tests are carried out regularly. If used intermittently the calibration checks should be made thoroughly before each set of tests.

PRINCIPLE OF OPERATION

The operation of an electrical pH meter is based on the principle that the solution to be tested can be considered as an electrolyte of a voltaic cell. One electrode, known as the reference electrode, remains at a constant voltage with respect to the solution and is unaffected by changes of pH. The voltage of the other electrode is affected by the conductivity, and indirectly by the pH, of the test solution, and the complex relationship between pH and voltage can be determined. In most instruments the voltage indicator is calibrated to read directly in pH units. The reference electrode most commonly used is the saturated calomel type. The other electrode may be of various types, of which the glass type is considered to be the most reliable. This consists of a thin-walled glass bulb, of a special kind of glass, enclosing a suitable electrolyte and electrode.

APPARATUS

(1) Electric pH meter of the type referred to above, covering the range pH 3.0 to 10.0, with a scale readable to 0.05 pH unit. One type of instrument is shown in Fig. 5.3.
(2) Three 100 ml glass beakers with cover-glasses and stirring rods.
(3) Two 500 ml volumetric flasks.
(4) Wash-bottle and distilled water.
(5) Pestle and mortar.
(6) 2 mm aperture BS sieve.
(7) Balance reading to 0.001 g.
(8) Tray (galvanised steel or plastics)

REAGENTS

(1) Buffer solution of pH 4.0.
(2) Buffer solution of pH 9.2.
      These are obtainable as powders ready for solution in water as directed by the manufacturers. Alternatively, they can be made up as follows:
      Solution of pH 4.0: Dissolve 5.106 g of potassium hydrogen phthalate in distilled water and make up to 500 ml.
      Solution of pH 9.2: Dissolve 9.54 g of sodium tetraborate (borax) in distilled water and make up to 500 ml.
(3) Potassium chloride, saturated solution (for maintenance of calomel electrode).

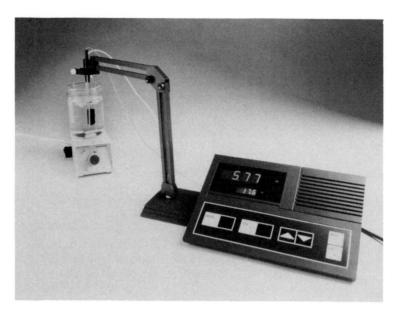

Fig. 5.3   Electric pH meter

PROCEDURE

(1)   From the initial soil sample prepare a sample passing the 2 mm sieve as described in Section 5.1.5, steps (1) to (7). In step (2), air-dry the soil by spreading it out on a tray exposed to air at laboratory temperature. In step (7), divide the material to give a representative sample of 60 g to 70 g to provide two test specimens.

(2)   From this sample weigh out 30 g of the soil, to within ± 0.1 g, for each test specimen. Place each specimen in a 100 ml beaker.

(3)   Add 75 ml of distilled water, stir for a few minutes.

(4)   Allow to stand overnight. Stir again immediately before testing.

(5)   After calibrating the pH meter (see below) wash the electrodes with distilled water and immerse them in the suspension.

(6)   Take two or three readings of pH, stirring briefly between each reading, when the meter reaches equilibrium. These readings should agree within +0.05 pH unit. About 1 min may be required to reach a constant value.

(7)   Remove the electrodes and wash with distilled water.

(8)   Recheck the calibration, using one of the buffer solutions. If out of adjustment by more than 0.05 pH unit, reset the instrument and repeat the test until consistent readings are obtained.

(9)   Leave the electrodes standing in distilled water when the instrument is not in use.

(10) If the pH of groundwater is to be measured, place about 80 ml in the beaker and follow stages (5)–(9).

The above procedure is a general guide but the detailed instructions provided by the manufacturer should be followed carefully.

REPORT RESULTS

Report the pH value of the soil suspension, or groundwater, to the nearest 0.1 pH unit, stating that the electrometric method was used.

CALIBRATION OF pH METER

The detailed procedure for calibration is given in the manufacturer's instructions accompanying each instrument and must be compensated for temperature. Calibration should be carried out at regular intervals, and the instrument, especially the electrodes, should be inspected frequently. The main stages for calibration are as follows.

Wash the electrodes with distilled water.
Set the electrical controls as directed.
Immerse the electrodes and a thermometer in the buffer solution of pH 4.0.
Adjust the buffer controls so that the pH scale reads 4.00. If the temperature of the solution is not 20°C, use the correction tables or graphs to determine the required corrected value.
Wash the electrodes with distilled water.
Immerse them in the buffer solution of pH 9.2, as a check on the reading before use.
Wash the electrodes with distilled water before using in the soil suspension or groundwater sample.

5.5.3   Colorimetric (Kuhn's Method)

This method is intended primarily for use in the field. The colour of the indicator is compared with the colours on a printed colour chart provided by the manufacturers. The method was first reported by S. Kuhn in 1930.

APPARATUS

(1)   A number of glass tubes, fitted with rubber bungs at each end. The tubes are about 200 mm long and 13 mm internal diameter, with two graduation marks at 115 mm and 140 mm from one end (see Fig. 5.4).
(2)   Wooden rack for the tubes.
(3)   Chattaway spatula with blade about 130 mm long and 10 mm wide.
(4)   Wash bottle with distilled water.
(5)   Colour chart.

REAGENTS

(1)(a) Indicator solution, obtainable specially for soil pH test.
   (b) Alternatively, an indicator solution may be made up using bromothymol blue, methyl red, thymol blue, and sodium hydroxide, as follows. Weigh out accurately 0.15 g bromothymol blue, 0.063 g methyl red and 0.013 g of thymol blue and transfer them to a 1000 ml beaker. Add 500 ml of distilled water. The beaker is heated gently and its contents stirred with a glass rod until the indicators have dissolved. The colour of the mixture of indicators in the beaker is adjusted by carefully adding drops of approximately 0.1 N sodium hydroxide solution until the colour approximately matches the colour corresponding to pH 7.0 on the colour chart. The mixture is allowed to cool, and is diluted to 1 litre with distilled water. The indicator should be stored in a stoppered bottle.
(2)   Barium sulphate of soil-testing grade.

The above apparatus and reagents are commercially available in a portable kit known as the BDH Soil Testing Outfit, as shown in Fig. 5.5.

PROCEDURE

(1)   Prepare an air-dried sample of 20–25 g of soil passing a 2 mm sieve, as described in Section 5.1.5, steps (1) to (7).

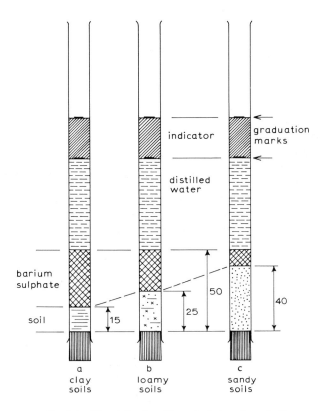

Fig. 5.4   *Glass tubes for pH test (colorimetric method)*

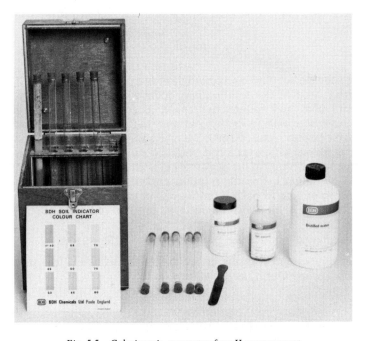

Fig. 5.5   *Colorimetric apparatus for pH measurement*

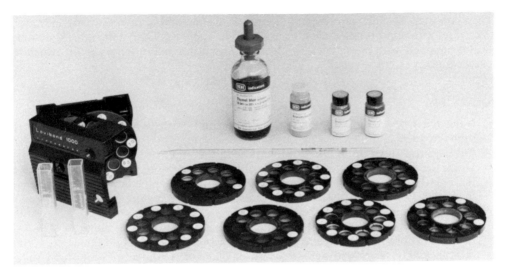

Fig. 5.6 Lovibond Comparator for pH measurement

(2) Place a stopper in the end of one of the glass tubes furthest from the graduation marks. Place the soil in the tube to a depth of (a) 15 mm for clay soil, (b) 25 mm for loamy (silty) soil, (c) 40 mm for sandy soil, as indicated in Fig. 5.4.

(3) Add barium sulphate on top of the soil so that the combined depth of soil and barium sulphate is 50 mm.

(4) Add distilled water to the tube up to the first graduation mark.

(5) Add indicator up to the second graduation mark.

(6) Place a rubber bung in the open end of the tube and shake the tube vigorously until all the soil and barium sulphate are in suspension.

(7) Place the tube in the rack to allow the solids to settle. The barium sulphate (which is insoluble in water) accelerates the settlement of the clay particles, which would otherwise remain as a turbid suspension, and leaves a clear-coloured supernatant liquid above the sediment.

(8) Compare the colour of the supernatant liquid in the tube with the colour chart. Record the pH value of the colour which matches most closely.

(9) Report the pH value to the nearest 0.5 pH unit, stating that the colorimetric method was used.

If at stage (7) the suspension clears only very slowly, too little water was added. Part of the suspension can be poured out, the remainder diluted with more water and more indicator, and the tube reshaken. However, as much soil as possible (consistent with obtaining a clear solution) should be used, because too little soil gives an unreliable result.

5.5.4 Lovibond comparator

This procedure can be used either with a full-range indicator, to obtain an approximate measure of pH (to the nearest unit), or with one of the many available narrow-range indicators, with which the pH can be read to 0.2 of a unit. It is intended for use with groundwater or with aqueous solutions.

APPARATUS

The Lovibond Comparator consists of a plastics holder in which two small test-tubes can be placed, and to which a rotatable comparator colour disc can be fitted (Fig. 5.6). A special pipette is provided.

There are over 30 different comparator discs available, and almost as many indicators. Sixteen discs with appropriate indicators are listed in Table 5.5. One of the three wide-range indicators marked with an asterisk would be sufficient for many purposes for which the greatest accuracy is not required.

PROCEDURE

(1)   If the water or solution to be tested contains solid particles in suspension, allow the solids to settle and filter if necessary.
(2)   Fill both test tubes up to the 10 ml mark with the water or solution to be tested.
(3)   Using the pipette provided, add the appropriate quantity of the selected indicator to the right-hand tube only. Do not immerse the tip of the pipette below the surface of the liquid. For most indicators 0.5 ml is used, but some require 0.2 or 0.1 ml. Follow the supplier's instructions.
(4)   Carefully mix the indicator into the liquid, using a clean glass stirring rod.
(5)   Insert the appropriate comparator disc in the recess provided. Hold the device so that the tubes may be viewed against north daylight or a source of white light (not a tungsten electric light bulb).
(6)   Rotate the comparator disc until the nearest colour match is obtained between the sample tube and the colour disc.
(7)   Read the pH number which appears in the indicator recess.
(8)   Report the result to the nearest whole number if a wide-range or universal indicator is used, or to the nearest 0.2 of a unit if a narrow-range indicator is used. State that the Lovibond comparator method was used.
(9)   Wash the tubes and pipette after use.

5.6 SULPHATE CONTENT TESTS

5.6.1 Scope of Tests

In this section methods of test for the determination of the following are described:

Total (acid-soluble) sulphates in soils, expressed as a percentage.
Water-soluble sulphates in soils, expressed as a percentage.
Sulphates in groundwater, expressed in grams per litre (in some references parts per 100 000, or parts per million (p.p.m.)).

There are two methods of analysis given in the British Standard for determining sulphates in soils and groundwater. These are:

(1)   The gravimetric method, in which the sulphates are precipitated as insoluble barium sulphate, which is collected and weighed.
(2)   A more rapid method using an ion-exchange column, which involves titration against a standardised sodium hydroxide solution.

For sulphate tests on soils, whether for total sulphates or for water-soluble sulphates, and whichever type of test procedure is used, it is first necessary to obtain a liquid extract containing the sulphates in solution. The gravimetric method of analysis of the extract is the same whether an acid extract or a water extract is obtained, and this procedure can also be used for the analysis of water samples.

The ion-exchange procedure can be used either for groundwater or for the water-soluble extract from soils, but not for the acid extract.

The ways in which these procedures are related to each other, and to the results obtained, are shown diagrammatically in Fig. 5.7.

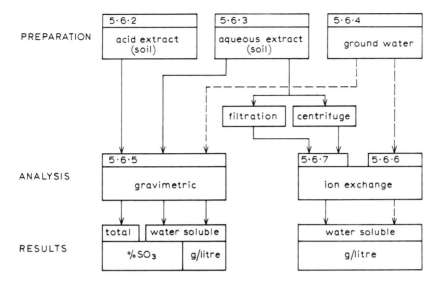

Fig. 5.7   Sulphate test procedures — flow diagram

For most of these tests the size of soil sample actually tested is very small, in some instances only about 2 g. It is, therefore, essential that the test sample be correctly prepared so as to be properly representative of the initial sample.

Another method which is available for determination of sulphate content is the analysis of the density of the precipitated insoluble sulphates by optical means, using a turbidimetric analysis. A nephelometer is an instrument of this type which can be calibrated to give a rapid indication of the amount of sulphates present if the concentration is below a certain level. This method is not discussed here, but further details are given in the paper by Bowley (1979). The equipment is expensive, and the accuracy is much lower than that obtainable by the standard methods, but it enables large numbers of tests to be carried out quickly.

5.6.2   Total Sulphates in Soils — Preparation of Acid Extract (BS1377: Part 3:1990:5.2)

This section describes the procedure for the extraction of a solution of acid-soluble sulphates in soil. For most practical purposes this includes all naturally occurring sulphates. The analysis of the extract is described in Section 5.6.5.

If the sulphate present in the soil is predominantly calcium sulphate, which has a low solubility in water, the total sulphate content as determined on the acid extract is likely to give a misleading and pessimistic impression of the danger to concrete or cement-stabilized material arising from the presence of sulphates. In cases where the total sulphate exceeds 0.5% the water soluble sulphate content of a 2 to 1 water-soil extract should be determined. If calcium sulphate is the only sulphate salt present its low solubility will ensure that the sulphate content of the aqueous extract does not exceed 1.2 g/litre. Sulphate contents in excess of this figure in a soil-water extract, or in ground water, as determined in this test therefore indicate the presence of other and more harmful sulphate salts.

The procedure can be divided into two main parts:

(A)   Physical preparation of soil sample.
(B)   Chemical treatment to prepare acid extract.

Fig. 5.8   Apparatus for determination of total sulphates in soil

APPARATUS

This list does not include the apparatus required for the analysis of the acid extract.

(1)   Analytical balance weighing to 0.001 g or better.
(2)   Glass weighing bottle 50 mm diameter and stopper.
(3)   Two conical beakers, 500 ml, with cover-glasses.
(4)   Glass filter funnel, 100 mm diameter.
(5)   Desiccator and desiccant.
(6)   Two glass stirring rods about 200 mm long.
(7)   Wash-bottle, containing distilled water.
(8)   Drying oven, set at 80°C.
(9)   Test sieves, 2 mm and 425 μm aperture, with receivers.
(10)   Riffle-box, small size, 300 cm³ capacity.
(11)   Pestle and mortar.

The above items are included in Fig. 5.8.

(12)* Buchner funnel 110 mm diameter and vacuum filtration flask, 500 ml or 1000 ml.
(13)* Vacuum pump and connecting vacuum tubing.
(14)* Dropping bottle with pipette.
(15)* Filter papers to fit the filtration funnel, hardened medium grade (e.g. Whatman no. 540) and fine grade (e.g. Whatman No. 42).
(16)* Red litmus paper.
(17)   Electric hotplate with controllable source of heat.

Items marked * are required only if the soil contains sesquioxides in appreciable quantity (steps (15) to (18) below).

REAGENTS

This list does not include the reagents needed for the analysis of the extract.

(1)   Dilute hydrochloric acid (10% HCl). Add 100 ml of concentrated hydrochloric acid (1.18 g/ml) to about 800 ml of distilled water, then add more distilled water to make 1 litre of solution.
(2)*   Dilute ammonia solution. Take 500 ml of ammonia (0.880 g/ml) and add distilled water to make 1 litre of solution.

(3)* Silver nitrate, 0.5% (v/v) solution. Dissolve 0.5 g of silver nitrate in 100 ml of distilled water. Store the solution in an amber-coloured glass bottle.

(4)* Concentrated nitric acid (density $1.42 \, g/cm^3$).

Items marked * are required only if the soil contains sesquioxides in appreciable quantity (steps (15) to (18) below).

All reagents must be of analytical reagent grade.

PROCEDURE

(A) *Preparation of sample*

The sample for test is prepared from the initial sample generally as described in Section 5.1.5. Only about 2 g of soil is used for the test, so it is essential that the test sample is properly representative of the soil. The actual mass of each test specimen (step (13) below) depends on the amount of sulphates present, and ideally should produce a precipitate of barium sulphate weighing about 0.2 g. An initial trial test may be desirable to ascertain the appropriate size of sample to use.

Details of the procedure, with reference to Section 5.1.5, are as follows.

(1) As step (1) of 5.1.5.
(2) As step (2) of 5.1.5, but using an oven drying temperature of 75° to 80°C.
(3) As step (3) of 5.1.5.
(4) As step (4) of 5.1.5. However, if the material retained on the 2 mm sieve includes lumps of gypsum, these should be removed by hand, crushed to pass the 2 mm sieve and added to the fraction passing the sieve.
(5) and (6) As steps (5) and (6) of 5.1.5.
(7) As step (7) of 5.1.5. The mass of material required is about 100 g.
(8) to (12) As steps (8) to (12) of 5.1.5. The mass of each test sample obtained in step (9) should be about 10 g. The oven drying temperature in step (10) is 75° to 80°C.
(13) Transfer at least two representative portions of the appropriate mass (e.g. about 2 g, but see above) from the weighing bottle to separate 500 ml conical beakers. Weigh the bottle after removal of each portion and calculate the mass of each test specimen transferred, to the nearest 0.001 g, by difference (m_3).

Each test specimen is then ready for treatment to obtain the acid extract.

(B) *Preparation of acid extract*

(1) Add 100 ml of the 10% hydrochloric acid to the sample in the 500 ml beaker. If frothing occurs, take care to ensure that no material is lost.

Soils containing sulphides will release hydrogen sulphide (H_2S), detectable by its smell, on acidification. Sulphides if present will give an over-estimate of the sulphate content due to sulphide oxidation. For these soils place 100 ml of the acid in a 500 ml beaker and heat to boiling. Remove from the heat source and, while stirring the acid solution, sprinkle the weighed soil sample on to the acid.

(2) Cover the beaker with a cover-glass, bring to the boil and simmer gently for 15 minutes in a fume cupboard. Rinse the underside of the cover glass with distilled water back into the beaker which is then normally ready for the gravimetric analysis described in Section 5.6.5.

However, if the soil contains sesquioxides in appreciable quantity (e.g. as in some tropical residual soils), these should be precipitated before proceeding with the analysis, as follows.

(3) Add a few drops of nitric acid while the suspension continues to boil.
(4) Add ammonia solution slowly (preferably from a burette), with constant stirring, to the boiling suspension until the sesquioxides are precipitated and red litmus is turned to blue by the liquid.

(5) Filter the suspension through a hardened filter paper into a 500 ml conical beaker. Wash the filter paper with distilled water until the washings are free from chloride, as indicated by an absence of turbidity when a drop is added to a small volume of silver nitrate solution.

(6) If a voluminous precipitate of sesquioxides forms when ammonia is added to neutralise the acid (step (4)) some sulphate may be entrapped which will not be removed by washing and could lead to low results. In this case a second precipitation is recommended. This is done by carefully removing the filter paper with the precipitate and replacing it in the original beaker. Add a 10% solution of hydrochloric acid and stir the contents until the sesquioxides have gone into solution (20 ml of 10% hydrochloric acid should be sufficient). Bring the contents to the boil, repeat steps (4) and (5) and combine the two filtrates.

5.6.3 Water-soluble Sulphates in Soils — Preparation of Aqueous Extract (BS1377: Part 3:1990:5.3)

This section gives the procedure for obtaining a 2:1 water–soil extract from a soil sample. Two procedures for measuring the sulphate content of the aqueous extract are described in Sections 5.6.5 and 5.6.7. The quantity of extract used for test depends upon which procedure is followed.

The procedure consists of two main parts:

(A) Physical preparation of soil sample.
(B) Preparation of aqueous extract.

APPARATUS

Items (1)–(17) of the apparatus listed in Section 5.6.2, with the addition of the following (see Fig. 5.9):

(18) Mechanical shaker, or stirrer, capable of keeping 50 g of soil in continuous suspension in 50 ml of water.
(19) Watch-glass 75 mm diameter.
(20) Extraction bottle of 200 ml capacity.
(21) Centrifuge and centrifuge tubes.
(22) Conical beaker 250 ml.
(23) Pipettes, 25 ml and 50 ml.
(24) Filter papers, Whatman Nos. 44 and 50, or Barcham Green Nos. 800 and 975.

PROCEDURE

(A) Preparation of sample

Stages (1)–(7). Prepare the soil sample from the initial bulk sample in the same way as for the total sulphates test (Section 5.6.2), except that a riffled sample of about 120 g instead of 100 g is prepared at step (A)(7).

(8) Weigh out two representative test specimens each of 50.00 g on watch-glasses and transfer each to a clean dry extraction bottle.

Each prepared test specimen is then ready for treatment to obtain the water extract.

(B) Preparation of water extract

(1) Add exactly 100 ml of distilled water to the extraction bottle or centrifuge tube, using the 50 ml pipette, and stopper tightly.
(2) Place in the shaker and agitate for 16 hours (i.e. overnight).

Fig. 5.9   Apparatus for determination of water-soluble sulphates in soil and groundwater

The final preparation stages (3–5, or 6–8, or 9–11) depend upon whether a gravimetric analysis (Section 5.6.5) or the ion-exchange method (Section 5.6.7) is to be followed, and for the latter whether filtration or a centrifuge is used.

Gravimetric method

(3)   Filter the soil suspension through a Whatman No. 50 or Barcham Green No. 975 filter paper into a clean, dry filter flask, using a Buchner funnel. Do not add any more water.

(4)   Transfer exactly 50 ml of the extract to a clean, dry 500 ml conical beaker, and add distilled water to make a total volume of about 300 ml.

(5)   The extract is now ready for the gravimetric analysis described in Section 5.6.5.

Ion-exchange method (filtration)

(6)   After stage (13) filter the soil suspension through a Whatman No. 50 or Barcham Green No. 975 filter paper into a clean, dry filter flask, using a Buchner funnel. Do *not* add any additional water, and do not wash the soil remaining on the filter paper.

(7)   Transfer exactly 25 ml of the water extract to a clean, dry 250 ml conical beaker, using the 25 ml pipette.

(8)   The extract is now ready for the ion-exchange method of analysis described in section 5.6.7.

Ion-exchange method (centrifuge)

(9)   After stage (13) centrifuge the suspension so that the solids settle out, leaving a clear, supernatant liquid above them.

(10)   Draw off exactly 25 ml of the supernatant liquid into the pipette and run into a 250 ml conical beaker.

(11)   The extract is now ready for the ion-exchange method of analysis described in Section 5.6.7.

5.6.4 Sulphates in Groundwater — Preparation of Water Sample (BS1377: Part 3:1990:5.4)

This section describes the simple procedure required for preparing a groundwater sample for the analysis of the sulphate content, which is expressed as grams per litre. The methods of analysis are described in Section 5.6.5 (gravimetric method) or Section 5.6.6 (ion-exchange procedure).

APPARATUS

(1)   Three conical beakers, 500 ml.
(2)   Two conical beakers, 250 ml.
(3)   Pipette, 50 ml.
(4)   Glass measuring cylinder, 100 ml.
(5)   Filtration funnel and stand.
(6)   Filter papers, Whatman No. 44 or Barcham Green No. 800, to suit the funnel.

PROCEDURE

(1)   A sample of groundwater of at least 500 ml should be collected in a clean bottle. The method of sampling is given in BRE Digest 250.
(2)   Filter the water sample through a Whatman No. 44 or Barcham Green No. 800 filter paper into a clean dry flask, to remove any particles in suspension.

   Continue with either stage (3) or stage (4), depending on which procedure is to be used.

(3)   *Gravimetric analysis* (Section 5.6.5). Transfer exactly 50 ml of the filtered water sample to each of two clean, dry 500 ml conical beakers, and add approximately 100 ml of distilled water.
(4)   *Ion-exchange method* (Section 5.6.6). After stage (2) transfer exactly 100 ml of the sample to each of two clean dry 250 ml conical beakers .

   The water samples are now ready for the analysis described in Section 5.6.5 or Section 5.6.6.

5.6.5 Sulphate Analysis on Prepared Solution — Gravimetric Method (BS1377: Part 3:1990:5.5)

The gravimetric procedure for the determination of the sulphate content of a liquid extract from soil (either an acid extract, or a 2:1 water–soil extract), and of a sample of groundwater, is described below. The analysis is the same in all cases, but the method of calculation differs slightly, depending on the purpose of the test.

APPARATUS

The following are additional to the apparatus required for preparation of the extract (items 1–24 of Sections 5.6.2 and 5.6.3). They are included in Fig. 5.8, except item (2).

(1)   Porcelain or silica ignition crucible, about 35 mm diameter and 40 mm high.
(2)   Electric muffle furnace capable of maintaining $800°C \pm 50°C$.
(3)   Bunsen burner, tripod and pipe clay triangle (if muffle furnace not available).
(4)   Glass stirring rod fitted with rubber policeman.
(5)   Balance, readable to 0.001 g.

REAGENTS

The following are additional to the reagents required for the preparation of the extract (items 1–4 of Section 5.6.2).

(5)   Barium chloride, 5% solution. Dissolve 50 g of barium chloride in 1 litre of distilled water. Filter before use if necessary.
   NOTE: BARIUM CHLORIDE IS POISONOUS.

(6)   Indicators: methyl red, or blue litmus paper.
(7)   Silver nitrate, 5% solution.

All reagents must be of analytical reagent grade.

PROCEDURAL STAGES

The test is carried out on the prepared solution in the conical flask, and follows on from the stage previously described, as under:

Total sulphates in soil — from stage (B)(2) or (5) of Section 5.6.2.
Water-soluble sulphates in soil — from stage (B)(5) of Section 5.6.3.
Sulphates in groundwater — from stage (3) of Section 5.6.4.

The procedural stages given below are divided into six parts:

(A)   Prepare precipitate.
(B)   Collect precipitate.
(C)   Ignite precipitate.
(D)   Weigh.
(E)   Calculate.
(F)   Report results.

PROCEDURE

(A)   *Preparation of precipitate*

(1)   Test the solution in the flask for acidity with blue litmus paper, and if necessary make slightly acid by adding about 20 drops of dilute hydrochloric acid. Alternatively, add two drops of methyl red indicator solution to the solution in the flask, and add dilute (10%) hydrochloric acid to acidify it (red coloration), plus a slight excess. An indicator solution is preferable to a paper indicator (as given in the British Standard) because it is easier to ensure thorough mixing. Methyl red is more sensitive than litmus.
(2)   Dilute to 300 ml (if necessary) and bring the solution to the boil.
(3)   Add 10 ml of the 5% barium chloride solution, drop by drop, while stirring the solution in the beaker. Continue boiling gently until the precipitate is properly formed.
(4)   Cover the solution and keep just below boiling point for at least 30 min. This digestion period is necessary to enable the precipitate to form in particles large enough to be retained by filtration.
(5)   Allow the suspension to settle, and add a drop or two of barium chloride solution to the supernatant liquid.
(6)   If any slight cloudiness is observed as the drops enter, precipitation is incomplete and stages (2)–(4) should be repeated.

(D)   *Collection of precipitate*

(7)   Transfer the precipitate with extreme care to a Whatman No. 42 filter paper in the glass funnel placed over a beaker, and filter. Wash several times with hot, distilled water until the washings are free from chloride.
(8)   To check for the presence of chloride, test a drop of the filtrate with a little of the silver nitrate solution. If there is no turbidity, the washings are free from chloride. Use a rubber policeman to remove the barium sulphate adhering to the walls of the beaker.
(9)   Transfer the filter paper and precipitate carefully to a porcelain or silica crucible, which has previously been ignited and weighed to 0.001 g (m_7).

(C)   Ignition of precipitate

If a muffle furnace is available, follow stages (10) and (11), omit stages (12*) and (13*), and continue from stage (14). Otherwise, continue from stage (12).

(10) Place the crucible and contents in an electric muffle furnace at room temperature.
(11) Raise the temperature of the furnace slowly to red heat (800°C), and maintain that temperature for 15 min. The filter paper should char slowly, not inflame.
(12*) Dry the filter paper slowly at first over a small bunsen flame. Do not allow the filter paper to inflame, but let it char slowly.
(13*) Ignite by heating to red heat, and maintain this heat for 15 min.
(14) Allow to cool to room temperature in a desiccator.

(D)   Weighing residue

(15) Weigh the crucible and contents to 0.001 g (m_8).
(16) Calculate the mass of precipitated (m_4) by difference:

$$m_4 = m_8 - m_7$$

(E)   Calculations

(17) *Total sulphates* (Acid extract procedure, Section 5.6.2). The percentage of total (acid-soluble) sulphates, expressed as SO_3, in the fraction of the soil sample finer than 2 mm is calculated from the equation

$$SO_3\,(\%) = \frac{34.3 m_4}{m_3}$$

(18) *Water-soluble sulphates* (Water extract procedure, Section 5.6.3). The percentage of water-soluble sulphates, expressed as SO_3, in the fraction of the soil sample finer than 2 mm is calculated from the equation

$$SO_3\,(\%) = 1.732 m_4$$

This is based on the mass of soil used (m_3) being exactly 50 g. If the mass differs from this, the equation given in stage (17) must be used instead.

(19) The water-soluble sulphates may alternatively be expressed as grams per litre in the 2:1 aqueous extract, as follows:

$$\text{Sulphates } (SO_3) \text{ in aqueous extract} = 6.86 \times m_4 \text{ g/litre}$$

(2)   *Sulphates in groundwater* (Section 5.6.4). The concentration of sulphates, expressed as SO_3, in the 50 ml groundwater sample is given by

$$SO_3 = 6.86 \times m_4 \text{ g/litre}$$

(21) If the results of individual determinations differ by no more than 0.2% (SO_3) or 0.2 g/litre, calculate the mean result. Otherwise the test should be repeated starting with two new test portions of soil.
(22) Calculate the percentage of the original soil sample passing the 2 mm sieve, which is equal to

$$\frac{m_2}{m_1} \times 100$$

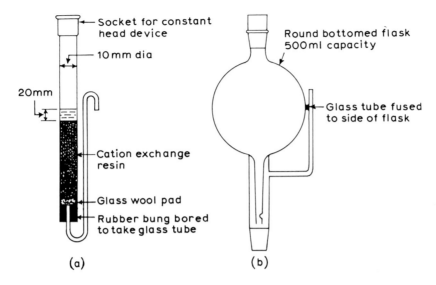

Fig. 5.10   Ion-exchange column and constant-head apparatus

where m_2 is the mass of material passing the 2 mm sieve (g) and m_1 is the initial dry mass of the sample (g).

(F)   Report results

(23) The sulphate content of soil, whether total (acid-soluble) or water-soluble, is reported as the percentage of SO_3 to the nearest 0.01%. It is important to state whether this relates to total sulphates or water-soluble sulphates. The sulphate content of ground-water, or of an aqueous solution, is reported as SO_3 in grams per litre, to the nearest 0.01 g/litre.

(24) The percentage by dry mass of the original sample passing the 2 mm sieve is reported to the nearest 1%.

5.6.6   Sulphates in Groundwater (Ion-exchange Method) (BS1377: Part3:1990:5.6)

This method is quicker and easier than the gravimetric procedure given above. However, it cannot be used if the groundwater contains chloride, nitrate or phosphate ions. If these or other anions are present, the gravimetric procedure (Section 5.6.5) should be used instead.

APPARATUS

(1) Glass tube 400 mm long and 10 mm diameter (the ion-exchange column), with swan-neck outlet, as detailed in Fig. 5.10(a) and shown on the left of Fig. 5.9.

(2) Constant-head device, made from a round-bottomed flask, as detailed in Fig. 5.10(b). If this is not available, the tap on a burette can be adjusted to provide a steady rate of flow into item (1), as shown in Fig. 5.9.

(3) 50 ml burette and burette stand

(4) Two 500 ml conical beakers.

(5) Beaker, 250 ml.

(6) Amber-coloured glass bottle with stopper.

REAGENTS

(1)   A strongly acidic cation-exchange resin. Suitable types are Zeo-Karb 225 or Amberlite IR-120.

(2)   Sodium hydroxide solution (c(NaOH)=approximately 0.1 mol/litre). Dissolve 2 g of sodium hydroxide in 500 ml of distilled water. Determine the exact concentration of this solution by titration against standardised potassium hydrogen phthalate solution (see below).

NOTE. SODIUM HYDROXIDE IS STRONGLY CAUSTIC — USE EYE PROTECTION

Keep the solution in an airtight plastics container.

(3)   Potassium hydrogen phthalate solution (c (KHC$_6$H$_4$(COOH)$_2$)=0.1 mol/litre). Weigh out 5.10 g of potassium hydrogen phthalate, which has been dried in an oven at 105° to 110°C for 2 hours, dissolve in distilled water and dilute to 250 ml in a volumetric flask.

Potassium hydrogen phthalate solution is used for standardisation of the NaOH solution in preference to hydrochloric acid solution because the concentration of the former is known to sufficient accuracy without the need for a standardisation check.

(4)   Silver nitrate solution. Dissolve 0.5 g of silver nitrate in 100 ml of distilled water. Store in an amber-coloured glass bottle.

(5)   Nitric acid, 5% v/v solution. Dilute 5 ml of concentrated nitric acid (density 1.42 g/ml) in about 60 ml of distilled water, then add more distilled water to make 100 ml.

(6)   Indicator solution, such as screened methyl orange, which gives a distinct colour change in the range pH 4 to pH 5.

(7)   Indicator solutions (if necessary): phenolphthalein or thymol blue.

(8)   Hydrochloric acid solution (cHCl = 4 mol/litre approximately). Dilute 360 ml of concentrated hydrochloric acid (1.18 g/ml) with distilled water to make 1 litre.

PROCEDURAL STAGES

(1)   Determine concentration of sodium hydroxide solution.
(2)   Prepare ion-exchange column.
(3)   Prepare groundwater sample.
(4)   Pass through ion-exchange column.
(5)   Add indicator.
(6)   Titrate.
(7)   Calculate.
(8)   Report result.

TEST PROCEDURE

(1)   *Determination of concentration of sodium hydroxide solution*

The exact concentration (B) of the sodium hydroxide solution (nominally 0.1 mol/litre) is obtained by titration against the potassium hydrogen phthalate solution prepared as described above, using phenolphthalein or thymol blue as indicator. Details are given in Vogel (1961).

Alternatively, a ready standardized solution of sodium hydroxide can be used, from which the specified concentration is prepared in accordance with the manufacturer's instructions.

(2)   *Preparation of ion-exchange column*

Take a quantity of ion-exchange resin sufficient to half-fill the column, place in a beaker and stir with distilled water. Empty the suspension of resin in water into the column. When the resin has settled, drain off surplus water so that about 20 mm depth of water remains above the resin (see Fig. 5.10a). Maintain this depth of water above the resin at all times.

The resin is activated by leaching with 100 ml of the hydrochloric acid solution followed by washing with distilled water. If the constant-head device (Fig. 5.10b) is used, place the acid in the round-bottomed flask, replace the stopper and leave the acid to pass through the column. Rinse the flask with distilled water and leave the water to percolate through the column until the liquid coming out shows no turbidity when tested with about 1 ml of silver nitrate solution acidified with a few drops of nitric acid.

If the constant-head device is not available, add the acid and water in increments but allow each increment to drain away before adding the next. Alternatively, use a burette as a substitute for the constant head device.

After four successive determinations of sulphate content, the resin must be re-activated as described above.

(3) *Preparation of sample*

As described in Section 5.6.4.

(4) *Use of ion-exchange column*

Place a 500 ml conical beaker under the outlet of the ion-exchange column. Pass the water through the column, followed by rinsing with two 75 ml increments of distilled water. Collect the water and washings together in the conical beaker.

(5) *Addition of indicator*

Add indicator to the collected liquid to impart sufficient colour for the detection of the end point of the titration.

(6) *Titration*

Titrate the liquid against the standardised sodium hydroxide solution. Record the volume of sodium hydroxide solution required to just neutralise the liquid to the nearest 0.05 ml (V). This has occurred when the solution retains a yellow colour after swirling the flask.

(7) *Calculation*

The sulphate content expressed as SO_3 of the groundwater is calculated from the equation

$$SO_3 = 0.4 \, BV \text{ g/litre}$$

where B is the concentration of the sodium hydroxide solution (mol/litre) (see step (1)); V is the volume of sodium hydroxide solution used (ml).

(8) If individual results differ by no more than 0.2 g/litre (SO_3), calculate the mean result. Otherwise repeat the test starting with two new 50 ml samples of filtered ground water.

(9) *Result*

The sulphate content of the groundwater, expressed as SO_3, is reported in grams per litre to the nearest 0.01 g/litre, as determined by the ion-exchange column method.

5.6.7 Sulphates in Aqueous Soil Extract (Ion-exchange Method) (BS1377: Part 3:1990:5.6)

This method is quicker and easier than the gravimetric method, but it cannot be used if the soil contains chloride, nitrate or phosphate ions, or other anions.

The procedure is similar to that described for groundwater samples in Section 5.6.6, except that a 2:1 water-soil extract must first be prepared as described in Section 5.6.3. The same apparatus and reagents as referred to in both these sections are required.

PROCEDURE

(1)   Determination of concentration of sodium hydroxide solution. As Stage (1) of 5.6.6.
(2)   *Preparation of ion-exchange column.* As Stage 2 of Section 5.6.6.
(3)   *Preparation of aqueous extract.* As described in Section 5.6.3. To the 25 ml of aqueous extract in the 250 ml beaker from Stage (B)(8) or (11), add distilled water to make up to 100 ml. Stages (4)–(6) are similar to Stages (4)–(6) in Section 5.6.6.
(7)   Calculation. The sulphate content, expressed as SO_3, of the soil–water extract, is calculated from the equation

$$SO_3 = 1.6\, BV \;\; \text{g/litre}$$

where B is the concentration of the sodium hydroxide solution (mol/litre).
(8)   If individual results differ by no more than 0.2 g/litre (SO_3), calculate the mean result. Otherwise repeat the test starting with two new test portions of soil.
(9)   *Results.* The sulphate content of the soil–water extract, expressed as SO_3, is reported in grams per litre to the nearest 0.01 g/litre, as determined by the ion-exchange column method. The percentage of the original soil sample passing the 2 mm sieve is reported to the nearest 1%.

5.7   ORGANIC CONTENT TESTS

5.7.1   Scope of Tests

Two procedures are given for the determination of the organic content of soils:

  Oxidation with hydrogen peroxide.
  Dichromate oxidation.

  The dichromate method is the one given in BS1377: Part 3:1990: as the standard procedure for soils. Hydrogen peroxide has only a limited action on undecomposed plant remains, such as roots and fibres.

5.7.2   Dichromate Oxidation Method (BS1377: Part 3:1990:3)

This method was first introduced by Walkley and Black in 1934, and was revised by Walkley in 1935. It has been found to give reproducible results. The accuracy is not absolute but is sufficient for most engineering purposes.
  Soils containing sulphides or chlorides have been found to give high results by this method. These substances, if present, can be removed at the sample preparation stage by the appropriate chemical treatment, as described under 'Procedure'. Methods for checking for their presence are included.

APPARATUS

(1)   Balances readable to 0.001 g and to 1 g.
(2)   Two 1 litre volumetric flasks.
(3)   Two 25 ml burettes, graduated to 0.1 ml, and burette stands.
(4)   10 ml pipette with rubber teat.
(5)   1 ml pipette with rubber teat.
(6)   Two 500 ml conical flasks.
(7)   Graduated measuring cylinders, 250 ml and 20 ml.
(8)   Glass weighing bottle. 25 mm diameter.
(9)   Sieves, 2 mm and 425 μm aperture.

Fig. 5.11 *Apparatus for organic content test (dichromate oxidation)*

(10) Pestle and mortar.
(11) Wash-bottle with distilled water.
(12) Drying oven set to give a temperature of $50° \pm 2.5°C$.
(13) Desiccator and desiccant.
(14) Riffle box, small ($300 \, cm^3$).
(15) Glass boiling tube.
(16) Filter funnel approximately 110 mm diameter.
(17) Filter papers, to suit the funnel; medium grade (e.g. Whatman No. 40) and fine grade (e.g. Whatman No. 42).
(18) Blue litmus paper.

The main items are shown in Fig. 5.11.

REAGENTS

(1) Potassium dichromate ($c \, (K_2Cr_2O_7) = 0.167 \, mol/litre$). Dissolved 49.035 g of potassium dichromate in distilled water to make 1 litre of solution.
(2) Hydrochloric acid, 25% v/v solution. Dilute 250 ml of concentrated hydrochloric acid (density 1.18 g/ml) with distilled water to make 1 litre of solution.
(3) Sulphuric acid, concentrated (density 1.84 g/ml).
(4) Sulphuric acid solution ($c \, (H_2SO_4) =$ approximately 1 mol/litre). Add 53 ml of concentrated sulphuric acid to about 500 ml of distilled water, then make up to 1 litre with distilled water.
(5) Ferrous sulphate solution. Dissolve 140 g of ferrous sulphate in sulphuric acid solution ($c \, (H_2SO_4) = 0.25 \, mol/litre$) to make 1 litre of solution. This solution is unstable in air and should be kept tightly stoppered. It should be standardised against the dichromate solution weekly. Prepare the sulphuric acid solution by adding 14 ml of concentrated sulphuric acid to about 800 ml of distilled water, then make up to 1 litre by adding

further distilled water. DO NOT ADD WATER TO THE CONCENTRATED ACID. Record the date the solution is made on the bottle.
(6)   Orthophosphoric acid, 85% v/v solution, density 1.70 to 1.75 g/ml.
(7)   Indicator solution. Dissolve 0.25 g of sodium diphenylamine sulphonate in 100 ml of distilled water.
(8)   Lead acetate paper. Filter paper which has been dipped in a solution of lead acetate.

PROCEDURAL STAGES

The procedure described below is divided into five parts:

(A)   Standardize ferrous sulphate solution.
(B)   Prepare sample.
(C)   Test for organic matter.
(D)   Calculate.
(E)   Report result.

PROCEDURE

(A) Standardization of ferrous sulphate solution

(1)   Run 10 ml of the potassium dichromate solution from the burette into a 500 ml conical flask.
(2)   Very carefully add 20 ml concentrated sulphuric acid. This will generate heat. Swirl the mixture and allow to cool, standing on a heat-insulating surface or pad. Protect from draughts.
(3)   Add 200 ml of distilled water very carefully.
(4)   Add 10 ml of orthophosphoric acid and 1 ml of indicator, and mix thoroughly.
(5)   Add ferrous sulphate from the burette in increments of 0.5 ml, swirling the flask, until the colour changes from blue to green.
(6)   Add a further 0.5 ml of potassium dichromate, changing the colour back to blue.
(7)   Add ferrous sulphate slowly drop by drop, with continued swirling, until the colour of the solution changes from blue to green after the addition of a single drop. Record the total volume of ferrous sulphate used, x, to the nearest 0.05 ml.

(B) Preparation of test specimen

The specimen for test is prepared from the original soil sample generally as described in Section 5.1.5. Details are as follows.

(1)   As step (1) of 5.1.5.
(2)   As step (2) of 5.1.5, but using an oven drying temperature of $50° \pm 2.5°C$.
(3)   As step (3) of 5.1.5.
(4)   As step (4) of 5.1.5. If the material retained on the 2 mm sieve is seen to contain organic matter, remove the pieces of organic matter by hand, crush them to pass the 2 mm sieve and add them to the fraction passing the sieve.
(5) and (6)   As steps (5) and (6) of 5.1.5.
(7)   As step (7) of 5.1.5. The mass of material required is 100 g.
(8)   As step (8) of 5.1.5.
(9)   Sub-divide this sample by riffling to obtain the following test specimen, as appropriate.

   (a) If sulphides and chlorides are known to be absent: two test samples each of about 5 g.
   (b) For a check test to determine whether sulphides are present: a sample of about 5 g.
   (c) For a check test to determine whether chlorides are present: a sample of about 50 g.
   (d) If sulphides and/or chlorides are present: a test sample of about 50 g.

(10) to (12)   As steps (10) to (12) of 5.1.5. The oven drying temperature in step (10) is $50° \pm 2.5°C$.

(C)   Qualitative checks for sulphides

The following procedure enables the presence of sulphides in the soil to be verified.

(1)   Place the 5 g check sample (which need not be weighed) in a boiling tube and add about 20 ml of hydrochloric acid, 25% solution.
(2)   Boil the acid, and check by smell (very carefully) whether hydrogen sulphide is evolved. If so the soil contains sulphides.
(3)   Alternatively, check for hydrogen sulphide by holding in the vapour a piece of filter paper that has been dipped in a 10% solution of lead acetate. This will turn black if hydrogen sulphide is present.
(4)   If the presence of sulphides is indicated they should be removed from the test specimen as described in (D) below before proceeding with the analysis for organic matter, otherwise a result that is too high will be obtained.
(5)   If the presence of sulphides is not indicated, omit procedure (D).

(D)   Elimination of sulphides

The procedure for the elimination of sulphides from the test sample is as follows.

(1)   Weigh about 50 g of the soil, after cooling as in (B) (11) above, to 0.01 g and place it in a 500 ml conical flask.
(2)   Add sulphuric acid (c (H_2SO_4) = 1.0 mol/litre) until no further evolution of hydrogen sulphide occurs. Check by testing with lead acetate paper.
(3)   Filter the contents of the conical flask on a medium grade filter paper, taking care to retain all solid particles. Wash several times with hot distilled water until the washings do not indicate acidity when tested with blue litmus.
(4)   Dry the soil retained on the filter paper to constant mass at a temperature of $50° \pm 2.5°C$ and cool in the desiccator.
(5)   Carefully remove all the soil from the filter paper and determine its mass to 0.01 g.
(6)   Sub-divide the treated sample, as in Section (B), step (9)(a), and dry and cool each test specimen as in steps (10) to (12).

(E)   Qualitative check for chlorides

The presence of chlorides in the soil can be verified by following the procedure given in Section 5.9.2.
    If the presence of chlorides is indicated they should be removed from the test sample as described in (F) below before proceeding with the analysis for organic matter, otherwise a result that is too high will be obtained.
    Alternatively, the effect of chlorides on the organic matter determination can be partly eliminated by using concentrated sulphuric acid in which silver sulphate has been dissolved, in place of the concentrated sulphuric acid. If the ratio of carbon to chloride does not exceed unity, 25 g of silver sulphate per litre of sulphuric acid will be sufficient to precipitate the chloride.
    If the presence of both sulphides and chlorides is indicated, the procedures given in Sections (D) and (F) should both be carried out on the sample of soil before determination of the organic content.
    If the presence of chlorides is not indicated, omit procedure (F).

(F)   Elimination of chlorides

The procedure for the elimination of chlorides from the test sample is as follows.

(1)   Weigh about 50 g of the soil, after cooling as in (B) (11) above, to 0.01 g.
(2)   Place the soil on a medium-grained filter paper (e.g. Whatman No. 40) in a funnel and wash with distilled water.

(3)   Continue washing until no turbidity is observed when a drop of the wash water is tested with silver nitrate solution.
(4)   Dry the soil retained on the filter paper to constant mass at a temperature of $50° \pm 2.5°C$ and cool in the desiccator.
(5)   Carefully remove all the soil from the filter paper and determine its mass to 0.01 g.
(6)   Sub-divide the treated sample, as in Section (B) step (9)(a) and dry and cool each test specimen as in steps (10) to (12).

(G)   *Analysis for organic matter*

(1)   Weigh each weighing bottle containing prepared soil, obtained as described in Section (B) or (D) or (F), to 0.001 g.
(2)   Transfer at least two representative portions of the appropriate mass (usually 5 g to 0.2 g, depending on the organic content) from the weighing bottle to separate dry 500 ml conical flasks. Weigh the bottle after removal of each portion and calculate the mass of each test specimen transferred, to the nearest 0.001 g, by difference (m_3).

The size of the sample for analysis depends on the amount of organic matter present in the soil. As much as 5 g may be required for soil low in organic matter, and as little as 0.2 g for a very peaty soil. After a number of determinations have been made experience will indicate the most suitable size of sample to be taken. Otherwise a series of samples of varying sizes should be tested, and the determination giving a total volume, V, of 5 ml to 8 ml of potassium dichromate solution reduced should be taken as the one giving the correct result.

(3)   Run 10 ml of the potassium dichromate solution into the conical flask from a burette, and add 20 ml concentrated sulphuric acid very carefully from a measuring cylinder. Swirl the mixture thoroughly for about 1 min, and then allow to stand on a heat-insulating surface for 30 min to allow oxidation of the organic matter to proceed. During this period protect the flask from cold air and draughts.
(4)   Add 200 ml of distilled water very carefully to the mixture, followed by 10 ml of orthophosphoric acid and 1 ml of the indicator, and shake the mixture thoroughly. If the indicator is absorbed by the soil add a further 1 ml of the solution.
(5)   Add ferrous sulphate solution from the second burette in 0.5 ml increments and swirl the contents of the flask until the colour of the solution changes from blue to green.
(6)   Add a further 0.5 ml of potassium dichromate solution, changing the colour back to blue.
(7)   Add ferrous sulphate solution slowly drop by drop with continued swirling until the colour of the solution changes from blue to green after the addition of a single drop.
(8)   Record the total volume of ferrous sulphate solution used, y, to the nearest 0.05 ml.

(H)   *Calculations*

Calculate the total volume, V ml, of potassium dichromate solution used to oxidise the organic matter in the soil sample from the equation

$$V = 10.5 \, (1 - y/x)$$

where y is the total volume of ferrous sulphate solution used in the test (ml); x is the total volume of ferrous sulphate solution used in the standardisation test (ml).

(2)   Calculate the percentage of the original soil sample passing the 2 mm BS test sieve from the equation

$$\text{Fraction finer than 2 mm} = \frac{m_2}{m_1} \times 100\%$$

where m_1 is the initial dry mass of sample (g); m_2 is the mass of the sample passing the 2 mm sieve (g).

(3) Calculate the percentage of organic matter present in the fraction of the soil specimen finer than 2 mm for each determination from the equation

$$\text{Percentage organic matter content} = \frac{0.67V}{m_3}$$

where m_3 is the mass of soil used in the test.

The test method is based on 'wet oxidation' of the organic content of the soil, and assumes that soil organic matter contains an average of 58% of carbon by mass. The method employed oxidises approximately 77% of the carbon in the organic matter, and these factors are included in the above equation. The factors will give correct results only for soil containing natural organic matter.

(4) If the individual results differ by no more than 2% organic matter, calculate the mean results. Otherwise repeat the test starting with two new test portions of soil.

(J)   Results

(1) Report the average percentage of organic matter content present in the soil fraction passing a 2 mm BS test sieve to the nearest 0.1%.
(2) Report the percentage by dry mass of the original sample passing the 2 mm sieve to the nearest 1%.
(3) If sulphides or chlorides have been identified in the soil, include a suitable statement in the report.

5.7.3   Peroxide Oxidation Method

This method is used as part of the pretreatment of soils before a fine particle size analysis, to eliminate colloidal organic matter (Section 4.8.1).

APPARATUS

(1) Balance accurate to 0.001 g.
(2) Porcelain evaporating dish, 150 mm diameter.
(3) Glass stirring rod.
(4) Thermometer, 0–100°C.
(5) Buchner funnel and filter flask.
(6) Source of vacuum.
(7) Filter papers, Whatman No. 50.
(8) Beaker, 400 ml.
(9) Drying oven 105–110°C and desiccator.

REAGENT

Hydrogen peroxide, 6% (20 volume) solution.

PROCEDURE

(1) Prepare a sample, oven dried at 50° ± 2.5°C, of 50 to 100 g of soil passing a 2 mm sieve, as described in Section 5.1.5, steps (1) to (7). Weigh the sample to 0.01 g (m_1) and place in a clean dry wide-mouth conical flask.
(2) Add 150 ml of hydrogen peroxide and stir gently with a glass rod. Cover, and allow to stand overnight.
(3) Heat gently to a temperature of about 60°C, stirring to release bubbles of gas. Avoid frothing over.
(4) Allow the reaction to continue until gas is no longer evolved at a very rapid rate.

(5) Boil the mixture, to reduce the volume to about 50 ml and to decompose excess peroxide.
(6) When cool, add more peroxide if necessary to complete the oxidation, and repeat stages (4) and (5). With very organic soils this process may take one or two days.
(7) Filter through a Whatman No. 50 filter paper, using a Buchner funnel and vacuum flask. Wash thoroughly with distilled water.
(8) Transfer the soil to a weighed and dried glass evaporating dish (mass m_2).
(9) Dry the dish and contents in the oven at 105–110°C, to constant mass.
(10) Weigh dish and contents (m_3).
(11) Calculate the loss due to hydrogen peroxide treatment from the equation

$$\text{loss} = \frac{m_3 - m_2}{m_1} \times 100\%$$

(12) Report the result as a percentage to the nearest 0.1% as the organic matter content determined by hydrogen peroxide treatment.

5.8 CARBONATE CONTENT TESTS

5.8.1 Scope of Tests

Four tests for the determination of the carbonate content of soils are described here. All depend upon the reaction between carbonates and hydrochloric acid, which liberates carbon dioxide. In all except the first method the amount of carbon dioxide evolved is measured. In all cases the test results are expressed as a percentage of CO_2. Initial preparation of soil is the same for all tests, and is described in Section 5.8.2.

The first procedure (Section 5.8.2) is a rapid titration method, in which the excess hydrochloric acid remaining after reaction with the carbonates is measured by titration against sodium hydroxide. This method is suitable when an approximate estimate only is required, for soils in which the carbonate content exceeds about 10%.

The second procedure (Section 5.8.3) is a gravimetric method in which the evolved carbon dioxide is absorbed and weighed. It is based on the procedure given in BS1881: Part 124:1988 for the determination of the carbonate content of hardened concrete. This method is best suited to a laboratory devoted to chemical testing.

In the third procedure, the amount of carbon dioxide is measured by volume, using the Collins apparatus. The fourth method is a simple version of the third. Both are relatively easy to carry out, once the necessary apparatus has been assembled, and the results are accurate enough for most engineering purposes (Sections 5.8.4 and 5.8.5).

In the ASTM Standards, Test Designation D-4373, the calcium carbonate content of soils is determined from the measurement of the pressure generated by the evolved carbon dioxide as a result of the reaction with hydrochloric acid. A purpose-made closed pressure vessel (the 'Rapid Carbonate Analyzer') is required. Calibration tests are first performed using a range of known masses of pure calcium carbonate. This test is not included here.

5.8.2 Rapid Titration Method (BS1377: Part 3:1990:6.3)

This is a quick test, once the test sample has been prepared and the acid solution standardized, and is suitable where an accuracy of about 1% is sufficient.

APPARATUS

(1) 250 ml tall-form beaker and watch glass cover.
(2) Two 100 ml burettes reading to 0.1 ml.

(3)   25 ml pipette.
(4)   250 ml conical flask.
(5)   1 litre volumetric flask.

REAGENTS

(1)   Hydrochloric acid (c (HCl)=approximately 1 mol/litre). Dissolve 88 ml of concentrated hydrochloric acid in distilled water to make one litre of solution.
(2)   Sodium hydroxide solution (c (NaOH)=approximately 1 mol/litre). Dissolve about 20 g of sodium hydroxide in 500 ml of distilled water and store in an airtight plastics container.

NOTE.   SODIUM HYDROXIDE IS STRONGLY CAUSTIC — USE EYE PROTECTION.

(3)   Screened methyl orange indicator. Screened methyl orange gives a more distinct end-point than the unscreened indicator but the latter may be used is preferred. Methyl red or bromocresol green are also suitable.

PROCEDURAL STAGES

(A)   Standardize hydrochloric acid.
(B)   Standardize sodium hydroxide.
(C)   Prepare sample.
(D)   Analyse for carbonates.
(E)   Calculate.
(F)   Report result.

PROCEDURE

(A)   *Standardization of sodium hydroxide solution*

(1)   Transfer 25 ml of the sodium hydroxide solution into a 250 ml volumetric flask, using a pipette.
(2)   Dilute with distilled water to make up to 250 ml. If the concentration of the original solution is B mol/litre, the concentration of this diluted solution is 0.1 B mol/litre.
(3)   Determine the concentration (0.1 B) of the diluted solution, using the method given in Section 5.6.6, step (1) of Test Procedure.
(4)   Multiply this value by 10 to obtain the concentration (B) of the concentrated solution.

(B)   *Standardization of hydrochloric acid solution*

(1)   Place 25 ml of the hydrochloric acid into a 250 ml conical flask using a pipette.
(2)   Place the conical flask on a white background and add sodium hydroxide solution slowly from a burette. During this operation, rotate the flask constantly with one hand while controlling the stop-cock on the burette with the other.
(3)   Continue adding sodium hydroxide until the end-point is reached and the hydrochloric acid is neutralised.
(4)   Record the volume of sodium hydroxide used.
(5)   Repeat steps (1) and (4) using two more 25 ml aliquots of acid solution. The volumes of sodium hydroxide used for each titration should not differ by more than 0.1 ml.
(6)   Calculate the mean volume of sodium hydroxide used (V_1 ml) and calculate the concentration (H) of the hydrochloric acid solution (mol/litre) from the equation

$$H = \frac{V_1}{25B}$$

where B is the concentration of the sodium hydroxide solution (mol/litre).

(C) Preparation of test specimen

The initial sample for testing is prepared from the original soil sample generally as described in Section 5.1.5, steps (1) to (7). The oven drying temperature in step (2) is 105° to 110°C, and the mass of material required in step (7) is about 50 g.

The test specimen is prepared as described in Section 5.1.5, steps (8) to (12). The mass required in step (9), to provide two test specimens, is about 12 g.

Take about 5 g of dry soil for each test specimen, and record the mass of each to the nearest 0.001 g (m).

(D) Analysis of test specimen

Each test specimen is analysed as follows.

(1) Place the weighed specimen into the 250 ml tall-form beaker.
(2) Add 100 ml of the hydrochloric acid solution slowly from the burette.
(3) Cover the beaker with the watch glass and allow to stand for 1 hour, stirring occasionally.
(4) When the soil has settled after the final stirring, remove 25 ml of the supernatant liquid with the pipette and transfer to a conical flask.
(5) Add six drops of the indicator solution and titrate with the sodium hydroxide solution as described in step (*B*)(3) until the same colour change as was observed in the standardization procedure occurs. Record the volume (V_2) of the sodium hydroxide solution used, to the nearest 0.1 ml.

(E) Calculations

Calculate the carbonate content of the soil, as a percentage of CO_2, from the equation

$$\text{Carbonate (as } CO_2) = \frac{8.8\,(25H - BV_2)}{m}\%$$

where H is the concentration of the hydrochloric acid solution (mol/litre); B is the concentration of the sodium hydroxide solution (mol/litre); m is the mass of the soil specimen (g); V_2 is the volume of sodium hydroxide used (ml).

If the individual results differ by no more than 2% of CO_2, calculate the mean result. Otherwise repeat the test starting with two new portions of soil.

(F) Results

(1) Report the average carbonate content of the soil sample, expressed as a percentage of CO_2, to two significant figures.
(2) If appropriate, report the percentage by dry mass of the original sample passing the 2 mm sieve, to the nearest 1%.

5.8.3   Gravimetric Method (BS1377: Part 3:1990:6.4)

This method refers to the method given in BS1881: Part 124:1988 for the determination of the carbonate content of hardened concrete. Its application to soils is described below.

APPARATUS

(1) Analytical balance, readable to 0.0001 g (i.e. to 0.1 mg).
(2) Muffle furnace, to provide controlled temperatures of 925° ± 25°C and 1200 ± 50°C.
(3) Desiccator with desiccant.

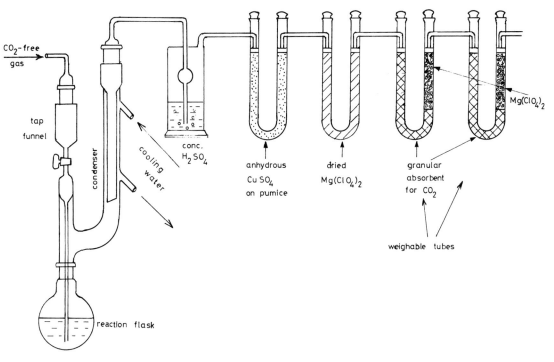

CO$_2$-free gas

tap funnel

condenser

cooling water

conc. H$_2$SO$_4$

anhydrous Cu SO$_4$ on pumice

dried Mg(ClO$_4$)$_2$

granular absorbent for CO$_2$

Mg(ClO$_4$)$_2$

weighable tubes

reaction flask

Fig. 5.12 Arrangement of carbon dioxide adsorption apparatus

(4) Apparatus for the determination of carbon dioxide by absorption. The assembled apparatus is shown diagrammatically in Fig. 5.12, and consists of the following components.

(a) Reaction flask, fitted with a tap funnel and connected to a water-cooled condenser (2 required).
(b) Drechsel bottle bubbler.
(c) Two absorption tubes for removal of atmospheric carbon dioxide.
(d) Two weighable absorption tubes for the absorption of carbon dioxide.

(5) Guard tubes apparatus for removal of carbon dioxide from a stream of air, i.e. to provide a source of gas free of carbon dioxide.

REAGENTS

(1) Concentrated sulphuric acid.
(2) Pumice coated with anhydrous copper sulphate.
(3) Dried magnesium perchlorate.
(4) Granular absorbent for carbon dioxide.
(5) Concentrated orthophosphoric acid, density 1.7 g/ml.
(6) Soda lime (sodium hydroxide and calcium hydroxide) for absorption of carbon dioxide from a stream of air.

PROCEDURAL STAGES

(A) Prepare sample.
(B) Prepare apparatus.
(C) Determine 'blank' measurement.
(D) Add acid.

(E)   Allow reaction to proceed.
(F)   Heat and cool.
(G)   Weigh absorption tubes.
(H)   Calculate.
(J)   Report.

TEST PROCEDURE

(A)   Preparation of test specimen

(1)   The test specimen is prepared and weighed as described in Section 5.8.2 (C), except that the mass required for test depends on the amount of carbonates present in the soil. The dry mass, m, will range from about 0.2 g for a pure limestone or chalk to 1 g for a relatively non-calcareous soil. If in doubt, ascertain the appropriate quantity by performing preliminary trial tests.
(2)   Place the specimen to be tested first in a reaction flask. Keep subsequent specimens in air-tight containers, such as glass weighing bottles, until required.

(B)   Preparation of apparatus

(1)   Assemble the apparatus shown in Fig. 5.12, but without the weighable absorption tubes. Ensure that the tap funnel is free of acid.
(2)   Connect the source of carbon dioxide-free gas to the tap funnel of an empty reaction flask and flush the apparatus with the gas for 15 minutes, applying a rate of flow of about 3 bubbles per second through the Drechsel flask.
(3)   Weigh the two absorption tubes to the nearest 0.0001 g and connect them to the apparatus.

(C)   Determination of 'blank' measurement

(1)   Disconnect the gas inlet, and place 30 ml of concentrated orthophosphoric acid into the tap funnel, with the tap closed.
(2)   Open the tap and re-connect the gas inlet, so that the gas pressure forces the acid into the reaction flask.
(3)   Slowly raise the temperature of the contents of the flask to boiling, and boil for 5 minutes.
(4)   Allow to cool for 15 minutes while maintaining the gas flow.
(5)   When cool, disconnect the absorption tubes and discontinue the flow of gas.
(6)   Weigh the absorption tubes to the nearest 0.0001 g.
(7)   Calculate the difference between the mass of the absorption tubes before and after the above operation, m_0 (g). This is known as the 'blank' for the apparatus which should be less than 1 mg.
(8)   If the blank is greater than 1 mg, check the joints and ensure that they are free of leaks, then repeat the blank determination. If the blank still exceeds 1 mg, replace the absorbents and repeat the whole procedure.

Analysis

(1)   Assemble the apparatus, and flush with carbon dioxide-free gas, as described in stage (B) (1) and (2) above.
(2)   Replace the empty reaction flask by the one containing the test sample, and continue to pass the carbon dioxide-free gas.
(3)   Weigh the two absorption tubes to the nearest 0.0001 g and connect them to the apparatus.
(4)   Disconnect the gas inlet, and place 30 ml of concentrated orthophosphoric acid into the tap funnel, with the tap closed.

(5)   Open the tap and re-connect the gas inlet, so that the gas pressure forces the acid into the reaction flask.
(6)   After effervescence has ceased, slowly raise the temperature of the contents of the flask to boiling, and boil for 5 minutes.
(7)   Allow to cool for 15 minutes while maintaining the gas flow.
(8)   When cool, disconnect the absorption tubes and discontinue the flow of gas.
(9)   Weigh the absorption tubes to the nearest 0.0001 g.

Calculations

(1)   Calculate the increase in mass in the absorption tubes, m_1 (g), during the test, to the nearest 0.0001 g.
(2)   Calculte the carbonate content, as CO_2, as a percentage to the nearest 0.1%, from the equation

$$CO_2 = \frac{m_1 - m_0}{m} \times 100\%$$

where m_1 is the increased in mass of the absorption tubes during the test (g); m_0 is the increase in mass of the absorption tubes during the 'blank' determination (g); m is the mass of the test sample (g).
(3)   Calculate the average percentage CO_2 from the two or more separate determinations.

Reporting

(1)   Report the average carbonate content of the soil sample, expressed as a percentage of CO_2, to two significant figures.
(2)   If appropriate, report the percentage by dry mass of the original sample passing the 2 mm sieve, to the nearest 1%.

5.8.4   Collins Calcimeter — Standard Method

This procedure for measuring the carbonate content of soils was first published by S. H. Collins in 1906. The apparatus which bears his name was developed from that known as Scheibler's apparatus.
   In this test a weighed amount of soil is treated with hydrochloric acid. The volume of carbon dioxide given off is measured, and is corrected for temperature and atmospheric pressure. The carbonate content of the soil is calculated from the corrected volume of carbon dioxide.

APPARATUS

(1)   Collins Calcimeter. This consists of several components (which are listed below) in a clear Perspex tank which can be filled with water. This is to maintain a uniform temperature in all components throughout the duration of the test. The results are very sensitive to temperature.
      The apparatus is illustrated in Fig. 5.13, and is shown diagrammatically in Fig. 5.14. It is no longer available commercially, but can be made up from the following components.

      Measuring cylinder for hydrochloric acid (A) (preferably of clear plastics)
      conical reaction flask (B), 150 ml, with rubber stopper (S)
      graduated burette reading to 0.1 cm³ (E)
      levelling tube (F)

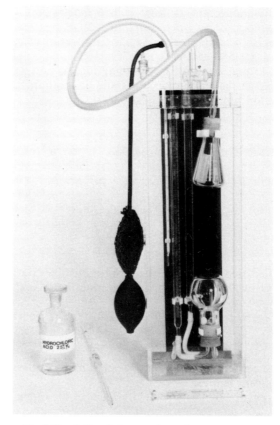

Fig. 5.13 Collins Calcimeter for carbonate content test

reservoir flask (R) with 3-hole stopper (J)
hand pressure bulb (D)
tap (G)
two-way tap (H) (see Fig. 5.15)
outlet (C)
rectangular Perspex tank (T)
mounting panel (P)
thermometer, 0–40°C reading to 0.2°C (Q)

(2)   Barometer, or means of ascertaining local barometric pressure.
(3)   Analytical balance reading to 0.001 g.
(4)   Drying oven and equipment.

REAGENT

Hydrochloric acid solution, 25% v/v. Dilute one volume of concentrated hydrochloric acid with three volumes of distilled water.

PROCEDURAL STAGES

(1)   Prepare test specimens.
(2)   Place in flask.
(3)   Prepare Calcimeter apparatus.
(4)   Insert acid.

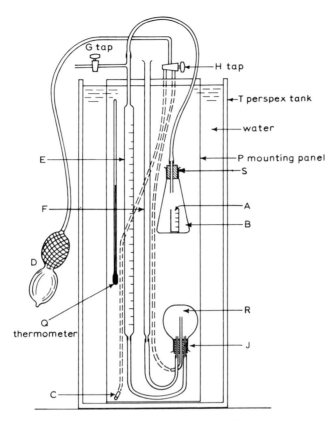

Fig. 5.14   Diagrammatic arrangement of Collins Calcimeter

(5)   Connect reaction flask.
(6)   Agitate water.
(7)   Adjust burette to zero.
(8)   Mix acid and soil.
(9)   Agitate water.
(10) Equalise burette.
(11) Repeat stages (7)–(10) until steady.
(12) Apply temperature correction.
(13) Read or obtain atmospheric pressure.
(14) Calculate.
(15) Report results.

TEST PROCEDURE

(1)   *Preparation of test specimen*

The test specimen is prepared and weighed as described in Section 5.8.2 (C), except that the mass required for test depends on the amount of carbonates present in the soil. The dry mass, m, will range from about 0.2 g for a pure limestone or chalk to about 20 g for a relatively non-calcareous soil. If in doubt, ascertain the appropriate quantity by performing preliminary trial tests.

(2)   *Placing in flask*

Place the specimen to be tested first in the reaction flask. Keep subsequent specimens in air-tight containers, such as glass weighing bottles, until required.

from pressure bulb

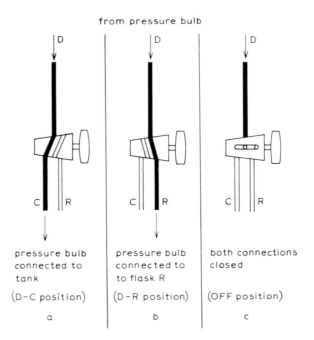

pressure bulb	pressure bulb	both connections
connected to	connected to	closed
tank	to flask R	
(D–C position)	(D–R position)	(OFF position)
a	b	c

Fig. 5.15   Two-way tap on Collins Calcimeter

(3)   Preparation of calcimeter

Fill the Perspex tank (T) with water at room temperature to within about 25 mm from the top. Use boiled or distilled water if the local mains water is hard. Before connecting the hand pressure bulb (D), and with tap G open to atmosphere, pour water into the levelling tube (F), using a funnel, so as to half-fill the reservoir flask (R) with water.

(4)   Insertion of acid

Add hydrochloric acid (25% v/v) to the graduated cylinder (A) from a pipette or burette, then carefully place it in the reaction flask (B). The amount of acid required will range from 10 ml for soils containing little calcareous matter to 15 ml for highly calcareous soils. Take care not to spill any acid on to the soil sample.

(5)   Connection of reaction flask

Connect the reaction flask to the apparatus by inserting the rubber stopper (S). Place the flask under the water in the tank and secure it into position with the spring clip on the mounting board.

(6)   Agitation of water

Set tap H as to connect the hand pressure bulb to the tube (C) (the D–C position, Fig. 5.15a). Gently blow air through the water in the tank by squeezing the hand bulb so as to obtain a uniform temperature. Record the temperature (T_1 °C) when it is steady.

(7)   Adjustment of burette

Open tap G and set tap H to the D–R position (Fig. 5.15b). Gently squeeze the bulb with one hand until the water level in burette E exactly coincides with the zero mark, then close tap G with the other hand. The water level in tube F should be at the same level as that in E. Release the pressure bulb, and the water level in tube F will fall. The level in E will also drop slightly.

(8) *Mixing acid and soil*

Carefully remove flask B from the water-bath, keeping it connected to the burette (E). Tilt the flask so that the acid in the cylinder (A) spills over the soil sample. Shake the flask well (being careful not to break or crack cylinder A), and replace it in the water-bath.

(9) *Agitation of water*

Set tap H to the D–C position, and gently blow air through the water in the tank until a uniform steady temperature $(T_2 \,°C)$ is established.

(10) *Equalisation of water levels*

Set tap H to the D–R position. Gently squeeze the pressure bulb until the water levels in tubes E and F are equal. Then close tap H (Fig. 5.15c). Read the volume indicated in the burette (E).

(11) *Check reaction complete*

Remove reaction flask B without disconnecting it, shake, return to the water-bath, and adjust the levels in tubes E and F (stages 7–10). Repeat this process until there is no further increase in the volume indicated by burette E. Record the final reading as the volume of CO_2 evolved $(V_g \,\text{ml})$.

(12) *Barometric pressure*

Read the barometric pressure at the time of the test. If an accurate barometer is not available, the local meteorological office can give this information, or it may be estimated approximately from the day's newspaper weather report.

(13) *Calculations*

(1) If the temperature at the end of the test $(T_2 \,C)$ differs by more than 0.2°C from that at the beginning $(T_1 \,°C)$, a correction to the measured volume is necessary. The correction (δV_1) is calculated from the equation

$$\delta V_1 = \frac{T_1 - T_2}{2} \,\text{ml}$$

If during the test the temperature falls from T_1 to T_2, δV_1 is positive; if the temperature rises, δV_2 is negative.

This correction is based on the fact that 136 ml of air expands by about 0.1 ml for a temperature rise of 0.2°C. The volume of air in flask B is equal to 150 ml less the volume of acid (10 to 15 ml), i.e. 135 to 140 ml.

(2) Calculate the means temperature $T_m \,°C$ during the test from the equation

$$T_m = \frac{1}{2}(T_1 + T_2)$$

(3) Determine the mass of carbon dioxide $(m_2 \,\text{mg})$, equivalent to 100 ml of gas measured in the apparatus corresponding to the mean temperature $T_m \,°C$, from table 5.7(a) under the volume of acid used $(V_a \,\text{ml})$.

(4) Determine the volume correction (δV_2) for barometric pressure, from table 5.7(b).

288

Table 5.7. DATA FOR COLLINS CALCIMETER TEST

(a) Mass m_2 (uncorrected)
Volume of acid used, V_a (ml)

T_m (°C)	10	12	14	16	18	20	30	40
12	198	201	204	208	211	214	230	249
14	196	199	202	204	207	210	226	244
16	193	196	198	201	203	206	221	238
18	191	193	196	198	201	203	216	233
20	188	190	193	195	198	200	212	228
22	186	188	190	192	194	196	208	223
24	183	185	187	189	191	193	204	218
26	181	183	185	186	188	190	200	213
28	178	180	182	183	185	187	196	208
30	176	178	179	181	182	184	192	203

(b) Corrections for barometric pressure (δV_2)

740 / 987	750 / 1000	760 / 1013	770 / 1027	780 / 1040	mmHg / mbar
−6	−3	0	+3	+6	
−5	−3	0	+3	+5	
−5	−3	0	+3	+5	
−5	−3	0	+3	+5	
−5	−3	0	+3	+5	
−5	−3	0	+3	+5	
−5	−3	0	+3	+5	
−5	−3	0	+3	+5	
−5	−3	0	+3	+5	
−5	−3	0	+3	+5	

(c) Corrections for volume V_g (ml) (δV_3)

0–9	10–19	20–29	30–40
+1	0	0	−1
+1	0	0	−1
+2	0	0	−2
+2	+1	−1	−2
+3	+1	−1	−3
+3	+1	−1	−3
+5	+3	−3	−5
+8	+4	−4	−8

Notes
1. Data are valid for a 150 ml reaction flask (B).
2. The main section of the table gives the mass of 100 ml CO_2 (mg) based upon a measured volume of 20 ml of CO_2 at 760 mmHg pressure, and at mean temperatures between 12 and 30°C (T_m).
3. The correction sections show values to be added or subtracted for different pressures and measured volumes of CO_2.

Example: $V_a = 20$ ml, $T_m = 22$°C, barometer reading 750 mmHg, $V_g = 10$ ml. Therefore 100 ml of CO_2 weighs: $196 - 3 + 1 = 194$ mg $= m_2$ (mg).
If the dry mass of sample used was 2.51 g $= m_1$ (g),

$$\text{carbonate content \% (as } CO_2) = \frac{194 \times 10}{1000 \times 2.51} = 0.77\%$$

4. 1 mmHg = 133.3 Pa = 1.333 mbar.

Table 5.8. DATA FOR SIMPLIFIED PROCEDURE Collins Calcimeter Test
(a) Volume (ml) of acid to be used

Temperature (°C)	Barometric pressure (mmHg)						
	730	740	750	760	770	780	790
12	19	17	15	13	11	9	7
14	21	19	17	15	13	11	9
16	23	21	19	17	15	13	11
18	25	23	21	19	17	15	13
20	27	25	23	21	19	17	15
22	29	27	25	23	21	19	17

Note: Where the quantity of acid indicated exceeds 15 ml, then half the stated quantity of 50% (v/v) HCl should be placed in tube A and the same amount of distilled water in flask B with the sample.

(b) Conversion to carbonate content

Mass of soil (g)	Carbonate content represented by each cm^3 of CO_2 (%)
0.2	1.0
2.0	0.1
20.0	0.01

(5) Determine the correction (δV_3) for the measured volume of gas, V_1, from Table 5.7(c).

(6) Calculate the corrected volume of carbon dioxide, V_g, from the equation

$$V_g = V_1 + \delta V_1 + \delta V_2 + \delta V_3$$

(7) Calculate the percentage carbonate in the soil sample, as CO_2, from the equation

$$\text{Carbonate (as } CO_2) = \frac{m_2 V_g}{1000\, m}\%$$

(8) Calculate the average percentage CO_2 from the two or more separate determinations.

Reporting

(1) Report the average carbonate content of the soil sample, expressed as a percentage of CO_2, to two significant figures.

(2) If appropriate, report the percentage by dry mass of the original sample passing the 2 mm sieve, to the nearest 1%.

5.8.5 Collins Calcimeter — Simplified method

This procedure is less accurate than the standard method described above, but it gives a direct reading of carbonate content without any calculations.

This is achieved by first reading the temperature of the water-bath and observing the barometric pressure. The volume of hydrochloric acid (25% v/v) required for the test is dependent on these two readings, and is read off from Table 5.8(a). This volume of acid is placed in the measuring cylinder (A, Fig. 5.14). The mass of soil used should be 0.2 g for limestone or chalk, 20.0 g for soil with little carbonte content or 2.0 g for an intermediate material.

The test is carried out exactly as described in Section 5.8.4. The volume of carbon dioxide evolved is read from burette E. This is converted to carbonate content of the soil (%) as indicated in Table 5.8(b).

Where the quantity of acid indicated in Table 5.8(a) exceeds 15 ml, use half the stated quantity of 50% (v/v) hydrochloric acid (instead of 25% HCl) in measuring cylinder A and the same amount of distilled water in flask B with the sample.

5.9   CHLORIDE CONTENT TESTS

5.9.1   Scope of Tests

The tests described here for the determination of the chloride content of soils comprise a rapid check to verify whether or not chlorides are present (Section 5.9.2); two tests for water-soluble chlorides; and one test for acid-soluble chlorides.

Of the tests for the determination of the water-soluble chloride content, the first (based on Volhard's method, Section 5.9.3) is given in BS1377: Part 3:1990. The second (Mohr's method, Section 5.9.4) is probably simpler, but both methods require careful observation and accurate weighing.

The acid-extract method (Section 5.9.6) (also due to Volhard) is included in BS1377: Part 3:1990.

5.9.2   Qualitative Check for Chlorides (BS1377: Part 3:1990:7.2.3.3)

The following is a quick check to verify whether or not chlorides are present in the soil. The reagents used are items (1) and (3) in Section 5.9.3.

(A)   Preparation of test specimen

Prepare a sample for the test from the original soil sample as described in Section 5.1.5, steps (1) to (7). The oven-drying temperature in step (2) is 105° to 110°, and the mass of soil required for the check test is about 50 g.

The check test specimen could alternatively be part of the sample prepared for either of the tests described in Sections 5.9.3 and 5.9.4.

Dry the specimen in an oven at 105° to 110°°C, to constant mass, and allow to cool to room temperature in the desiccator.

(B)   Check test

(1)   Place the 50 g check specimen in a 500 ml conical flask and add to it an approximately equal mass of distilled water.
(2)   Agitate the contents intermittently for 4 hours, allow to settle and pour off some of the supernatant solution into a beaker.
(3)   If necessary filter the solution through a medium grade filter paper (e.g. Whatman No. 40) until about 25 ml of clear solution is obtained.
(4)   Acidify the liquid with the nitric acid solution, add about five drops of the silver nitrate solution and allow to stand for 10 minutes.

If no turbidity is apparent after this time the soluble chloride ion content of the soil is negligible and the test for chloride content is not necessary. These observations should be reported.

5.9.3   Chloride Content — (BS1377: Part 3:1990:7.2)

The principle used in this method is that due to Volhard, in which an excess of silver nitrate solution is added to the acidified chloride solution and the unreacted portion is back-titrated with potassium thiocyanate, with ferric alum used as an indicator.

APPARATUS

(1) Analytical balance readable to 0.001 g.
(2) Two 1000 ml volumetric flasks.
(3) Graduated measuring cylinders, 10 ml and 500 ml.
(4) Pipettes, 100 ml and 25 ml.
(5) Two 50 ml burettes and burette stands.
(6) At least four conical flasks, 250 ml.
(7) Wash-bottle and distilled water.
(8) Amber-coloured glass reagent bottle.
(9) Three plastics bottles with wide mouth and watertight screw tops, 2 litres capacity.
(10) Filter funnel, about 100 mm diameter.
(11) Filter papers, to suit the funnel, medium grade (e.g. Whatman No. 40) and fine grade (e.g. Whatman No. 42).
(12) Mechanical shaker, either a vibratory type or one which rotates the watertight containers end-over-end at 30 to 60 rpm.
(13) Drying oven, to give controlled temperatures of $105° \pm 5°C$ and $150° \pm 5°C$.
(14) Desiccator and desiccant.
(15) 500 ml volumetric flask.
(16) Two beakers, about 250 ml capacity.

REAGENTS

(1) Silver nitrate $(c\,(AgNO_3) = 0.100\,mol/litre)$. Dry about 20 g of silver nitrate in the oven at 110°C for 1–2 h, and cool in the desiccator. Weigh out 16.987 g of the dried silver nitrate, dissolve in distilled water, and make up to exactly 1000 ml with distilled water in the volumetric flask. Store the solution in the amber-coloured bottle away from sunlight.
(2) Thiocyanate solution. Either potassium thiocyanate $(c\,(KSCN) = $ approximately 0.1 mol/litre) or ammonium thiocyanate $(c\,(NH_4SCN) = $ approximately 0.1 mol/litre) may be used. Dissolve 10.5 g of potassium thiocyanate, or 8.5 g of ammonium thiocyanate, in distilled water and make up to 1000 ml in a volumetric flask. This solution will be a little stronger than 0.1 N, and the exact normality is determined as described in steps (1)–(5) of the procedure below.
(3) Nitric acid $(c\,(HNO_3) = $ approximately 6 mol/litre). Dilute 100 ml of nitric acid (density 1.42 g/ml) with distilled water to 250 ml, and boil until it is colourless.
(4) 3,5,5-Trimethylhexan-1-ol.
(5) Ferric alum indicator solution. Add 60 g of water to 75 g of ammonium ferric sulphate, warm to dissolve, and add 10 ml of the nitric acid (item (3)). Allow to cool and store in a glass bottle.

PROCEDURE STAGES

The procedure described below is divided into five main sections:

(A) Standardise thiocyanate solution.
(B) Prepare sample.
(C) Prepare solution and titrate.
(D) Calculate.
(E) Report result.

PROCEDURE

(A) Standardisation of thiocyanate solution

(1) Use a 25 ml pipette to place 25 ml of the silver nitrate solution into a 250 ml conical flask.
(2) Add 5 ml of the nitric acid solution and 1 ml of the ferric alum indicator solution.
(3) Add thiocyanate solution from a burette until the colour first permanently changes from colourless to brick-red.
(4) Record the volume of thiocyanate solution used (V_1 ml).
(5) Calculate the concentration (T) of the thiocyanate solution from the equation

$$T = \frac{2.5}{V_1} \text{ mol/litre}$$

(B) Preparation of test specimen

The sample is prepared for testing as described in Section 5.9.2 (*A*), except that the mass of soil required is about 500 g for each test specimen.

(C) Preparation of water-soluble chloride extract

The water-soluble chloride extract is obtained from each test portion as follows.

(1) Weigh a clean and dry screw-capped bottle to 1 g, and record its mass.
(2) Place the dried test specimen in the bottle and weigh the bottle and contents to 1 g.
(3) Calculate the mass of soil by difference.
(4) Add to the bottle a mass of distilled water equal to twice the mass of the test specimen. A portion of the same 2:1 water–soil extract as prepared for sulphate analysis (Section 5.6.3) may be used. However for non-cohesive soils a 1:1 extract may be more convenient, in which case a different factor is used for the calculation below. Fasten the water-tight cap securely.
(5) Secure the bottle to the shaking apparatus and shake for at least 16 hours. When convenient the soil can be left shaking overnight.
(6) Filter the water through a medium-grade filter paper into a clean beaker until at least 100 ml of clear filtrate has been collected. If the filtrate is not completely clear, filter through a fine-grade paper. If the solids settle quickly and the supernatant liquid is clear it can be carefully poured off instead of filtering.

(D) Analysis of extract

Each water extract sample is analysed as follows.

(1) Take 100 ml of the filtered extract by means of the pipette and transfer to the 250 ml conical flask.
(2) Add 5 ml of the nitric acid solution to the flask followed by silver nitrate solution from a burette until all the chloride has been precipitated, and then add a little excess silver nitrate.
(3) Record the total volume (V_2 ml) of silver nitrate solution added.
(4) Add 2 ml of 3, 5, 5 trimethylhexan-1-ol, fit the stopper, and shake the flask vigorously to coagulate the precipitate.
(5) Carefully loosen the stopper, avoiding loss of solution, rinse with distilled water, and collect the washings in the solution.
(6) Add 5 ml of the ferric alum indicator solution, followed by the standardised thiocyanate solution from a burette until the first permanent colour change occurs, that is from colourless to brick-red, and is the same depth of colour as was used for the standardisation described in (*A*) above.
(7) Record the volume (V_3 ml) of thiocyanate solution added.

(*E*)   *Calculations*

Calculate the amount of chloride ions present in each water extract, as a percentage by dry mass of soil, from the equation

$$\text{Chloride ion content} = 0.007092 \, (V_2 - 10 \, T V_3)\%$$

where V_2 is the volume of the silver nitrate solution added (ml); V_3 is the volume of standardised thiocyanate solution added (ml); T is the molarity of the standardised thiocyanate solution.

If the water-soil ratio is 1:1 instead of 2:1 (see (*C*) (4) above) the constant 0.007092 in this equation is replaced by 0.003546.

If more than one specimen has been tested, and if the individual results differ by no more than 0.1% of chloride ion content, calculate the mean result. If they differ by more than 0.1%, repeat the test starting with new test specimens.

(*F*)   *Report*

(1)   Report the average percentage of chloride ions in the soil sample, to the nearest 0.01%.
(2)   The water to soil ratio used for preparing the soluble extract should also be reported.

5.9.4   Chloride Content — Mohr's Method

This method for determination of water-soluble chlorides in soil is based on Mohr's procedure, and is outlined by M. J. Bowley (1979). The test solution and a blank for comparison are each titrated with silver nitrate solution, potassium chromate being used as an indicator.

APPARATUS

(1)   Analytical balance readable to 0.0001 g.
(2)   Three conical flasks, 250 ml.
(3)   Burette, 50 ml, and burette stand.
(4)   Filtration funnel.
(5)   Filter papers, Whatman Nos. 541 and 44.
(6)   Ashless tablets.
(7)   Glass bottles, about 500 ml, with stopper.
(8)   Pipette, 25 ml.
(9)   Drying oven, desiccator.
(10) Desiccator and desiccant.

REAGENTS

(1)   Sulphuric acid solution ($c \, (H_2SO_4) = 0.005$ mol/litre).
(2)   Silver nitrate solution ($c \, (AgNO_3) = 0.02$ mol/litre).
(3)   Potassium chromate saturated solution.
(4)   Indicator papers, narrow-range (pH 6.0–7.0).

PROCEDURAL STAGES

The procedure described below is divided into five main sections:

(A)   Prepare sample.
(B)   Prepare solutions.
(C)   Titrate.
(D)   Calculate.
(E)   Report result.

PROCEDURE

(A) *Preparation of test specimen*

(1) Initial preparation of the sample is as described in Section 5.9.2 (A), except that the mass of soil required is about 1 kg.
(2) Sieve the sample on a 425 μm sieve and crush all retained particles to pass the sieve. Mix the crushed material thoroughly with the material already passing the sieve.
(3) Dry the sample in an oven at 105° to 110°C to constant mass and cool in the desiccator.
(4) Weigh out exactly 100 g (to ±0.01 g) of the dried soil for each test specimen, and place in a 500 ml bottle.

(B) *Preparation of solutions*

(4) To the soil sample add 200 ml of distilled water. Insert the stopper and shake frequently over a period of 24 h so as to dissolve all the water-soluble chlorides. Do not apply heat.
(5) Filter a portion of the liquid, if necessary, to remove any solids in suspension, and transfer 25 ml by means of a pipette into a 250 ml conical flask.
(6) Acidify with sulphuric acid solution, just sufficiently to bring the pH to between 6 and 7, checking with a narrow-range pH paper.
(7) Add two drops of saturated potassium chromate solution. Add the same to two similar conical flasks each containing 25 ml distilled water. These are for colour comparison and blank determination.

(C) *Titration*

(8) Titrate the blank with the silver nitrate solution until a blood-red tinge is just obtained and remains permanent. Record the volume (V_1 ml) of silver nitrate solution used.
(9) Titrate the test solution in the same way, until the same colour as in (8) is obtained. Record the volume (V_2 ml) of silver nitrate solution used. The volume (V ml) of silver nitrate solution required for the chloride reaction is the difference between the measured volumes, i.e. $V = (V_2 - V_1)$ ml.

(D) *Calculation*

(10) Calculate the amount of soluble chloride ions present in the water extract, as the equivalent sodium chloride, as a percentage by dry mass of soil, from the equation

$$\text{chloride ion content} = 0.00568\ V\%$$

If the mass of dry soil is m g instead of 100 g, this equation becomes

$$\text{chloride ion content} = \frac{0.568\ V}{m}\%$$

Notes

(11) If the chloride content is very high, it is more convenient to use a more concentrated solution of silver nitrate (c ($AgNO_3$) = 0.1 mol/litre) in steps (8) and (9). The calculation given in step (10) then becomes

$$\text{chloride ion content} = 0.0284\ V\%$$

if 100 g of dry soil is used or

$$\text{chloride ion content} = \frac{2.84\ V}{m}\%$$

if a mass m g of dry soil is used.

(12) The volume of water extract taken for the titration analysis is given in step (5) as 25 ml, but this is suitable only when the concentration of chlorides is fairly low. If the concentration is high, take a lesser quantity, say 5 ml (measured accurately with a pipette), and dilute with distilled water to make up to 25 ml for a working solution. The aim should be to use approximately 10 ml of silver nitrate solution during titration. An amount greatly in excess of this is wasteful, and using very much less (say less than 5 ml) reduces the accuracy of the measurement by burette.

If 5 ml of extract is used instead of 25 ml, the calculations for chloride content given in step (10) must be amended by a factor of 5, so that

$$\text{chloride ion content} = 0.0284 \, V \%$$

or

$$\text{chloride ion content} = \frac{2.84 \, V}{m} \%$$

Similarly, if any other volume (x ml) of extract is used, the calculation must be proportionately adjusted by multiplying by the factor $25/x$.

(E) *Report results*

(13) The result is reported as a percentage to two significant figures as the soluble chloride ion content.

5.9.5 Acid-Soluble Chloride Content (BS1377:Part 3:1990:7.3)

This procedure is based on Volhard's method, and is used for the determination of the acid-soluble chlorides in soil, including chlorides that are not extracted by water. The chlorides are extracted from dry soil with dilute nitric acid.

APPARATUS

(1) Analytical balance readable to 0.001 g.
(2) 1000 ml volumetric flask.
(3) 10 ml graduated glass measuring cylinder.
(4) 50 ml graduated glass measuring cylinder.
(5) 15 ml pipette.
(6) 500 ml beaker.
(7) Two 50 ml burettes.
(8) At least two stoppered conical flasks, 250 ml capacity.
(9) Filter funnel of approximately 100 mm diameter.
(10) Filter papers of a diameter appropriate to the size of the funnel: coarse grade (e.g. Whatman No. 541).
(11) Support for filter funnel (e.g. burette stand).

REAGENTS

The reagents required are the same as those listed in Section 5.9.4.

PROCEDURAL STAGES

The procedure described below is divided into six main sections.

(A) Standardise thiocyanate solution.
(B) Prepare sample.

(C)   Prepare acid extract.
(D)   Analyse extract.
(E)   Calculate.
(F)   Report result.

PROCEDURE

(A) *Standardisation of thiocyanate solution*

The procedure is the same as that given in Section 5.9.4.

(B)   *Preparation of test specimens*

(1)   Initial preparation of the soil for test is as described in Section 5.1.5, steps (1) and (2). The oven drying temperature is 105° to 110°C.
(2)   Sieve the dried sample on a 150 μm BS test sieve (if appropriate, guarded by a sieve of larger aperture.)
(3)   Crush all retained particles to pass the 150 μm sieve and mix thoroughly with the material already passing the sieve.
(4)   Divide the material by successive riffling to produce representative samples each of about 10 g.
(5)   Dry the samples in the oven at 105° to 110°C and allow to cool in the desiccator.

(C)   *Preparation of acid extract*

The acid-soluble chloride extract is obtained from each prepared sample as follows.

(1)   Weigh out 5 g (to an accuracy within 0.005 g) as the test specimen and place it in a 500 ml beaker.
(2)   Add 50 ml of distilled water to disperse the particles, followed by 15 ml of nitric acid.
(3)   Heat to near boiling point and keep warm for 10 to 15 minutes.
(4)   Filter through a coarse-grade filter paper into a conical flask, wash with hot water and collect the washings with the filtrate. Some cloudiness in the filtrate is permissible. Allow to cool.

(D)   *Analysis of extract*

(1)   Add silver nitrate solution from a burette until all the chloride has been precipitated, then add a little excess silver nitrate.
(2)   Record the total volume (V_2 ml) of silver nitrate solution added.
(3)   Add 2 ml of 3,5,5 trimethylhexan-1-ol, fit the stopper, and shake the flask vigorously to coagulate the precipitate.
(4)   Carefully loosen the stopper, avoiding loss of solution, rinse with distilled water, and collect the washings in the solution.
(5)   Add 5 ml of the ferric alum indicator solution, followed by the standardised thiocyanate solution from a burette until the first permanent colour change occurs, that is from colourless to pink, and is the same depth of colour as was used for the standardisation described in Section 5.9.3 (A).
(6)   Record the volume (V_3 ml) of thiocyanate solution added.

(E)   *Calculations*

Calculate the chloride content as a percentage by dry mass of soil from the equation

$$\text{Chloride content} = 0.07092\,(V_2 - 10\,T V_3)\%$$

where V_2 is the volume of the silver nitrate solution added (ml); V_3 is the volume of standardised thiocyanate solution added (ml); T is the molarity of the standardised thiocyanate solution.

If more than one specimen has been tested, and if the results differ by no more than 0.1% of chloride content, calculate the mean result. If they differ by more than 0.1%, repeat the test starting with two new test specimens.

(F) *Report*

(1) Report the percentage of chlorides in the soil sample, to the nearest 0.01%.
(2) Report that the acid extraction method was used.

5.10 MISCELLANEOUS TESTS

5.10.1 Scope

The tests described below are procedures which have not been covered in the foregoing sections. The first is a test for the determination of the total solids dissolved in groundwater, irrespective of the actual substances. The second is the determination of the mass lost from a soil by ignition at a specified temperature. The third is an outline of the use of indicator papers for assessing the approximate concentration of certain dissolved salts.

5.10.2 Total Dissolved Solids

This test enables the total amount of dissolved solids in a sample of water, e.g. ground water, to be determined.

The result obtained by this method may not be exact, especially if ammonium salts are present, but it is sufficiently accurate for practical purposes where an indication of the amount of dissolved salts is required.

APPARATUS

The following apparatus is required.

(1) Buchner funnel of about 100 mm diameter.
(2) Vacuum filtration flask of about 500 ml capacity, to take the funnel.
(3) Source of vacuum, and vacuum tubing.
(4) Filter papers to fit the funnels, e.g. Whatman No. 40.
(5) Evaporating dish.
(6) Drying oven capable of maintaining a temperature of 180°C to within $\pm 10°C$.
(7) Desiccator with desiccant.
(8) Volumetric flasks, of a size appropriate to the amount of water required for each test determination.
(9) Balance readable to 0.5 mg.
(10) Electric hotplate, or Bunsen burner and tripod.
(11) Shallow container for holding boiling water, for use as a boiling water bath.
(12) Wash-bottle containing distilled water.

PROCEDURE

(1) Filter the sample of groundwater, using the Buchner funnel and flask so as to remove any suspended solids.
(2) Collect a known volume (V ml) of filtered water in each volumetric flask. The volume of water tested should be such as to yield between 2.5 mg and 1000 mg of dissolved solids.
(3) For each determination, heat the evaporating dish to 180°C for 30 min, allow to cool in a desiccator and weigh to 0.5 mg (m_1 grams).

(4)   Pour a portion of the filtered water sample into the dish and evaporate over a bunsen burner (or electric hot-plate) on a boiling water-bath. This must be done in a clean atmosphere to prevent contamination with airborne solids.
(5)   Add further portions of the water sample to the dish as evaporation proceeds. When the flask is empty, rinse it twice with 10 ml of distilled water and add the rinsings to the evaporating dish.
(6)   Allow to evaporate to dryness and wipe dry the outside of the dish.
(7)   Heat the dish and contents in an oven at 180°C for 1 h.
(8)   Allow to cool in a desiccator, and weigh to 0.5 mg (m_2 grams).
(9)   Repeat stages (7) and (8), but heating for 30 min, until the difference between successive weighings does not exceed 1 mg.

CALCULATIONS

Calculate the total dissolved solids (TDS) in each measured sample of water from the equation

$$TDS = \frac{m_2 - m_1}{V} \times 10^6 \text{ parts per million}$$

where m_1 is the mass of the dried evaporating dish (g); m_2 is the mass of the dish with dissolved solids after drying at 180°C (g); V is the measured volume of the sample of water used (ml).

   Calculate the mean of the individual results if more than one sample has been tested. If they differ by more than 10% of the mean value, repeat the test starting with two new water samples.

REPORT RESULT

The result expressed in parts per million is reported to two significant figures as the total dissolved solids in the water sample when dried at 180°C.

   If it was not possible to filter the sample free of all traces of turbidity, this fact should be reported.

5.10.3   Loss on Ignition (BS1377: Part 3:1990:4)

The mass lost from a soil on ignition is related to the organic content of soils such as sandy soils containing little or no clay, chalky material, peats, and clays containing more than about 10% of organic matter. However, in some soils, material other than organic matter could contribute a major proportion of the mass loss on ignition.

APPARATUS

(1)   Balance, readable to 0.001 g.
(2)   Silica or porcelain crucible or similar container of about 30 ml capacity.
(3)   (a) Electric muffle furnace capable of maintaining a temperature of 440°C to within
          ±25°C, or
       (b) Meker gas burner, pipeclay triangle and tripod.
(4)   Drying oven capable of maintaining a temperature of 50° ± 2.5°C.
(5)   Desiccator with desiccant.

PROCEDURAL STAGES

(A)   Prepare crucible.
(B)   Prepare sample.
(C)   Ignite.

(D)   Calculate.
(E)   Report.

PROCEDURE

(A)   *Preparation of crucible*

Before starting each series of tests, place the crucible in the unheated muffle furnace and heat to 440°C±25°C, or heat to red heat over the Meker burner. Maintain the temperature for one hour. Remove the crucible from the furnace or burner and allow it to cool to room temperature in the desiccator. Weigh the crucible to the nearest 0.001 g.

Repeat the above to verify that the mass remains constant to within 0.01 g. If not repeat these steps again until constant mass is achieved.

Record the crucible mass to the nearest 0.001 g (m_c).

(B)   *Preparation of test specimen*

Initial preparation of the sample is generally as described in Section 5.1.5. Details are as follows.

(1) to (6)   As steps (1) to (6) of 5.1.5. The oven drying temperature in step (2) is 50° ± 2.5°C.
(7)   As step (7) of 5.1.5. A prepared sample of about 20 g is required for two determinations.
(8) and (9)   As steps (8) and (9) of Section 5.1.5. The mass of each test specimen should be about 5 g.
(10)   Place each specimen in a prepared crucible, dry in the oven at 50° ± 2.5°C to constant mass and cool in the desiccator.
(11)   Weigh the crucible and contents to the nearest 0.001 g (m_3).

(C)   *Ignition of soil*

Each test specimen is ignited as follows.

(1)   Place the crucible with soil in the unheated muffle furnace, heat to 440°C±25°C, and maintain this temperature for a period of not less than three hours, or until constant mass is achieved. The period required for ignition will vary with the type of soil and size of sample.
(2)   Remove the crucible and contents from the furnace and allow to cool to room temperature in the desiccator.
(3)   Weigh the crucible and contents to the nearest 0.001 g (m_4).

(D)   *Calculations*

(1)   Calculate the loss on ignition, LOI, as a percentage of the dry mass of soil passing a 2 mm BS test sieve from the equation

$$\text{LOI} = \frac{m_3 - m_4}{m_3 - m_c} \times 100\%$$

where m_3 is the mass of the crucible and oven-dry soil specimen (g); m_4 is the mass of the crucible and specimen after ignition (g); m_c is the mass of the prepared crucible (g).
(2)   Calculate the average LOI from the separate determinations.
(3)   Calculate the percentage of the original soil sample passing the 2 mm BS test sieve from the equation

$$\text{Fraction finer than 2 mm} = \frac{m_2}{m_1} \times 100\%$$

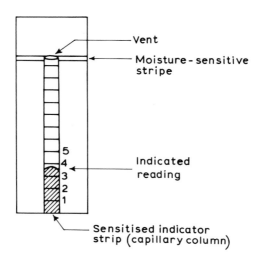

Fig. 5.16   Typical indicator strip

where m_1 is the original dry mass of sample (g); m_2 is the mass of sample passing the
2 mm sieve (g).

(E)   *Report*

(1)   Report the average loss on ignition as a percentage of the soil fraction passing the
2 mm BS test sieve, to two significant figures.
(2)   Report the furnace temperature and period of ignition.
(3)   Also report the percentage by dry mass of the original sample passing the 2 mm sieve,
to the nearest 1%.

5.10.4   Use of Indicator Papers

Litmus papers, blue and red, for indicating acidity or alkalinity are the most familiar type
of indicator papers used in chemical testing. Indicator papers for the measurement of pH
are described in Section 5.5.1. Sensitised papers of a more complex type have become
available in recent years for the rapid quantitative determination of many substances present
in solution in water. The most important of these in soils work are sulphate content and
chloride content. One brand of these indicating papers is known as 'Quantabs'. They are
available in sealed bottles containing 50 test strips.

The test strip carries a scale marked from 0 to 10, at the top end of which is a thin
coloured horizontal stripe (see Fig. 5.16). The test strip is immersed in the solution to be
tested, and held there for a few seconds until the horizontal indicator stripe turns colour.
This confirms that water has made its way to the top of the test strip by capillary action,
so that the whole length of the strip has been wetted. The length of the strip, from the
bottom upwards, which has turned colour after the lapse of 30 s is dependent on the
concentration of the substance in the solution. The reading on the scale where the
colour-change ends is referred to a table supplied with the indicator strips, from which the
percentage (e.g. of sulphates) is read off.

Full instructions are provided by the manufacturer.

These indicators are not intended to be a substitute for chemical analysis, but they are
useful for providing an approximate indication. They can be used in the first instance to
assess whether or not a full analysis is necessary.

BIBLIOGRAPHY

Bowley, M. J. (1979). 'Analysis of sulphate-bearing soils'. Building Research Eastablishment Current Paper CP 2/79, March 1979. Building research Establishment, Garston, Watford

BS812: Part 117:1990, 'Testing aggregates — Methods for determination of water-soluble chloride salts'. British Standards Institution, London

BS1881: Part 124:1988, 'Testing concrete — Methods for analysis of hardened concrete. British Standards Institution, London

BS3978: 'Specification for water for laboratory use'. British Standards Institution, London

Building Research Establishment (1981). Digest No. 250. 'Concrete in sulphate-bearing soils and groundwaters'. Building research Establishment, Garston, Watford

Collins, S. H. (1906). 'Scheibler's apparatus for the determination of carbonic acid in carbonates; an improved construction and use for accurate analysis'. *J. Soc. Chem. Ind.*, Vol. 25

Cumming, A. C. and Kay, S. A. (1948). *Quantitative Chemical Analysis.* Gurney and Jackson, London

Gutt, W. H. and Harrison, W. H. (1977). 'Chemical resistance of concrete'. BRE Current Paper CP 23/77, May 1977. Building Research Establishment, Garston, Watford. (Reprint from *Concrete* 1977, Vol. 11, No. 5)

Kuhn, S. (1930). 'Eine neue kolorimetrische Schnellmethode zur Bestimmung des pH von Boden'. *Z. Pflernahr. Dung.*, Vol. 18A

Vogel, A. I. (1961), *A Textbook of Qualitative Inorganic Analysis*, 3rd edition. Longmans, London

Walkley, A. and Black, I. A. (1934). 'An examination of the Degtjareff method of determining soil organic matter, and a proposed modification of the chromic acid titration method'. *Soil Sci.*, Vol. 37(1)

Wilson, C. L. and Wilson, D. W. (1959). *Comprehensive Analytical Chemistry.* Elsevier, Amsterdam

Chapter 6

Compaction tests

6.1 INTRODUCTION

6.1.1 Scope

Many civil engineering projects require the use of soils as 'fill' material. Whenever soil is placed as an engineering fill, it is nearly always necessary to compact it to a dense state, so as to obtain satisfactory engineering properties which would not be achieved with loosely placed material. Compaction on site is usually effected by mechanical means such as rolling, ramming or vibrating. Control of the degree of compaction is necessary to achieve a satisfactory result at reasonable cost. Laboratory compaction tests provide the basis for control procedures used on site.

Compaction tests furnish the following basic data for soils:

(1)   The relationship between dry density and moisture content for a given degree of compactive effort.
(2)   The moisture content for the most efficient compaction — that is, at which the maximum dry density is achieved under that compactive effort.
(3)   The value of the maximum dry density so achieved.

Item (1) is expressed as a graphical relationship from which items (2) and (3) can be derived. The latter are the moisture and density criteria, against which the compacted fill can be judged if *in situ* measurements of moisture content and density are made.

There are several different standard laboratory compaction tests. The test selected for use as the basis for comparison will depend upon the nature of the works, the type of soil and the type of compaction equipment used on site. This chapter describes the tests accepted in Britain as standard practice, and two tests of American origin which have special applications.

Tests which are carried out on site to determine the density and other characteristics of the compacted fill are not described here.

6.1.2 Development of Test Procedures

A test to provide data on the compaction characteristics of soil was first introduced by R. R. Proctor in the USA in 1933, in order to determine a satisfactory state of compaction for soils being used in the construction of large dams, and to provide a means for controlling the degree of compaction during construction. The test made use of a hand rammer and a cylindrical mould with a volume of $1/30\,\text{ft}^3$, and became known as the standard 'Proctor' compaction test (Proctor, 1933; Taylor, 1948). The test now known as the British Standard 'light' compaction test is very similar, although the equipment used differs in some details.

At that time it was believed that the Proctor test represented in the laboratory the state of compaction which could be reasonably achieved in the field. But with the subsequent introduction of heavier earth-moving and compaction machinery, especially for the construction of large dams, higher densities became obtainable in practice. A laboratory test using increased energy of compaction was then necessary to reproduce these higher compacted densities, so a test was introduced which used a heavier rammer with the same mould. This intensified procedure became known as the 'modified AAHSHO' test. It is similar to the British Standard 'heavy' compaction test. The 'Proctor' mould of $1/30\,\text{ft}^3$ ($944\,\text{cm}^3$) is used in ASTM standards. When the British Standard changed to SI units in 1975 the volume of the mould was rounded up to $1000\,\text{cm}^3$, and this is known as the 'one-litre compaction mould'. The dimensions and masses of rammers were rationalised to metric units at the same time. It is essential to appreciate that the BS and ASTM tests, although similar in principle, require different apparatus and use procedures which differ in some details.

Details of the BS and ASTM compaction moulds are summarised in Table 6.2, and data for compaction rammers are included in Table 6.3.

Granular soils, especially gravels, are most effectively compacted by vibration. A laboratory test using a vibrating hammer was introduced into British Standards in 1967 to establish the compaction characteristics for these conditions. Because particles up to coarse gravel size are necessary so as to represent these materials as closely as possible in the test, a large mould (the CBR mould) is used. This procedure is known in this country as the British Standard vibrating hammer compaction test.

Dry densities measured on compacted soils *in situ* are still often expressed as a percentage of the maximum dry density for a specified degree of compaction. This percentage is called the 'relative compaction' of the soil. When the dry density required on site is greater than the BS 'light' maximum dry density, the field density is more usually related to the BS 'heavy' maximum dry density, rather than quoting values of relative compaction in excess of 100%.

6.2 DEFINITIONS

COMPACTION   The process of packing soil particles more closely together, usually by rolling or mechanical means, thus increasing the dry density of the soil.
OPTIMUM MOISTURE CONTENT (OMC)   The moisture content of a soil at which a specified amount of compaction will produce the maximum dry density.
MAXIMUM DRY DENSITY   The dry density obtained using a specified amount of compaction at the optimum moisture content.
RELATIVE COMPACTION   The percentage ratio of the dry density of the soil to its maximum compacted dry density determined by using a specified amount of compaction.
DRY DENSITY–MOISTURE CONTENT RELATIONSHIP   The relationship between dry density and moisture content of a soil when a specified amount of compaction is applied.
PERCENTAGE AIR VOIDS (V_a)   The volume of air voids in a soil expressed as a percentage of the total volume of the soil.
AIR VOIDS LINE   A line showing the dry density–moisture content relationship for a soil containing a constant percentage of air voids.
SATURATION LINE (ZERO AIR VOIDS LINE)   The line on a graph showing the dry density–moisture content relationship for a soil containing no air voids.

6.3 THEORY

6.3.1 Process of Compaction

Compaction of soil is the process by which the solid soil particles are packed more closely together by mechanical means, thus increasing the dry density (Markwick, 1944). It is achieved through the reduction of the air voids in the soil, with little or no reduction in the water content. This process must not be confused with consolidation, in which water is

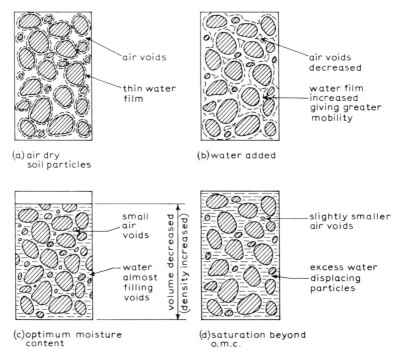

Fig. 6.1   *Representation of compaction of soil grains*

squeezed out under the action of a continuous static load. The air voids cannot be eliminated altogether by compaction, but with proper control they can be reduced to a minimum. The effect of the amount of water present in a fine-grained soil on its compaction characteristics, when subjected to a given compactive effort, is discussed below.

At low moisture content the soil grains are surrounded by a thin film of water, which tends to keep the grains apart even when compacted, (Fig. 6.1a). The finer the soil grains, the more significant is this effect. If the moisture content is increased, the additional water enables the grains to be more easily compacted together (Fig. 6.1b). Some of the air is displaced and the dry density is increased. The addition of more water, up to a certain point, enables more air to be expelled during compaction. At that point the soil grains become as closely packed together as they can be (i.e. the dry density is at the maximum) under the application of this compactive effort (Fig. 6.1c). When the amount of water exceeds that required to achieve this condition, the excess water begins to push the particles apart (Fig. 6.1d), so that the dry density is reduced. At higher moisture contents little or no more air is displaced by compaction, and the resulting dry density continues to decrease.

If at each stage the compacted dry density is calculated, and plotted against moisture content, a graph similar to curve A in Fig. 6.2 is obtained. This graph is the 'moisture–density relationship' curve. The moisture content at which the greatest value of dry density is reached for the given amount of compaction is the optimum moisture content (OMC), and the corresponding dry density is the maximum dry density. At this moisture content the soil can be compacted most efficiently under the given compactive effort. The relationship between bulk (wet) density and moisture content is shown by the dotted curve (W) in Fig. 6.2. This curve is not generally plotted, except perhaps as a guide during a compaction test before the moisture contents are measured.

A typical compaction curve obtained from the British Standard 'light' compaction test (Section 6.5.3) is shown in Fig. 6.3 as curve A. If a heavier degree of compaction, corresponding to the BS 'heavy' compaction test (Section 6.5.4), is applied at each moisture

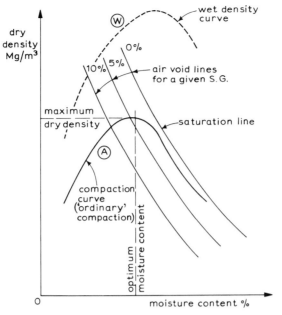

Fig. 6.2   Dry density–moisture content relationship for soils

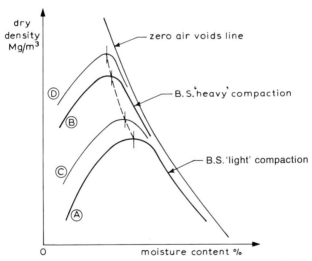

Fig. 6.3   Dry density–moisture curves for various compactive efforts

content, higher values of density, and therefore of dry density, will be obtained. The resulting moisture–density relationship will be a graph such as curve B in Fig. 6.3. The maximum dry density is greater, but the optimum moisture content at which this occurs is lower, than in the 'light' test.

Every different degree of compaction on a particular soil results in a different compaction curve, each with unique values of optimum moisture content and maximum dry density. For instance a compaction test similar to the BS 'light' test but using, say, 50 blows per layer instead of 27 would give a graph similar to that shown by curve C in Fig. 6.3. A test similar to the 'heavy' compaction test but using a greater number of blows would give a graph similar to curve D. It can be seen that increasing the compactive effort increases the maximum dry density, but decreases the optimum moisture content.

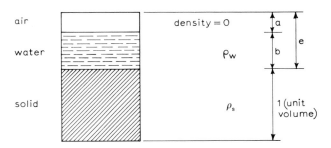

Fig. 6.4   Representation of soil with air voids

6.3.2   Air Voids Lines

A compaction curve is not complete without the addition of air voids lines. An air voids line is a (curved) line showing the dry density–moisture content relation for soil containing a constant percentage of air voids. A set of air voids lines can be drawn from calculated data if the particle density of the soil grains is known, and three are indicated in Fig. 6.2. The derivation of the equation relating dry density to moisture content for a given percentage of air voids, V_a, is given below. Note that V_a is the volume of air voids in the soil expressed as a percentage of the *total* volume of soil, as in BS1377:1990: Part 1:2.2.37, and not as a percentage of the voids. V_a is *not* the same as $(100-S)$, where S is the saturation expressed as a percentage of the total voids.

If all the air voids are removed, so that the total voids between solid particles are filled with water, the soil reaches the fully saturated condition. The equation relating the saturated dry density to moisture content, from which the zero air voids line can be drawn, can be derived by putting V_a equal to zero.

The notation is the same as that used in Section 3.3.2, with some additional symbols.

Volume of solids	=	1
Volume of air voids	=	a
Volume of water in voids	=	b
Total volume $= V$	$= 1+a+b$	
Mass of solids $= 1 \times \rho_s$	$= \rho_s$	
Mass of air $= a \times 0$	$= 0$	
Total dry mass	$= \rho_s$	

Mass of water, from moisture content $= \dfrac{w}{100} \times \rho_s$

Therefore,

$$\text{Volume of water} = \frac{w\rho_s}{100\rho_w} = b \qquad (6.1)$$

Volume of air voids, a, expressed as a percentage of the total volume, is denoted by V_a.

i.e.

$$V_a = \frac{a}{V} \times 100 = \frac{100a}{1+a+b} \qquad (6.2)$$

Hence

$$a = \frac{V_a(1+b)}{100 - V_a}$$

therefore

$$V = 1 + \frac{V_a(1+b)}{100 - V_a} + b$$

$$= \frac{(1+b)(100 - V_a) + V_a(1+b)}{100 - V_a}$$

$$= \frac{100(1+b)}{100 - V_a} \tag{6.3}$$

Substituting for b from Eq. (6.1),

$$V = \frac{100\left(1 + \dfrac{w\rho_s}{100\rho_w}\right)}{100 - V_a}$$

$$= \frac{1 + \dfrac{w\rho_s}{100\rho_w}}{1 - \dfrac{V_a}{100}}$$

Dry density $\rho_D = \dfrac{\text{dry mass}}{\text{volume}} = \dfrac{\rho_s}{V}$

$$= \rho_s \cdot \frac{1 - \dfrac{V_a}{100}}{1 + \dfrac{w\rho_s}{100\rho_w}}$$

i.e.

$$\rho_D = \frac{\left(1 - \dfrac{V_a}{100}\right)\rho_w}{\dfrac{\rho_w}{\rho_s} + \dfrac{w}{100}} \tag{6.5}$$

Using SI units, and putting $\rho_w = 1\ \text{Mg/m}^3$, gives the equation

$$\rho_D = \frac{1 - \dfrac{V_a}{100}}{\dfrac{1}{\rho_s} + \dfrac{w}{100}} \quad \text{Mg/m}^3 \tag{6.6}$$

Table 6.1. DATA FOR CONSTRUCTING AIR VOIDS LINES*

Moisture content, w (%)	Air voids, V_a (%)	Particle density ρ_s Mg/m³				
		2.60	2.65	2.70	2.75	2.80
0	0	2.60	2.65	2.70	2.75	2.80
	5	2.47	2.52	2.57	2.61	2.66
	10	2.34	2.39	2.43	2.48	2.52
5	0	2.30	2.34	2.38	2.42	2.46
	5	2.19	2.22	2.26	2.30	2.33
	10	2.07	2.11	2.14	2.18	2.21
10	0	2.06	2.09	2.13	2.16	2.19
	5	1.96	1.99	2.02	2.05	2.08
	10	1.86	1.89	1.91	1.94	1.97
15	0	1.87	1.90	1.92	1.95	1.97
	5	1.78	1.80	1.83	1.85	1.87
	10	1.68	1.71	1.73	1.75	1.77
20	0	1.71	1.73	1.75	1.77	1.79
	5	1.63	1.65	1.67	1.69	1.71
	10	1.54	1.56	1.58	1.60	1.62
25	0	1.58	1.59	1.61	1.63	1.65
	5	1.50	1.51	1.53	1.55	1.56
	10	1.42	1.43	1.45	1.47	1.48
30	0	1.46	1.48	1.49	1.51	1.52
	5	1.39	1.40	1.42	1.43	1.45
	10	1.31	1.33	1.34	1.36	1.37
35	0	1.36	1.37	1.39	1.40	1.41
	5	1.29	1.31	1.32	1.33	1.34
	10	1.23	1.24	1.25	1.26	1.27

*Dry densities (Mg/m³) corresponding to various moisture contents for soils of different particle densities

For the fully saturated condition (no air voids), $V_a = 0$. Therefore,

$$\rho_{D(sat)} = \frac{1}{\dfrac{1}{\rho_s} + \dfrac{w}{100}} \, \rho_w \quad \text{Mg/m}^3 \tag{6.7}$$

This equation defines the zero air voids line, or the saturation line. It is impossible for a point on a compaction curve (in terms of dry density) to lie to the right of this line, whatever degree of compactive effort is applied.

Curve for 0, 5 and 10% air voids (i.e. $V_a = 0, 5, 10\%$) are shown in Fig. 6.2. These curves are defined only by the particle density of the soil grains. Sets of standard curves can be drawn up for various particle densities, so that the set applicable to a particular soil can be selected, either by use of the data given in Table 6.1 or direct from eqs. (6.6) and (6.7). The air voids lines do not apply to the wet density curve (W) in Fig. 6.2.

6.3.3 Compactive Efforts

The procedures used for various types of BS and ASTM compaction test are summarised in Table 6.3.

Table 6.2. DETAILS OF COMPACTION MOULDS

	Internal dimensions			Internal volume	
	Diameter		Height		
Type of mould	(mm)	(in)	(mm)	(cm³)	(cu ft)
BS 1 litre	105		115.5	1000	
CBR	152		127	2305	
ASTM 4 in	101.6	(4)	116.4	944	(1/30)
6 in	152.4	(6)	116.4	2124	(0.075)

Table 6.3. COMPACTION PROCEDURES

		Rammer		No.	Blows	Refer
		Mass	Drop	of	per	to
Type of test	Mould	(kg)	(mm)	layers	layer	section
BS 'light'	One litre	2.5	300	3	27	6.5.3
	CBR	2.5	300	3	62	6.5.5
ASTM (5.5 lb)	4 in	2.49	305	3	25 ⎫	6.5.7
	6 in	2.49	305	3	56 ⎭	
BS 'heavy'	One litre	4.5	450	5	27	6.5.4
	CBR	4.5	450	5	62	6.5.5
ASTM (10 lb)	4 in	4.54	457	5	25 ⎫	6.5.7
	6 in	4.54	457	5	56 ⎭	
BS Vibrating hammer	CBR	32 to 41*	(vibration)	3	(1 min)	6.5.9

*Downward force to be applied.

The mechanical energy applied in each type of BS test, in terms of the work done in operating the rammer, is derived and compared below.

BS 'LIGHT' COMPACTION TEST

$$(2.5\,\text{kg}) \times \frac{(300\,\text{mm})}{1000} \times 27 \times 3 = 60.75\,\text{kg m}$$

$$= 60.75 \times 9.81\,\text{Nm} = 596\,\text{J}$$

$(\text{kg m} \times 9.81 = \text{newton metres} = \text{joules})$.
Volumes of soil used $= 1000\,\text{cm}^3 = 0.001\,\text{m}^3$. Therefore,

$$\text{work done per unit volume of soil} = \frac{596}{1000}\,\text{J/cm}^3 = 596\,\text{kJ/m}^3$$

BS 'HEAVY' COMPACTION TEST

$$4.5 \times \frac{450}{1000} \times 27 \times 5 \times 9.81 = 2682\,\text{J, or } 2682\,\text{kJ/m}^3$$

'LIGHT' COMPACTION IN CBR (CALIFORNIA BEARING RATIO) MOULD

$$\text{Volume} = 2305 \text{ cm}^3$$

$$2.5 \times 0.3 \times 62 \times 3 \times 9.81 = 1368 \text{ J}$$

$$\frac{1368}{2305} \times 1000 = 594 \text{ kJ/m}^3$$

'HEAVY' COMPACTION IN CBR MOULD

$$4.5 \times 0.45 \times 62 \times 5 \times 9.81 = 6158 \text{ J}$$

$$\frac{6158}{2300} \times 1000 = 2672 \text{ kJ/m}^3$$

The calculations verify that for the 'light' compaction tests, whether carried out with the one-litre mould or the CBR mould, the compactive energy per unit volume of soil is about the same.

For the 'heavy' test the energy is similar with both procedures. The energy applied per unit volume in the 'heavy' test is 4.5 times as much as that used in the 'light' test ($2682/596 = 4.5$ exactly).

VIBRATING HAMMER COMPACTION (see Section 6.5.9)

Assuming a 600 W motor, and that 50% of the electrical input is converted to mechanical energy, half of which is absorbed by the soil sample (the other half being taken mainly by the operator).

$$\text{energy applied to sample} = 600 \times \tfrac{1}{2} \times \tfrac{1}{2} \times 60 \times 3 \text{ J}$$

$$= 27\,000 \text{ J}$$

$$\frac{27\,000}{2300} \times 1000 = 11\,739 \text{ kJ/m}^3$$

The ratio of the calculated energy applied by the vibrating hammer to that applied by the 'heavy' compaction test is $11\,739/2672 = 4.39$, which is of the same order of magnitude as the ratio (4.5) of the 'heavy' to the 'light' compactive effort.

6.3.4   Effect of Stone Content

In the laboratory compaction tests using the one-litre mould only the fraction of soil passing a 20 mm sieve is used. Particles larger than 20 mm which are removed before test may consist of gravel, fragments of rock, shale, brick or other hard material, and are collectively referred to below as 'stones'. The soil actually tested is called the 'matrix' material.

The density achieved on site for the total material cannot be compared directly with the results of laboratory compaction tests on the matrix material only. If the matrix material is compacted to reach a particular density, the presence of stones will give the total material a higher density, because the stones have a greater density than the matrix material they displace. The resulting *in situ* density of the whole material can be calculated from the equation derived below, provided that (a) the proportion of stones in the total material is known; (b) this proportion is not large (i.e. not more than about 25% of the total by dry mass), so that the stones are distributed within the matrix in such a way that they are not in contact with each other (Maddison, 1944; McLeod, 1970).

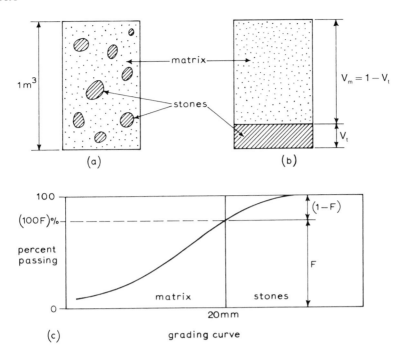

Fig. 6.5   Representation of stony soil

Table 6.4.   SYMBOLS FOR STONE CONTENT EQUATIONS

	Symbols used		
Soil properties	Matrix material	Stones	Total material
Dry density	ρ_{mD}		$\rho = ?$
Particle density	ρ_s	ρ_t	
Volume	V_m	V_t	1
Mass in a unit volume of soil	m_m	m_t	$(m_m + m_t)$

In practice, the presence of stones requires additional compactive effort to achieve the same degree of compaction of the matrix as when the matrix is compacted alone. However, this effect is not great for small percentages of stones, and does not affect these calculations. If the percentage of stones is quite large, there may not be sufficient matrix materials completely to fill the voids between the stones, and this could be an unsatisfactory fill material for many purposes.

A unit volume of 'stony' soil is represented diagramatically in Fig. 6.5(a), and the stones are imagined to be fused together in one piece occupying a volume V_t, as in Fig. 6.5(b). The idealised grading curve of the whole material is shown in Fig. 6.5(c). The proportion of material finer than 20 mm, expressed as a decimal fraction, is denoted by F.

The symbols used in the expressions below are summarised in Table 6.4. Four relationships can be written in the form of equations, as follows.

The total dry mass in a unit volume is equal to $\rho_D = m_m + m_t$ \hfill (6.8)

The mass of matrix materials is equal to its density multiplied by its volume — that is,

$$m_m = (1 - V_t)\rho_{mD} \hfill (6.9)$$

The mass of stones is equal to the volume of solid material multiplied by the density of that material — that is,

$$m_t = V_t \rho_t \qquad (6.10)$$

From the grading curve, the fraction of the matrix material to the whole is equal to the ratio of its dry mass to the total dry mass — that is,

$$F = \frac{m_m}{m_m + m_t} \qquad (6.11)$$

From these equations the relationship between the dry density of the material containing stones, ρ_D, and the dry density of the matrix material measured in the laboratory, ρ_{mD}, can be derived, and is as follows:

$$\rho_D = \frac{\rho_t}{(1 - F) + F\left(\dfrac{\rho_t}{\rho_{mD}}\right)} \qquad (6.12)$$

Using customary SI units, and putting $\rho_w = 1\,\mathrm{Mg/m^3}$, this equation becomes

$$\rho_D = \left[\frac{\rho_t}{(1 - F)\rho_{mD} + F\rho_t}\right]\rho_{mD} \qquad (6.13)$$

This is the theoretical dry density to be expected *in situ*, derived from the dry density, ρ_{mD}, of the matrix material measured in the laboratory.

The overall moisture content of the total material will differ from that of the matrix, owing to presence of the stones. The stones themselves may absorb a certain amount of moisture, which will be removed by the normal oven drying procedure. Let w_m = moisture content of matrix, and w_t = moisture content (absorbed moisture) of stones. This absorbed moisture does not alter the volume of the stones. Moisture contents are expressed as decimal fractions. Other notation is as before.

Mass of water contained in matrix

$$= w_m m_m = w_m F(m_m + m_t)$$
$$= w_m F \rho_D$$

Mass of water contained in stones

$$= w_t m_t = w_s (1 - F)\rho_D$$

Therefore, total mass of water contained in unit volume of combined material

$$= W = w_m F + w_t (1 - F)\rho_D$$

Moisture content of total material

$$= w = \frac{W}{\text{dry mass}} = \frac{W}{\rho_D}$$

Therefore,

$$w = Fw_m + (1 - F)w_t \qquad (6.14)$$

If the stones contain no absorbed water (e.g. if they consist of pieces of quartz gravel), the value of w_t is zero and w is simply equal to $F \times w_m$.

Relationships similar to the above are given in ASTM Standards, D4718.

6.4 APPLICATIONS

6.4.1 Objectives of Proper Compaction

Soils may be used as fill for many purposes, the most usual being:

(1)   To refill an excavation, or a void adjacent to a structure (such as behind a retaining wall).
(2)   To provide made-up ground to support a structure.
(3)   As a sub-base for a road, railway or airfield runway.
(4)   As a structure in itself, such as an embankment or earth dam, including reinforced earth.

Compaction, by increasing the density, improves the engineering properties of soils. The most significant improvements, and the result effects on the mass of fill as a whole, are summarised in Table 6.5.

Table 6.5. EFFECTS OF PROPER COMPACTION OF SOILS

Improvement	*Effect on mass of fill*
Higher shear strength	Greater stability
Lower compressibility	Less settlement under static load
Higher CBR value	Less deformation under repeated loads
Lower permeability	Less tendency to absorb water
Lower frost susceptibility	Less likelihood of frost heave

6.4.2 Construction Control

The relationship between dry density and moisture content for soil subjected to a given compactive effort, established by laboratory compaction tests, provides reference data for the specification and control of soil placed as fill. On many projects the laboratory compaction tests are supplemented by field compaction trials by using the actual placing and compacting equipment which is to be employed for construction (Williams, 1949).

Sometimes it is necessary to adjust the natural moisture content of a soil to a value at which it can be most effectively compacted, or at which it has the highest strength. The required moisture content, and the dry density to be achieved, can be assessed on the basis of the dry density — moisture content relationship derived from laboratory compaction tests on samples taken from the borrow area.

While compaction *in situ* needs to be of a sufficient degree to obtain the required density, it is equally important not to overcompact fine-grained soils. Overcompaction not only is wasteful of effort, but should be avoided because overcompacted soil, if not confined by overburden, can readily absorb water, resulting in swelling, lower shear strength and greater compressibility. Tops and sides of embankments are particularly sensitive to this effect.

6.4.3 Design Parameters

When the compaction characteristics of a soil are known, it is possible to prepare samples in the laboratory at the same dry density and moisture content as that likely to be attained

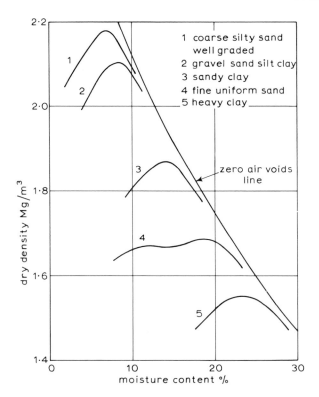

Fig. 6.6 Compaction curves for some typical soils

after compaction in the field. These samples can be subjected to laboratory tests for the determination of their shear strength, compressibility and other engineering properties. Design parameters derived from these tests enable the stability, deformation and other characteristics of the fill to be assessed. They can also provide the basis for the initial design of an embankment or earth dam.

More elaborate tests can be carried out on compacted samples to measure the changes of pore pressure due to changing conditions of applied stress. During construction, pore pressures can be monitored so as to ensure that they do not at any time exceed certain limiting values established by the tests.

A specification for compacted fill may require a certain 'relative compaction' (measured in terms of dry density) to be achieved, within specified limits of moisture content. More usually a specification defines the maximum air voids permitted in the compacted soil within the required dry density range. For this reason it is necessary to determine the density of soil particles so that air voids lines can be added to the compaction test graphs.

6.4.4   Types of Compaction Curve

The form of compaction curve for five typical materials is shown in Fig. 6.6. For ease of comparison they have been related to a common zero air voids line by adjusting the curves to the same particle density. These curves relate to BS or ASTM 'light' compactive effort.

In general, clay soils, and well-graded sandy or silty soils, show a clearly defined peak to the compaction curve. Uniformly graded free-draining soils, consisting of a narrow range of particle sizes, give a flatter compaction curve from which the optimum condition is not easy to define. A 'double peak' is often obtained from uniformly graded fine sands. For these materials a moisture content for optimum compaction is not easy to define. The results of laboratory tests can be meaningless or misleading, and provide a poor guide to field compaction behaviour. A higher dry density can often be obtained in the field, and a maximum density test (Section 3.7.2 or 3.7.4) might be more appropriate.

6.4.5 Compaction of Chalk

Chalk is a very variable material which in its natural state exists as a virtually saturated porous rock. When excavated and re-compacted its properties and behaviour can range from those of rock to those of soil, depending on the proportion of 'putty chalk' formed as a result of breakdown of the natural material (see Section 2.4.3).

If the proportion of putty chalk is high enough to control the behaviour of the mass, the fill material will be weak and unstable, and may be difficult to compact at all.

The extent to which chalk is likely to break down during earth-works construction processes can be assessed from the chalk crushing value (CCV). The test to obtain this value was developed at TRRL when it was realised that other methods of test for soil and rock could not realistically represent the susceptibility of chalk to crushing. The CCV, together with the saturation moisture content (see Sections 2.4.3 and 2.5.4) enables the chalk to be classified as to whether it is suitable for use as fill, and if so to assess the appropriate construction methods to use. (Ingoldby and Parsons, 1977; Ingoldby, 1977).

6.5 COMPACTION TEST PROCEDURES

6.5.1 Types of Test

The tests described in this section are those given in BS1377:1990: Part 3 as the recognised tests for the determination of the moisture-density relationship of soils. Very similar tests, except for the vibrating hammer method, are also given in ASTM Standards, and their differing features are outlined in separate sections.

These tests are described in the following Sections.

(1) 'Light' compaction (BS clause 3.3)—Section 6.5.3. (ASTM D698)—Section 6.5.7.
(2) 'Heavy' compaction (BS clause 3.5)—Section 6.5.4. (ASTM D1577)—Section 6.5.7.
(3) Compaction of soils containing large particles, in CBR mould (BS clauses 3.4 & 3.6, and ASTM clauses D698 & D1577)—Section 6.5.5.
(4) Compaction using vibrating hammer (BS clause 3.7)—Section 6.5.9.

The British Standard describes these tests under the title 'Determination of dry density/moisture content relationships'. The ASTM title is 'Moisture-density relations'.

Preparation of soil for the BS tests is given in Section 6.5.2, and for the ASTM tests in Section 6.5.6.

The use of an automatic compaction apparatus as an alternative to hand compaction (Section 6.5.8) is included.

It is important to refer to the test designation in full when reporting results or when quoting the tests, including whether reference is made to British or ASTM Standards.

6.5.2 Preparation of Soil for BS Compaction Tests

GENERAL

The method of preparation of test samples from the original soil sample depends upon

(a) the largest size of particles present in the original sample;
(b) whether or not the soil particles are susceptible to crushing during compaction.

Criterion (a) is assessed by inspection, or by passing the soil through sieves in the gravel-size range. The amount of coarse material determines the size of mould to be used, i.e. whether the one litre (4 in) or the CBR (6 in) mould should be used.

Criterion (b) can sometimes be assessed by inspection and handling, but trial compaction may be desirable, with sieving tests on the soil before and after compaction to determine whether any particles break down during the process. Breakdown of particles results in a change in the soil characteristics, and if a single batch of soil is compacted several times that change will be progressive during the test. A separate batch of susceptible soil is needed for each determination of compacted dry density, consequently a much larger sample is required.

Cohesive soils need to be broken down into small pieces before adjusting the moisture content for compaction. These soils should not be dried first, but should be chopped with a suitable knife, or shredded using a cheese-grater, while at natural moisture content. The extent of chopping or shredding should be consistent, because the results of a compaction test depend on the size of pieces. In any case the results will not necessarily relate directly to results obtained *in-situ* because there the extent of breaking down is quite different from that obtained in the laboratory. Typical methods are to chop the soil into pieces to pass a 20 mm sieve; or to shred it to pass a 5 mm sieve. The method used should be recorded.

GRADING CRITERIA

For the purpose of compaction tests, soil is divided into six zones on the particle size chart, depending on the percentages retained on the 20 mm and 37.5 mm sieves. The six grading zones are designated and defined as follows.

Zone (1): No particles retained on (i.e. 100% passing) the 20 mm sieve.

Zone (2): 100% passing the 37.5 mm sieve, and not more than 5% retained on the 20 mm sieve.

Zone (3): 100% passing the 37.5 mm sieve, and between 5% and 30% retained on the 20 mm sieve.

Zone (4): 100% passing the 63 mm sieve, and not more than 5% retained on the 37.5 mm sieve, and not more than 30% retained on the 20 mm sieve.

Zone (5): 100% passing the 63 mm sieve, and between 5% and 10% retained on the 37.5 mm sieve, and not more than 30% retained on the 20 mm sieve.

Zone (X): More than 10% retained on the 37.5 mm sieve, or more than 30% retained on the 20 mm sieve.

The criteria (in terms of percentages retained on each sieve) are summarised in Table 6.6. These zones are also shown diagrammatically in Fig. 6.7, which represents the relevant portion of a particle size distribution chart. If the grading curve passes through more than one zone, the highest-numbered zone applies. If the grading curve passes through zone (X), the soil is not suitable for these tests unless the coarse material is removed.

If appropriate, soils in grading zones (1) and (2), which would normally be compacted in the one litre mould, may be compacted in the CBR mould, provided that there is enough material. This procedure is useful when CBR tests (see Volume 2) are to be performed on the compacted soil over a range of moisture contents.

METHOD OF PREPARATION (BS1377: Part 4:1990:3.2)

(1)   *Grading zone*

Determine the grading zone to which the soil belongs, by sieving on the 37.5 mm and 20 mm sieves as appropriate. Use undried soil for this assessment, and determine the dry mass of soil passing the 20 mm sieve from the moisture content measured on a representative portion.

The amount of soil used for this preliminary sieving should be not less than the mass indicated in Table 4.4. If enough soil is available a separate representative portion may be used for sieving, and that portion may be dried if it is not to be used for the compaction test.

Table 6.6. GRADING CRITERIA FOR BS COMPACTION TESTS

Grading zone	Retained on sieves (%)		Minimum mass required (kg)		Mould used for test	Mass for each determination (kf)
	37.5 mm	20 mm	(a)	(b)		
(1)	0	0	6	15	one litre	2.5
(2)	0	0–5				
(3)	0	5–30	15	40	CBR	6
(4)	0–5	0–30				
(5)	5–10	5–30				
(X)	> 10 or > 30		Tests not applicable			

(a) Single batch — soil particles not susceptible to crushing.
(b) Multiple batches — soil particles susceptible to crushing.

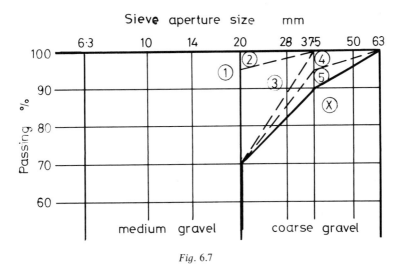

Fig. 6.7

The sample is dealt with as follows, according to the grading zone to which it is allocated.

Grading zone (1) — Can be compacted in the one-litre mould.

Grading zone (2) — Either remove the material retained on the 20 mm sieve and compact in the one-litre mould; or compact in the CBR mould.

Grading zone (3) — Compact in the CBR mould.

Grading zone (4) — Remove the material retained on the 37.5 mm sieve and compact in the CBR mould.

Grading zone (5) — Remove and weigh the material retained on the 37.5 mm sieve. Replace this material by the same mass of similar material passing the 37.5 mm sieve and retained on the 20 mm sieve. Compact in the CBR mould.

Grading zone (X) — Not suitable for these tests.

Any coarse material removed should be weighed and the mass recorded.

(2) Susceptibility to crushing

Assess whether the soil particles are susceptible to crushing under the degree of compaction to be applied in the test. Soils susceptible to crushing contain granular material of a soft

Table 6.7. MOISTURE CONTENTS FOR COMPACTION TESTS

	Moisture content (%)		
	Suggested lowest value		Increments for subsequent stages
Type of soil	2.5 kg test	4.5 kg test	
Sandy and gravelly	4 to 6	3 to 5	1 to 2
Cohesive	$(w_p - 10)$ to $(w_p - 8)$	about $(w_p - 15)$	2 to 4

w_p is the plastic limit of the fraction finer than 425 μm.

nature, e.g. soft limestone, sandstone, chalk, or other minerals likely to be broken down by compaction. If necessary compact a portion of the soil by the appropriate method. If in doubt assume that soil particles are susceptible to crushing.

Whenever practicable the procedures described for susceptible material should be followed for all soils.

(3)   Mass of soil for test

The mass of soil to be prepared for the test is obtained from Table 6.6 when the grading zone and susceptibility to crushing of particles have been established. Obtain the required representative mass from the original sample (after removing coarse material if necessary) by riffling or quartering, as described in Section 1.5.5.

(4)   Adjustment of moisture content

The lower end of the moisture content range for a test, and suitable increments of moisture content for each stage, should be judged from experience. The values given in Table 6.7 provide a general guide.

(5)   Single batch of soil (BS1377: Part 4:1990:3.2.4 & 3.2.5)

It is often convenient to make the first determination with the soil at the moisture content 'as received'. For subsequent determinations, adjust the moisture content as follows.

(a)   To obtain a lower moisture content, allow the soil to partially air-dry to the moisture content at which it is to be compacted. Do not allow the soil to dry more than necessary, and mix frequently to prevent local over-drying. Estimate the moisture content by inspection, or by weighing at intervals.

(b)   To obtain a higher moisture content, mix additional water thoroughly into the soil as described in (6) below.

Place the soil in an airtight container if it is not to be used immediately. For a cohesive soil, leave it in the container for a maturing period of at least 24 hours to allow for a uniform distribution of water in the sample.

(6)   Multiple soil batches (BS1377: Part 4:1990:3.2.6 & 3.2.7)

Subdivide the prepared soil sample to give five or more representative specimens for test. Each specimen should be of about 2.5 kg for the one-litre mould, or 6 kg for the CBR mould (see Table 6.6).

Add a different amount of water to each specimen, so as to cover the required range of moisture contents (see Table 6.7). The range should provide at least two values on either side of the optimum moisture content at which maximum dry density occurs.

Thorough mixing in of the water is especially with cohesive soils. After mixing a cohesive soil should be allowed to mature for at least 24 hours in a sealed container.

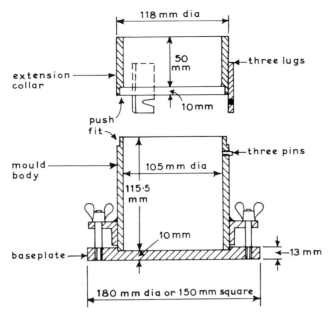

Fig. 6.8   British Standard one-litre compaction mould

6.5.3 'Light' Compaction Test (2.5 kg rammer method) (BS1377: Part 4:1990:3.3)

This test is suitable for soils containing particles no larger than 20 mm. The detailed procedure depends on whether or not the granular material is susceptible to crushing during compaction.

If the soil contains particles larger than 20 mm, refer to Section 6.5.5.

APPARATUS

(1) Cylindrical metal mould, internal dimensions 105 mm diameter and 115.5 mm high. This gives a volume of 1000 cm³. The mould is fitted with a detachable baseplate and removable extension collar (see Fig. 6.8).
(2) Metal rammer with 50 mm diameter face, weighing 2.5 kg, sliding freely in a tube which controls the height of drop to 300 mm (see Fig. 6.9).
(3) Measuring cylinder, 200 ml or 500 ml (plastics).
(4) 20 mm British Standard sieve and receiver.
(5) Large metal tray, e.g. 600 × 600 × 60 mm deep.
(6) Balance, 10 kg capacity reading to 1 g.
(7) Jacking apparatus for extracting compacted material from the mould.
(8) Small tools: palette knife; steel straight-edge, 300 mm long; steel rule; scoop or garden trowel.
(9) Drying oven, 105–110°C, and other equipment for moisture content determination.

Compaction test equipment is shown in Fig. 6.10. If a mechanical compaction apparatus is available, refer to Section 6.5.8. The procedure described below is the same in principle whether compaction is effected by the hand rammer or by the machine.

PROCEDURAL STAGES

(1) Prepare apparatus.
(2) Prepare test sample or samples.
(3) Place soil in mould.

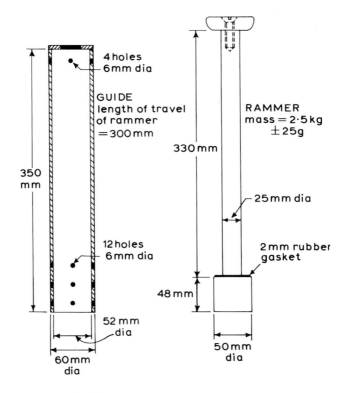

Fig. 6.9   Rammer for BS 'light' compaction test

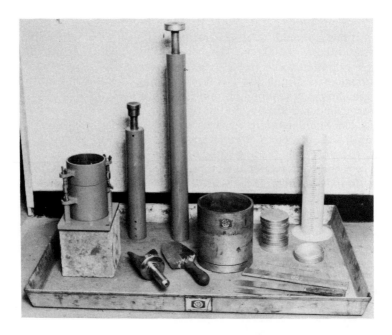

Fig. 6.10   Equipment for compaction tests

(4)   Compact soil into mould.
(5)   Trim off.
(6)   Weigh.
(7)   Remove soil from mould.
(8)   Measure moisture content.
(9)   Break up and remix.
(10) Repeat stages (4)–(8) *either* (A) after adding more water to sample *or* (B) using next batch (total of at least five compactions).
(11) Calculate.
(12) Plot graph.
(13) Read off optimum values.
(14) Report results.

TEST PROCEDURE — PARTICLES NOT SUSCEPTIBLE TO CRUSHING (Clause 3.3.4.1)

(1) *Prepare apparatus*

Verify that the mould, baseplate, extension collar and rammer to be used are those that conform to BS1377. Check that the mould, extension collar and baseplate are clean and dry. Weigh the mould body to the nearest 1 g (m_1). Measure its internal diameter (D mm) and length (L mm) in several places to 0.1 mm using vernier calipers, and calculate the mean dimensions. Calculate the internal volume of the mould (V cm^3) from the equation

$$V = \frac{\pi \times D^2 \times L}{4000}$$

The mould is designed to give a volume $V = 1000$ cm^3, but this may change slightly with wear.

Check that the lugs or clamps hold the extension collar and baseplate securely to the mould, and assemble them together. A wipe with a slightly oily cloth on the internal surfaces will assist removal of soil afterwards. A disc of thin filter paper may be placed on the baseplate for the same purpose.

Check the rammer to ensure that it falls freely through the correct height of drop, and that the lifting knob is secure.

(2) *Preparation of sample*

Prepare the soil as described in Section 6.5.2, to provide the single sample of about 6 kg (step (3)), and adjust the moisture content to the desired starting value (step (5)).

(3) *Place into mould*

Place the mould assembly on a solid base, such as a concrete floor or plinth, or a concrete cube. A resilient base may result in inadequate compaction.

Add loose soil to the mould so that after compaction the mould will be one-third filled.

(4) *Compaction in mould*

Compact the soil by applying 27 blows of the rammer dropping from the controlled height of 300 mm (Fig. 6.11).

Take care to see that the rammer is properly in place before releasing. The hand which holds the tube must be kept well clear of the handle of the falling rammer. Do not attempt to grab the lifting knob before the rammer has come to rest; a finger or thumb trapped between knob and tube can sustain a nasty injury.

The first few blows of the rammer, which are applied to soil in a very loose state, should be applied in a systematic manner to ensure the most efficient compaction and maximum

Fig. 6.11   *Compacting soil into mould*

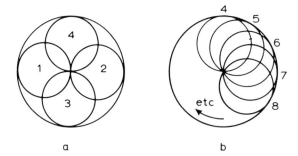

Fig. 6.12   *Sequence of blows using hand rammer*

reproducibility of results. The sequence shown in Fig. 6.12(a) should be followed for the first four blows. By this means the effort dissipated in displacing loose material is kept to a minimum. After that the rammer should be moved progressively around the edge of the mould between successive blows, as indicated in Fig. 6.12(b), so that the blows are uniformly distributed over the whole area. Soil must not be allowed to collect inside the tube of the rammer, because this will impede the free fall of the rammer. Make sure that the end of the tube is resting on the soil surface, and does not catch on the edge of the mould, before releasing the rammer. The guide tube must be held vertically. Place the tube gently on the soil surface; the rammer does the compaction, not the tube.

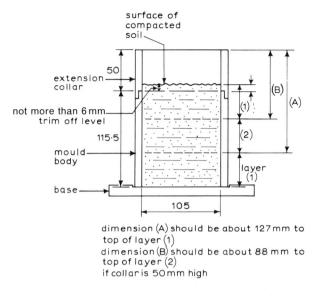

dimension (A) should be about 127 mm to
top of layer (1)
dimension (B) should be about 88 mm to
top of layer (2)
if collar is 50 mm high

Fig. 6.13 Soil in mould after compaction

If the correct amount of soil has been used, the compacted surface should be at about one-third of the height of the mould body — that is, about 77 mm below the top of the mould body, or 127 mm below the top of the extension collar. If the level differs significantly (by more than, say, 5 mm) from this, remove the soil, break it up, mix it with the remainder of the prepared material and start this stage again.

Lightly scarify the surface of the compacted soil with the tip of a spatula or point of a knife. Place a second, approximately equal, layer of soil in the mould, and compact with 27 blows as before. Repeat with a third layer, which should then bring the compacted surface in the extension collar to not more than 6 mm above the level of the mould body (see Fig. 6.13). If the soil level is higher than this, the result will be inaccurate, so the soil should be removed, broken up and remixed, and the test repeated with slightly less soil in each layer.

(5)   Trim off
Remove the extension collar carefully. Cut away the excess soil and level off to the top of the mould, checking with the straight-edge. Any small cavities resulting from removal of stones at the surface should be filled with fine material, well pressed in.

(6)   Weigh

Remove the baseplate carefully, and trim the soil at the lower end of the mould if necessary. Weigh soil and mould to the nearest 1 g (m_2).

The British Standard procedure does not call for the removal of the baseplate before weighing. If the soil is granular and will not hold together well, the baseplate is best left on. In this case the mass m_1 refers to the mould with baseplate. If the soil is cohesive enough to hold together, it is preferable not to include the baseplate in the weighings, because the mould with baseplate weighs substantially more than the soil it contains.

(7)   Remove soil

Fit the mould on to the extruder and jack out the soil (Fig. 6.14). Alternatively, remove the soil by hand, but this can be difficult with gravelly soils containing a clay binder. Break up the sample on the tray.

Fig. 6.14   Jacking soil out of mould

(8)   *Measure moisture content*

Take up to three representative samples in moisture content containers for measurement of moisture content, using the standard procedure described in Section 2.5.2. This must be done immediately, before the soil begins to dry out. The average of the three measurements is denoted by $w\%$.

Alternatively, moisture content samples may be taken, one from each layer, as the soil is placed in the mould for compaction.

(9)   *Break up and remix*

Break up the material on the tray, by rubbing through a 20 mm sieve if necessary, and mix with the remainder of the prepared sample. Add an increment of water, approximately as follows:

Sandy and gravelly soils: 1–2% (50–100 ml of water to 5 kg of soil).
Cohesive soils: 2–4% (100–200 ml of water to 5 kg of soil).

Mix in the water thoroughly.

(10)   *Repeat with added water*

Repeat stages (4)–(9) for each increment of water added, so that at least five compactions are made. The range of moisture contents should be such that the optimum moisture content (at which the dry density is maximum) is near the middle of that range. If necessary to define the optimum value clearly, carry out one or more additional tests at suitable moisture contents. Keep a running plot of dry density against moisture content so as to see when the optimum condition has been passed.

Above a certain moisture content the material may be extremely difficult to compact. For instance, a granular soil may by then contain excessive free water, or a clay soil may be very soft and sticky. In either event the optimum condition has been passed and there is no point in proceeding further.

TEST PROCEDURE — PARTICLES SUSCEPTIBLE TO CRUSHING (Clause 3.3.4.2)

(1)   As step (1) above.

(2)   *Preparation of sample*

Prepare the soil as described in Section 6.5.2, to provide a sample of about 15 kg (step (3)), from which five (or more) separate batches of about 2.5 kg are obtained and made up to different moisture contents (step (6)).

(3) to (8)   Treat the first batch as described insteps (3) to (8) of the 'non-susceptible' procedure. The whole compacted sample can be used for the moisture content determination if it is not required for further tests.

(9)   Repeat stages (3)–(8) for each batch in turn. If necessary, make up another batch or batches and test them if other points are required on the compaction curve.

CALCULATION, PLOTTING AND REPORTING

The following stages refer to the above procedures.

(1)   *Calculate*

Calculate the bulk density of each compacted specimen from the equation

$$\rho = \frac{m_2 - m_1}{1000} \ \text{Mg/m}^3$$

where m_1 = mass of mould (and base if included) and m_2 = mass of soil and mould (and base if included). If the volume of the mould is not 1000 cm^3 but is V cm^3, then

$$\rho = \frac{m_2 - m_1}{V} \ \text{Mg/m}^3$$

Calculate the average moisture content, $w\%$, for each compacted specimen.
Calculate the corresponding dry density from the equation

$$\rho_D = \left(\frac{100}{100 + w}\right) \rho \ \text{Mg/m}^3$$

Typical density and moisture content data and calculations are given in Fig. 6.15.
Calculate the percentage of stones retained on the 20 mm sieve.

(2)   *Plot graph*

Plot each dry density, ρ_D, against the corresponding moisture content, w. Draw a smooth curve through the points. The curves for 0, 5 and 10% air voids may be plotted as well.

A typical graph, together with other test data, is shown in Fig. 6.16, which includes three air voids lines.

**Compaction Test
Work sheet**

B.S. 1377:1975		Test 12			Location:	Easthampstead		Loc No.	1998
No. of layers	3	Rammer	2·5	kg	Soil description Brown sandy clay with a little fine gravel			Sample No.	27/4
Blows per layer	27	Drop	300	mm	Sample type	Bulk bag	Operator C.B.A	Date started	10.3.78
Compacted by hand/machine					Sample preparation	Air dried and riffled			
One-litre/CBR/cylinder no.		7			No. of separate batches	5	Special techniques Separate batches used		

DENSITY		Volume of cylinder (V)	1002	cm^3				
Measurement no.			(1)	(2)	(3)	(4)	(5)
Cylinder & soil	A	g	3786	3907	3999	3962	3908
Cylinder	B	g	1917	1917	1917	1917	1917
Wet soil	A − B	g	1869	1990	2082	2045	1991
Wet density	ρ	Mg/m^3	1·865	1·986	2·078	2·041	1·987

MOISTURE CONTENT

		64	44	18								
Container no.		64	44	18								
Wet soil & container	g	104·12	97·48	89·67								
Dry soil & container	g	96·02	90·42	82·76								
Container	g	9·36	16·58	9·51								
Dry soil	g	88·66	73·84	73·25								
Moisture loss	g	8·10	7·06	6·91								
Moisture content $w_{1,2,3}$	%	9·35	9·56	9·44								
AVERAGE MOISTURE	%		9·45		12·55		15·95		18·71		21·47	
DRY DENSITY ρ_D	Mg/m^3		1·704		1·765		1·792		1·719		1·636	

$$\rho = \frac{A - B}{V}$$

$$W = \frac{W_1 + W_2 + W_3}{3} \qquad \rho_D = \rho \times \frac{100}{100 + w}$$

Fig. 6.15   Compaction test data and calculations

(3)   Read off optimum values

Ascertain the point of maximum dry density on this curve, and read off the maximum dry density value. The maximum value may lie between two plotted points, but the peak should not be exaggerated when drawing the curve. Read off the corresponding moisture content, which is the optimum moisture content for this degree of compaction.

(4)   Report results

The report should state that the test was carried out in accordance with Clause 3.3 of BS1377: Part 4:1990, and should include the following.

The graphical plot, showing the experimental points and giving a description of the soil.

Method of preparation of the sample, and whether a single sample or separate batches were used; and if relevant the size of lumps or pieces to which a cohesive soil was broken down.

The percentage by dry mass (to the nearest 1%) of the original material retained on the 20 mm and 37.5 mm sieves.

The maximum dry density for the degree of compaction used, to the nearest 0.01 Mg/m^3.

The optimum moisture content, to two significant figures.

The particle density used for constructing the air voids lines, and whether measured (if so the method used) or assumed.

6.5.4   'Heavy' Compaction Test (4.5 kg rammer method) BS1377: Part 4:1990:3.5)

This test gives the dry density — moisture content relationship for a soil compacted in five layers in the same mould as used in the 'light' compaction test, using 27 blows per layer with a 4.5 kg rammer falling 450 mm. The total compactive energy applied is 4.5 times greater than in the 'light' test. From the density–moisture curve the optimum moisture content, and the maximum dry density, for this heavier degree of compaction can be determined.

As with the 'light' test, this test is suitable for soils containing particles no larger than 20 mm, and the detailed procedure depends on whether or not the particles are susceptible to crushing.

If the soil contains particles larger than 20 mm, refer to Section 6.5.5.

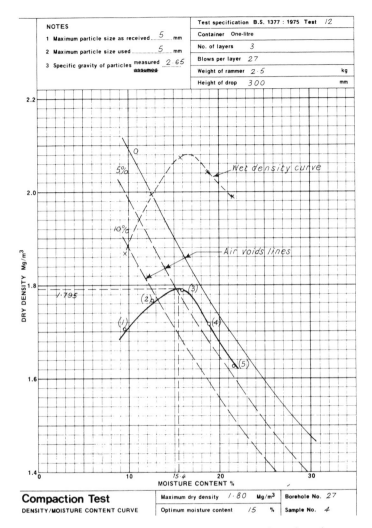

Fig. 6.16 Dry density–moisture content test results and graph
(The wet density curve is not normally plotted)

APPARATUS

(1) Mould — as for the 'light' compaction test (Section 6.5.3).
(2) Metal rammer with 50 mm diameter face, weighing 4.5 kg, and a controlled height drop of 450 mm (see Fig. 6.17). Otherwise, it is similar to item (2) of Section 6.5.3.

Items (3)–(9) are as for the 'light' test (Section 6.5.3).

PROCEDURAL STAGES

The stages are similar to those given in Section 6.5.3 for the 'light' test.

TEST PROCEDURE

The procedure is similar to that described in Section 6.5.3, with the exception of the detailed modifications referred to below. As in Section 6.5.3, the test is carried out either on a single sample of soil, or on separate batches, depending on the nature of the soil particles.

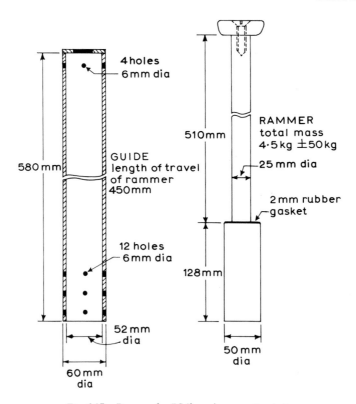

Fig. 6.17   Rammer for BS 'heavy' compaction test

Stages (1) *and* (2).   Sample preparation, as for the 'light' test, depends upon whether the soil particles are susceptible to crushing. The quantity of water to be added to the sample initially, or to the first batch is a little less than for the 'light' test (see Table 6.7).

(3) & (4)   *Compaction.*   Compaction is carried out in five layers instead of three, the 4.5 kg rammer with a drop of 450 mm being used, with 27 blows for each layer. Take extra care when using this rammer to ensure that it is properly in place before releasing. See stage (4) of Section 6.5.3.

If the correct amount of soil has been used for compacting the first layer, the compacted surface should be at about one-fifth of the height of the mould body — that is, about 92 mm below the top of the mould body, or 142 mm below the top of the extension collar. If significantly different from this, remove the soil and start this stage again.

Compact four more equal layers into the mould as before. The final compacted surface should be not more than 6 mm above the top of the mould body (see Fig. 6.13). If it is higher than this, remove the soil, break it up and repeat this stage, using slightly less soil in each layer.

Stages (5)–(10).   As for Section 6.5.3. Moisture content increments are similar to those suggested in stage (9).

CALCULATION, PLOTTING AND REPORTING

These are as described in Section 6.5.3, except that the reported procedure is in accordance with Clause 3.5 of BS1377: Part 4:1990.

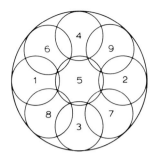

Fig. 6.18   Sequence of blows using hand rammer in CBR mould

6.5.5   Compaction of Stony Soils

For soils containing gravel-size fragments larger than 20 mm, a calculated correction can be applied to the maximum dry density to estimate the corresponding maximum dry density in the field. The principle is explained in Section 6.3.4, but applies only if the 'stone' content does not exceed about 25%.

For soils containing larger proportions of coarse material, the only satisfactory method of obtaining the compaction characteristics is to carry out a test in a larger container so that a larger maximum particle size can be used. A CBR mould, as used in the vibrating hammer test (Section 6.5.9), is used for this purpose. The nominal volume of this mould is 2305 cm³, but this may change slightly with wear and the dimensions should be checked as described in Section 6.5.3. When this mould is used for compaction tests, up to 30% of particles retained on a 20 mm sieve can be included in the test sample. Verify that the BS mould is used.

Either the equivalent 'light' or the equivalent 'heavy' standard compaction test may be carried out with the CBR mould. The total quantity of material passing the 37.5 mm sieve required is 25 kg, or five batches each of 8 kg if the particles are susceptible to crushing. The procedures are the same as those described in Sections 6.5.3 and 6.5.4, except that 62 blows are required in each layer instead of 27. This is because of the increased volume of soil compared with the smaller mould (see Table 6.2). Weighings are made to the nearest 5 g instead of 1 g.

Application of the first few blows should be done systematically, but the pattern differs from that used for the 'one-litre' mould because of the larger size. The first two blows should be applied at the edge and diametrically opposite each other; the next two half-way between; and the fifth at the centre (see Fig. 6.18). The next four (numbered 6, 7, 8, 9) are placed between those already applied. After that, work systematically around the mould and across the middle so that the whole area is uniformly compacted.

The test report should state that the procedure is in accordance with Clause 3.4 or 3.6 (as appropriate) of BS1377: Part 4:1990.

6.5.6   Preparation of Soil for ASTM Tests

In ASTM test designations D698 (5.5 lb rammer method) and D1557 (10 lb rammer method), four categories of soil are recognised, depending on the largest sizes of particles remaining after initial preparation. These categories relate to the following test methods.

Methods A and B: Material retained on a No. 4 (4.75 mm) sieve is removed and no correction is made. However if the amount of retained material is 7% or more by mass, Method C is recommended instead.

Method C: Material retained on a $\frac{3}{4}$ in (19.0 mm) sieve is removed and no correc-
                         tion is made. However if the amount of retained material is 10% or more
                         by mass, Method D is recommended instead.

Method D: This method applies only when the amount of material retained on the
                         $\frac{3}{4}$ in (19.0 mm) is from 10% to 30%. Sieve the retained material on a
                         3 in (75 mm) sieve and discard the material retained on that sieve.
                         Replace the material between 3 ins and $\frac{3}{4}$ in sizes by an equal mass
                         of similar material, taken from an unused portion of the sample, passing
                         a $\frac{3}{4}$ in sieve and retained on a No. 4 (4.75 mm) sieve. Mix in the replaced
                         material thoroughly.

If the amount of material retained in the $\frac{3}{4}$ in (19.0 mm) is 30% or more, the test methods
for the determination of maximum dry density or optimum moisture content are not
applicable.

For method A, the 4 in compaction mould is used, and a test sample of about 11 kg is
required. For methods B, C and D the 6 in mould is used, and test samples of about 23 kg
are required. In all cases the test sample is divided into at least four portions for compaction,
each of which is brought to a different moisture content so as to bracket the optimum
moisture content.

Otherwise the method of preparation for test is generally similar to that described in
Section 6.5.2.

6.5.7 ASTM Compaction Test Procedures (ASTM D698 and D1557)

The ASTM compaction test procedures are similar in principle to the corresponding British
Standard procedures described in Sections 6.5.3, 6.5.4 and 6.5.5.

Verify that the mould, baseplate, extension collar and rammer are those that conform to
ASTM D698 or D1557, as appropriate, whether the 4 in or 6 in mould is used.

In Test Designation D698, the 5.5 lb (2.49 kg) rammer with a 12 in (305 mm) drop is used.
Soil prepared by Method A (Section 6.5.6) is compacted into the 4 in (101.6 mm) diameter
mould (944 cm^3) in 3 layers applying 25 blows of the rammer on each layer. Soil prepared
by Methods B, C or D requires the 6 in (152.4 mm) diameter mould (2124 cm^3), and is
compacted in 3 layers with 56 blows of the rammer on each layer.

In Test Designation D1557, the 10 lb (4.54 kg) rammer with a drop of 18 ins (457 mm) is
used. Soil prepared by Method A (Section 6.5.6) is compacted into the 4 in (101.6 mm)
diameter mould (944 cm^3) in 5 layers applying 25 blows of the rammer on each layer. Soil
prepared by methods B, C or D requires the 6 in (152.4 mm) diameter mould (2124 cm^3),
and is compacted in 5 layers with 56 blows of the rammer on each layer.

An automatic compaction device as described in Section 6.5.8, suitably designed to give
the compactive efforts required for the ASTM procedures, may be used in place of the hand
rammer in any of these tests.

In all cases the whole compacted sample, after removal from the mould, should be used
for the determination of moisture content if the soil is of high permeability, such that the
moisture content is not uniformly distributed throughout the sample.

Calculations, plotting and reporting are similar to the requirements for the BS tests.

6.5.8 Use of Automatic Compactor

An automatic compaction apparatus eliminates much of the physical effort required for
carrying out compaction tests. However, it has been found that the densities achieved by
machine are often less than those obtained by hand compaction. This is partly because the
blow pattern differs from that recommended in Sections 6.5.3 and 6.5.5, and partly because
the base not only rotates but also has to provide horizontal movement when a CBR mould

Fig. 6.19   *Automatic compaction apparatus*

is being used so that its whole area may be covered by the rammer. This results in the mould support being less rigid than a concrete base.

A machine of the type shown in Fig. 6.19 incorporates the following features:

(1) The blow pattern follows closely the recommended pattern, by applying widely spaced blows at first, which flatten the soil surface, followed by overlapping blows.
(2) The area of a CBR mould is covered by shifting the position of the rammer assembly instead of by moving the base.
(3) The rotating base is supported by an inertia block offering a machined annular surface of large area, which makes for a very rigid support.

Separate machines are designed specifically for the BS and the ASTM compaction tests, and the machine used must be to the correct specification.

The performance of an automatic compaction machine can be assessed by performing parallel tests on duplicate samples using a hand rammer and the machine with the appropriate setting. If the density obtained by the machine is within $\pm 2\%$ of the density obtained by using the hand rammer, the machine is satisfactory and meets the requirements of BS1377.

6.5.9   Compaction by Vibration (BS1377: Part 4:1990:3.7)

This test is applicable to granular soils passing the 37.5 mm sieve. The principle is similar to that of the rammer procedures, except that a vibrating hammer is used instead of a drop-weight rammer, and a larger mould (the standard CBR mould) is necessary.

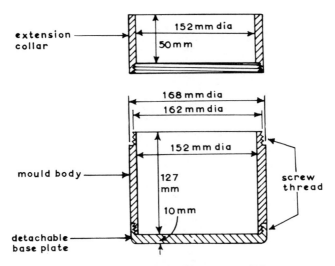

Fig. 6.20   CBR mould, screw type (BS)

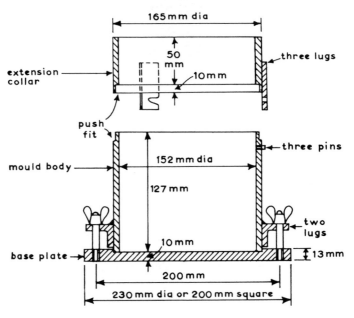

Fig. 6.21   CBR mould, clamp type (BS)

APPARATUS

(1)   Cylindrical metal mould, internal dimensions 152 mm diameter and 127 mm high (CBR mould). The mould can be fitted with an extension collar and baseplate. Details of two types of mould are shown in Figs. 6.20 and 6.21. (Note: The mould shown in Fig. 6.21 must not be confused with the similar ASTM compaction mould which is 116.4 mm high.)

(2)   Electric vibrating hammer, power consumption 600–700 W, operating at a frequency in the range 25–45 Hz. For safety reasons the hammer should operate on 110 V and an earth-leakage circuit breaker (ELCB) should be included in the line between the hammer and the mains supply. A check test to determine whether the hammer meets

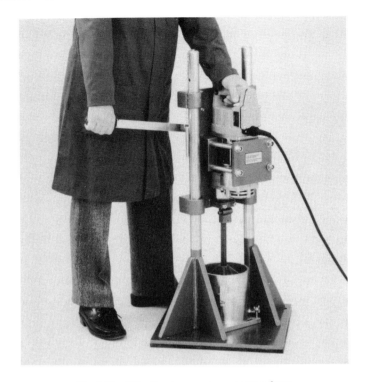

Fig. 6.22   Vibrating hammer in supporting frame

the requirements of the British Standard is described below. A special supporting frame for the hammer may be used for easier operation, as shown in Fig. 6.22.

(3)   Steel tamper for attaching to the vibrating hammer, with a circular foot 145 mm diameter (see Fig. 6.23).
(4)   37.5 mm BS sieve and receiver.
(5)   Depth gauge or steel rule reading to 0.5 mm.
(6)   Laboratory stop-clock reading to 1 s.

Other items required are items (3)–(9) as listed in Section 6.5.3, except that a balance of higher capacity is required, for which readability to 5 g is adequate.

PROCEDURAL STAGES

(1)   Prepare apparatus.
(2)   Obtain soil samples.
(3)   (A) Prepare test sample, *or* (B) prepare five batches.
(4)   Compact into mould in layers.
(5)   Measure height.
(6)   Weigh.
(7)   Remove soil from mould.
(8)   Measure moisture content.
(9)   Break up and remix.
(10A)Add more water, and repeat stages (4)–(9)
        or
(10B)Repeat stages (4)–(8), using the next batch.
        In either case the total number of compactions should be at least five.
(11)   Calculate.

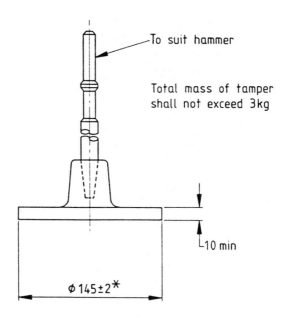

Fig. 6.23   Tamper for vibrating hammer (Courtesy of BSI, London)

(12) Plot graph.
(13) Read off optimum values.
(14) Report results.

TEST PROCEDURE — PARTICLES NOT SUSCEPTIBLE TO CRUSHING (Clause 3.7.5.1)

(1) *Prepare apparatus*

See that the component parts of the mould are clean and dry. Assemble the mould, baseplate and collar securely, and weigh to the nearest 5 g (m_1). Measure the internal dimensions of the assembly, and calculate the internal volume, as described in Section 6.5.3. The nominal dimensions of the mould give an area of cross-section of 18 146 mm^2 and a volume of 2304.5 cm^3 (say 2305 cm^3), but these may change slightly with wear. The inside height of the mould with collar is recorded (h_1 mm).

The comments regarding preparation of the compaction mould given in Section 6.5.3 apply equally to the CBR mould. It is particularly important to ensure that the lugs and clamps holding the mould assembly together are secure and in good condition, in order to withstand the effects of vibration. If the mould has screw-on fittings (Fig. 6.20), the threads must be kept clean and undamaged. Avoid cross-threading when fitting the baseplate and extension collar, and make sure that they are tightened securely as far as they will go without leaving any threads exposed. Screw threads and mating surfaces should be lightly oiled before tightening.

Ensure that the vibrating hammer is working properly, in accordance with the manufacturer's instructions. See that it is properly connected to the mains supply, and that the connecting cable is in sound condition. The supporting frame, if used, must move freely without sticking. The hammer should have been verified as described below.

The tamper stem must fit properly into the hammer adaptor, and the foot must fit inside the CBR mould with the necessary clearance (3.5 mm all round).

(2)   *Preparation of sample*

Prepare the soil as described in Section 6.5.2, to provide a single sample of about 15 kg (step (3)), and adjust the moisture content to the desired starting value (step (5)). A typical moisture content for a sandy and gravelly soil would be about 3% to 5% but the actual value should be judged from experience.

(3)   *Place into mould*

Place the mould assembly on a solid base, such as a concrete floor or plinth. If the test is performed out of doors because of noise and vibration problems, place the mould on a concrete paved area, not on unpaved ground or on thin tarmac. Any resilience in the base results in inadequate compaction.

Add a quantity of soil to the mould, such that after compaction the mould is one-third filled A preliminary trial may be necessary to ascertain the correct amount of soil. A disc of polyethylene sheet, of a diameter equal to the internal diameter of the mould, may be placed on top of the layer of soil. This will help to prevent sand particles moving up through the annular gap between the tamper and the mould.

(4)   *Compaction into mould*

Compact the layer with the vibrating hammer, fitted with the tamper, for 60 s, applying a firm pressure vertically downwards throughout. The downward force, including that resulting from the mass of the hammer and tamper, should be 300–400 N. This force is sufficient to prevent the hammer bouncing up and down on the soil. The correct force can be determined by standing the hammer, without vibration, on a platform scale and pressing down until a mass of 30–40 kg is indicated. With experience the pressure to be applied can be judged, but an occasional check on the platform scale is advisable. If the hammer-supporting frame is used, the hand pressure required is much less but should be carefully checked.

Repeat the above compaction procedure with a second layer of soil, and then with a third layer. The final thickness of the compacted specimen should be between 127 mm and 133 mm; if it is not, remove the soil and repeat the test.

(5)   *Measure height*

After compaction remove any loose material from the surface of the specimen around the edge, so that the surface is reasonably flat. Clean off the top edge of the mould collar. Lay the straight-edge across the top of the collar, and measure down to the surface of the specimen with the steel rule or depth gauge, to an accuracy of 0.5 mm. Take readings at four points spread evenly over the surface, and 15 mm from the side of the mould. Calculate the average depth (h_2 mm). The mean height of the compacted specimen, h, is given by

$$h = (h_1 - h_2)\,\text{mm}$$

(6)   *Weigh*

Weigh the mould with the compacted soil, collar and baseplate to the nearest 5 g (m_2).

(7)   *Remove soil*

Remove the soil from the mould and place on the tray. A jacking extruder makes this operation easy if fittings to suit the CBR mould are available. But sandy and gravelly (non-cohesive) soil should not be too difficult to break up and remove by hand.

(8)   *Measure moisture content*

Take two representative samples in large moisture content containers for measurement of moisture content. This must be done immediately after removal from the mould, before the soil begins to dry out. The moisture content samples must be large enough to give results representative of the maximum particle size of the soil (see Section 2.5.2). The average of the two moisture content determinations is denoted by $w\%$.

(9)   *Break up and remix*

Break up the material on the tray and rub it through the 20 mm or the 37.5 mm sieve if necessary. Mix in with the remainder of the sample. Add an increment of water so as to raise the moisture content by 1 or 2% (150–300 ml of water for 15 kg of soil). As the optimum moisture content is approached it is preferable to add water in smaller increments.

(10)   *Repeat with added water*

Repeat stages (3)–(9) for each increment of water added. At least five compactions should be made, and the range of moisture contents should be such that the optimum moisture content is within that range. If necessary, carry out one or more additional tests at suitable moisture contents.

Above a certain moisture content the soil may contain an excessive amount of free water, which indicates that the optimum condition has been passed.

TEST PROCEDURE — PARTICLES SUSCEPTIBLE TO CRUSHING (Clause 3.7.5.2)

(1)   As step (1) above.
(2)   Prepare the soil as described in Section 6.5.2, to provide a sample of about 40 kg (step (3)), from which five (or more) separate batches of about 8 kg are obtained and made up to different moisture contents (step (6)).
(3) to (8)   Treat the first batch of soil as described in steps (3) to (8) above.
(9)   Repeat steps (3) to (8) on each batch of soil in turn. If additional points are required to define the optimum condition on the compaction curve, make up additional 8 kg batches at appropriate moisture contents and compact each batch as above.

CALCULATION, PLOTTING AND REPORTING

The following stages apply to both the above procedures.

(1)   *Calculate*

Calculate the bulk density of each compacted specimen from the equation

$$\rho = \frac{m_2 - m_1}{18.15 \times h} \quad \text{Mg/m}^3$$

where m_1 = mass of mould, collar and baseplate; m_2 = mass of mould, collar and baseplate with soil; h = height of compacted soil specimen = $h_1 - h_2$.

The above equation applies only if the average diameter of the mould is 152 mm. If it is not, and is represented by D mm, use the area of cross-section $A(=\pi D^2/4)$ in the equation

$$\rho = \frac{m_2 - m_1}{A \times h} \times 100 \quad \text{Mg/m}^3$$

Calculate each dry density from the corresponding moisture content, $w\%$, from the equation

$$\rho_D = \frac{100}{100 + w} \times \rho \ \text{Mg/m}^3$$

Calculate the percentage of coarse material retained on the 37.5 mm sieve.

(2) Plot graph

Plot the values of dry density, ρ_D, against moisture content, w, and draw a smooth curve through the points. The curves corresponding to 0, 5 and 10% air voids may be plotted as well.

(3) Read off optimum values

Read off the maximum dry density, and the corresponding moisture content, from the compaction curve.

(4) Report results

The report should state that the test was carried out in accordance with Clause 3.7 of BS1377: Part 4:1990, and should include the following.

The graphical plot, showing the experimental points and giving a description of the soil.

Method of preparation of the sample, and whether a single sample or separate batches were used.

The percentage by dry mass (to the nearest 1%) of the original material retained on the 37.5 mm sieve.

The maximum dry density for the degree of compaction used, to the nearest 0.01 Mg/m^3.

The optimum moisture content, to two significant figures.

The particle density used for constructing the air voids lines, and whether measured (if so the method used) or assumed.

VERIFICATION OF VIBRATING HAMMER (BS1377: Part 4:1990:3.7.3)

The following procedure may be used to ascertain whether the vibrating hammer used for the above test complies with the requirements of BS1377:1990 and is in satisfactory working order.

About 5 kg of an unused sample of clean, dry silica sand from the Woburn Beds of the Lower Greensand in the Leighton Buzzard district is required. The specified grading requires 100% passing the 850 μm sieve, at least 75% passing 600 μm, at least 75% retained on 425 μm, and 100% retained on 300 μm. The sand must be dry, and free from flakey particles, silt, clay and organic matter.

Add water to the sieved sand to bring its moisture content to 2.5% (125 ml of water to 5 kg of dry soil). Mix the water in thoroughly, and check the actual moisture content, which should not differ from the stated value by more than 0.5%.

Compact the sand into the CBR mould in three layers with the vibrating hammer. Measure the height of the compacted sample, weigh and determine the compacted dry density to the nearest 0.002 Mg/m^3, as described above. Repeat twice on the same sample of sand, making three tests in all.

If the range of values of dry density exceeds 0.01 Mg/m^3, repeat the above procedure. The vibrating hammer is satisfactory for the vibrating compaction test if the mean dry density achieved exceeds 1.74 Mg/m^3.

This test is valid only for the sand specified above. Other types of sand will give different results.

Fig. 6.24   Harvard compaction apparatus

6.5.10   Harvard Compaction Method

The Harvard compaction test procedure is given in *ASTM STP* 479 (Wilson, 1970) as a suggested method for determining the compaction characteristics of fine-grained soil when only a small quantity of material is available. The action of the apparatus differs from the drop-weight principle of the conventional compaction tests in that the soil is subjected to kneading rather than impact. Results from the Harvard test may not be directly comparable with the BS or ASTM tests, and is not intended as a substitute for them.

This small-scale procedure can be useful in the laboratory for the preparation of small recompacted specimens for use in other tests. The controlled degree of compaction which it provides gives results which are more reproducible than those obtainable by arbitrary hand tamping methods.

APPARATUS

The compaction device consists of a hand-held spring-loaded tamper and special mould, which are shown in Fig. 6.24. The spring is compressed by means of the adjusting nut to a compression of 40 lb (18.2 kgf or 178 N), so that a small increase of force above that value will compress the spring further. Springs of different stiffnesses can be substituted. The metal tamper rod is 0.5 in (12.7 mm) diameter.

The mould used has an internal diameter of $1\frac{5}{16}$ ins (33.34 mm) and is 2.816 ins (71.5 mm) high, giving a volume of $\frac{1}{454}$ ft^3 (62.4 cm^3). This volume was selected because the mass of soil, in grams, is equal to its density in pounds per cubic foot. An extension collar about $1\frac{1}{2}$ ins (38 mm) high may be added to the mould, both of which can be fitted to a detachable baseplate.

The Harvard compaction procedure can be modified to provide additional or lesser degrees of compaction, but the relationship to the BS or ASTM compaction efforts can only be determined experimentally for a particular soil.

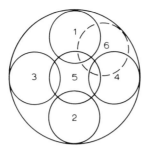

Fig. 6.25   Sequence of tamps using Harvard apparatus

A specially designed jig (the collar remover) enables the compacted soil to be held in place and kept intact while the extension collar is being removed.

A sample ejector quickly and easily removes the compacted specimen from the mould.

SOIL SAMPLE

Soil for use with this apparatus should contain particles no larger than 2 mm. The usual procedure for sub-dividing, sieving, mixing and curing should be followed.

If a complete moisture–density relationship test is to be done, separate batches should be used for each moisture content, and a compacted specimen should *not* be remixed and reused.

COMPACTION PROCEDURE

Compaction is effected by placing the plunger on the soil surface and pressing down with the hand grip until it can be felt that the spring is just starting to compress, and then releasing and moving to the next position. The first four tamps should be applied in opposite quadrants touching the edge of the mould, followed by one at the centre (see Fig. 6.25). The next four should be in a similar pattern but spaced between the first four, then at the centre. This sequence is repeated until the required number of tamps have been applied, at a rate of about one tamp every $1\frac{1}{2}$ s.

With a 40 lb (178 N) spring, compaction in three layers with 25 tamps per layer is roughly equivalent to BS 'light' compaction, but this is given only as a general guide and not an established relationship.

Measurement of density and moisture content, and calculations, are done in the same way as for the compaction tests described above. Results of a moisture–density relationship test should include a note reporting the type of test, size of mould and compression spring used.

Apart from its use for determining the moisture–density relationship, the Harvard tamping device provides a convenient means of preparing small reproducible test specimens for shear strength and other tests on recompacted soil.

6.6 MOISTURE CONDITION TESTS

6.6.1 Scope

PURPOSE OF TESTS

The procedure for determining the moisture condition value (MCV) of a soil was developed at the Transport & Road Research Laboratory as a rapid means of assessing the suitability of soil for use in earthworks construction in relation to the specified limits of moisture content or strength (see TRRL Reports by Parsons (1976); Parsons and Boden (1979);

Parsons and Toombs (1987)). Because of the variability of materials encountered on a typical earthworks construction site it is not usually possible to assign unique values of soil parameters such as moisture content, plastic limit and optimum moisture content. This causes difficulties in the control of the quality of earthworks, and it is these difficulties which the MCV test attempts to overcome.

Some of the merits of the MCV test are summarised as follows.

(1)   It provides an immediate result, without having to wait for the determination of moisture content or other parameters.
(2)   The test is applicable to a wide range of soil types except granular soils with no fines.
(3)   Some variation within a given soil type are not critical.
(4)   The test can be performed on site as well as in the laboratory.
(5)   Test results show good reproducibility.
(6)   The likelihood of operator error is small.
(7)   Variability associated with sampling is not excessive because a reasonably large sample is used.
(8)   The test can provide a useful indicator of the engineering quality, and of some aspects of behaviour, of the soil.

It is possible that relationships can be derived between MCV and laboratory-measured soil parameters such as undrained shear strength and CBR value, as well as soil classification. The MCV could also be related to the performance of earthmoving plant, and could indicate where high degrees of compaction are likely to be difficult to achieve, or where excess water pressures might be produced by over-compaction.

TYPES OF TEST

The procedures given in clause 5 of BS1377: Part 4:1990 comprise the following.

(1)   Determination of the MCV of a sample of soil at the moisture content 'as received'. (Section 6.6.3.)
(2)   Determination of the relationship between MCV and moisture content — the Moisture Condition Calibration (MCC). (Section 6.6.4).
(3)   Rapid assessment of whether or not a sample of soil is suitable for compaction by comparison with a previously determined criterion. (Section 6.6.5.).

In the MCV test the soil is repeatedly compacted into a rigid mould under the blows of a falling rammer. The apparatus used is a modification of the machine used for the determination of the aggregate impact value, described in BS812: Part 112:1990. The minimum compactive effort required to produce near-full compaction of the soil fraction passing a 20 mm sieve is determined.

The relationship between MCV and moisture content for a particular soil type can be determined. A criterion can then be established against which a rapid assessment test can determine whether or not a similar soil complies with the pre-calibrated standard.

6.6.2   Principles

The MCV test is based on the principles of compaction of soil described in Section 6.3.1. If a soil is subjected to compaction tests using different compactive efforts, as the moisture content increases the curves relating dry density to moisture content tend to converge, and they lie close to the zero air voids line, as shown in Fig. 6.3. Compaction curves for a soil when three different compactive efforts are used, denoted by A, B and C (A being the lightest, C the heaviest) are shown diagrammatically in Fig. 6.26.

When the soil is at moisture content m_1, compactive effort A gives a dry density corresponding to point a; effort B to point b; and effort C to point c. Increasing the compactive effort results in corresponding increases in dry density at this moisture content.

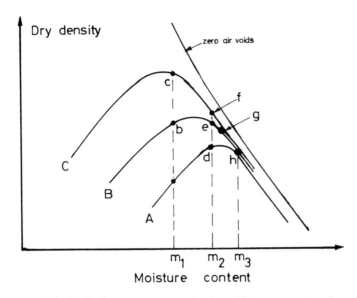

Fig. 6.26   Idealised compaction curves for three different compactive efforts

At the higher moisture content m_2, compactive effort B still gives a higher dry density than effort A (points e and d respectively). However effort C (point f) gives a relatively insignificant increase compared with effort B because point e is already close to the zero air voids line. Thus at moisture content m_2, compactive effort B is sufficient to produce very nearly full compaction of the soil. Increasing the moisture content a little more (point g), curves B and C virtually coincide and effort B gives full compaction.

Increasing the moisture content further to m_3 and beyond, no significant increase in dry density can be achieved by using compactive efforts B or C compared with effort A. At this moisture content (point h), effort A is sufficient to produce full compaction.

It can be seen that the higher the moisture content of the soil, the lower is the compactive effort required beyond which no significant increase in dry density occurs, i.e. the lower is the effort required to give full compaction. A measure of the 'moisture condition' can be obtained by determining the lowest compactive effort beyond which the increase in dry density is not significant.

In the test the change in height of the sample (which is related to change in density) in the mould is determined by measuring the penetration of the rammer. The change is considered to be 'not significant' when the change in penetration due to additional compaction is 5 mm or less.

The bulk density and dry density are not required to be calculated in the BS test. However the determination of these values, although liable to some error, requires little additional effort and they provide further useful data for comparison with other test results.

6.6.3   Moisture Condition Value (MCV) Test (BS1377: Part 4:1990:5.4)

The test is carried out on soil containing particles passing a 20 mm sieve. It is particularly relevant to cohesive soils, but for non-cohesive (granular) soils there may be difficulties in interpretation of results, especially if particles are susceptible to crushing.

APPARATUS

(1)   Moisture condition apparatus complete with mould, separating disc and a means of measuring the penetration or protrusion of the rammer. A typical apparatus is shown

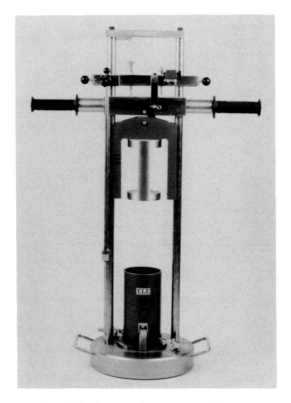

Fig. 6.27   Apparatus for moisture condition test

in Fig. 6.27, and the main features are shown diagrammatically in Fig. 6.28. Full specification details are given in BS1377: Part 4:1990:5.2.

Essential requirements are as follows.

Mass of base of frame: at least 31 kg
Mould with detachable permeable base:
   Internal diameter 100 mm
   Internal height at least 200 mm
   Internal surface has a protective coating
   Permeability of base to allow water discharge of 4 to 7 litres per minute when the water
     level in the mould is maintained at a head of 175 mm above the base.
Rammer: falling mass 7 kg
        diameter of face 97 mm
        height of drop 250 mm
Scale and vernier to measure penetration or protrusion of rammer to 0.1 mm.
Fibre disc to separate soil from rammer: 99.2 mm diameter.
Lifting system to release rammer at pre-set level, fitted with automatic counter.
Drop height control to regulate the height of drop in the range 100 mm to 260 mm, to
   within ± 5 mm.

It is not necessary to stand the apparatus on a plinth or inertia block because the specified mass of the base provides enough inertia.

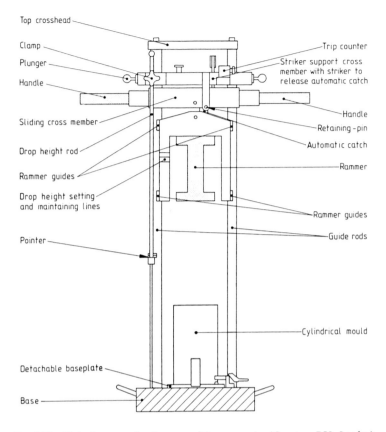

Top crosshead

Clamp

Plunger

Handle

Sliding cross member

Drop height rod

Rammer guides

Drop height setting
and maintaining lines

Pointer

Detachable baseplate

Base

Trip counter

Striker support cross
member with striker to
release automatic catch

Handle

Retaining-pin

Automatic catch

Rammer

Rammer guides

Guide rods

Cylindrical mould

Fig. 6.28   Main features of moisture condition apparatus (Courtesy BSI, London)

The energy per blow delivered by the rammer is $2\frac{1}{3}$ times that delivered by the BS 2.5 kg rammer. The energy delivered per unit volume of soil per blow is about 3 times that of the 2.5 kg rammer, or about 11% more than that of the BS 4.5 kg rammer.

(2)   Balance, 2 kg capacity reading to 1 g.
(3)   20 mm sieve and receiver.
(4)   Large metal tray, e.g. $600 \times 600 \times 60$ mm deep.
(5)   Drying oven, 105°–110°C, and other equipment for moisture content determination.
(6)   Jacking apparatus for extracting compacted soil from the mould.

PROCEDURAL STAGES

(1)   Check apparatus.
(2)   Prepare test sample.
(3)   Place soil in mould.
(4)   Fit mould.
(5)   Apply blows and measure penetrations.
(6)   Remove sample.
(7)   Calculate.
(8)   Plot graph.

Derivation of the MCV, and reporting of results, are described under separate headings.

TEST PROCEDURE

(1) *Checking apparatus*

Ensure that the rammer drops freely.

Adjust the height of drop of the rammer to give a fall of 250 mm to the top of the rigid disc when placed in the mould on the machine base, in accordance with the manufacturer's instructions. (A smaller height of drop may be appropriate in some tests in which higher moisture contents are used; if so the height of drop should be clearly stated.)

Ensure that all components of the apparatus are secure.

Check that the mould and its components are clean and dry, and that the internal protective coating has not been worn away by abrasion.

Measure the internal dimensions of the mould.

For safety, when checking or adjusting the apparatus or placing the mould and fittings with the rammer in the raised position, ensure that it is securely held by the retaining pin.

(2) *Preparation of test sample*

Sieve the original sample of soil on a 20 mm sieve, break down any aggregations of retained particles, and remove individual particles retained on the sieve.

Weigh the sample and the removed material to 1 g if the proportion of coarse particles is to be reported.

Take a representative portion of soil passing the 20 mm sieve for determination of moisture content.

Subdivide the soil passing the sieve to give a representative test sample of about 1.5 kg (to within ± 20 g). Do not break down any aggregation of particles in this sample.

If compacted densities are to be measured (not a requirement of the BS, but useful for comparisons), weigh the test sample to the nearest 1 g.

(3) *Placing soil in the mould*

Place the 1.5 kg sample of soil as loosely as possible in the mould. If necessary, push the soil in if it would otherwise overflow the rim, but push only enough for the surface of the loose soil to finish within about 5 mm of the top of the mould.

The loose condition can be achieved by pouring the soil into the mould, through a funnel if necessary. If the soil is not placed in its loosest condition the reproducibility of the result can be affected.

(4) *Fitting mould*

Secure the rammer of the apparatus in the raised position with the retaining pin.

Place the mould in position on the base of the apparatus and place the fibre disc on top of the sample.

Adjust the automatic counter to zero.

Holding the rammer steady, remove the retaining pin and lower the rammer gently on to the disc covering the loose sample. Allow the rammer to penetrate into the mould under its own weight until it comes to rest.

Set the height of drop to 250 mm.

(5) *Application of blows*

Raise the rammer until it is released by the automatic catch, so that one blow is applied to the sample. Measure the depth of penetration of the rammer inside the mould, or the distance to the top of the rammer from the rim of the mould, i.e. the protrusion, to 0.1 mm using the depth gauge and vernier.

It is immaterial whether penetration (which will increase with further blows) or protrusion (which will decrease) is measured, because the plotting is based on changes in measurements.

Measurement of protrusion using a depth gauge is usually easier, and the test data given below are based on that type of measurement, but reference is also made to penetration.

Re-set the height of drop of the rammer to 250 mm. Apply further blows as above, and re-set the height of drop to 250 mm as necessary, until there is no further significant increase in penetration or until 256 blows have been applied.

Measure the penetration or protrusion after the blows numbered as follows.

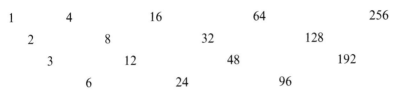

In each line of this sequence, every number is 4 times the previous number. The criterion for 'no significant increase in penetration' has been arbitrarily set at a change of 5 mm between the application of n blows and $4n$ blows (see Section 6.6.2 above).

Enter the readings opposite the blow number in the second column of the form shown in Fig. 6.29. If the material is very dry and more than 256 blows would be required, report the MCV as 'more than 18'.

When all the required blows have been applied, raise the rammer carefully and secure it with the retaining pin.

(6) Removal of sample

Remove the mould from the base of the apparatus, take off the base and extract the sample. Clean and dry the mould ready for its next use.

Although not required by the British Standard, a representative sample may be taken for determination of moisture content.

(7) Calculation

Calculate the change in penetration between a given number of blows, n, and four times as many blows, $4n$. (For example, between 1 and 4; 2 and 8; 3 and 12, etc.).

Enter the difference in the third column of the table in Fig. 6.29 opposite n blows. For example, enter the difference between 4 and 16 blows on the same line as 4 blows.

If desired, calculate the approximate density of the compacted soil from the known mass of soil used. The height of the compacted sample can be determined from the mould and rammer measurements and the final penetration or protrusion measurement. If the height is denoted by H (mm), and the mass of soil used is 1.5 kg, the density is equal to $191/H$ Mg/m^3.

From the density and moisture content calculate the dry density using equation (3.12) in Section 3.3.2. This value, although not a requirement of the BS, enables the tested sample to be related to the moisture/density relationship obtained from a compaction test.

(8) Plotting graph

Plot each change of penetration, on the linear scale, against each initial number of blows, n, on the logarithmic scale, using a form similar to that shown in Fig. 6.30. Use the value of penetration change which is on the same line as the number n.

Draw a smooth continuous curve through the plotted points. Interpretation of this graph differs according to the requirements of different authorities, as described below.

DERIVATION OF MCV

(a) BS1377 method

This method of derivation of the MCV from the graph is given in Clause 5.4.2.3 of BS1377:

Moisture condition

Single sample/separate batches*	
Initial sample mass	g
Moisture content	g
Dry mass	g
Mass retained on 20 mm sieve	g
	%

Location	
Job ref.	
Borehole/ Pit no.	
Soil description	
Sample no.	
Depth	
Test method BS 1377 : Part 4 : 1990 : **5**	
Date	

*Delete as appropriate

Total number of blows n	Penetration or protrusion mm	Change in penetration n to 4 n mm
1		
2		
3		
4		
6		
8		
12		
16		
24		
32		
48		
64		
96		
128		
192		
256		

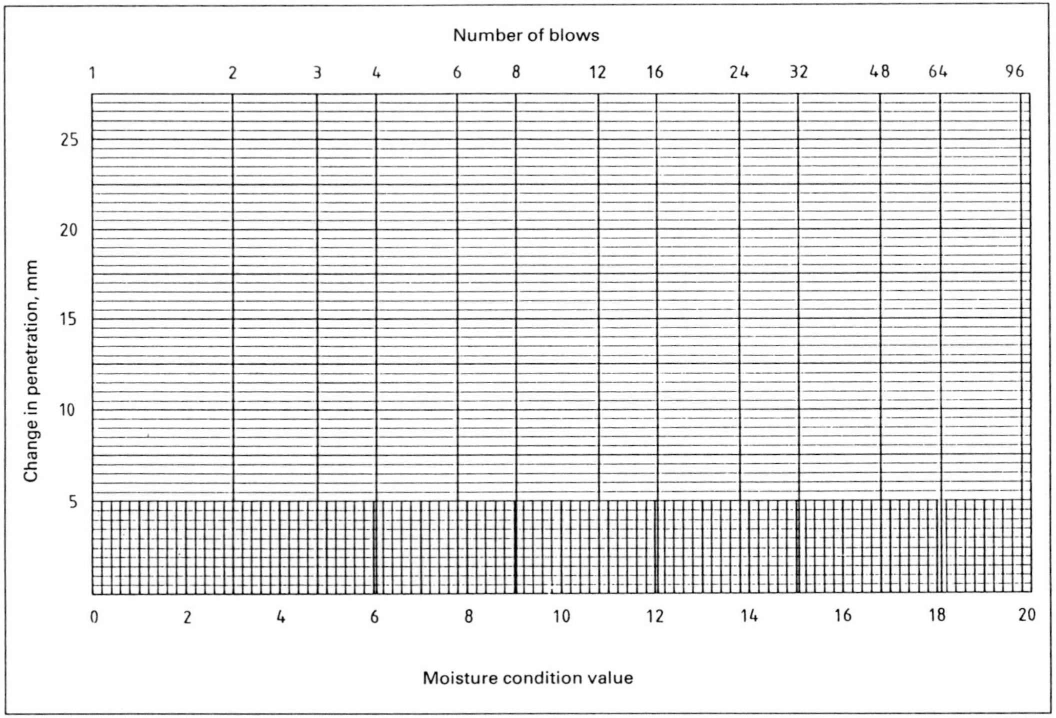

Fig. 6.29   Form for recording moisture condition value test (Courtesy BSI, London)

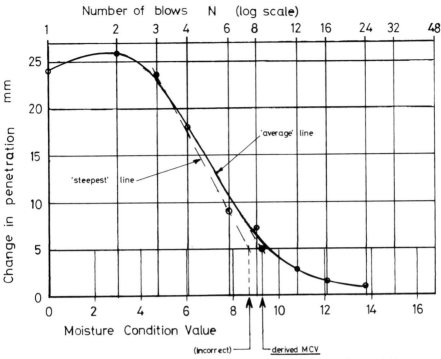

Fig. 6.30 *Typical plot of change in penetration against number of applied blows (logarithmic scale) (see paragraph (c))*

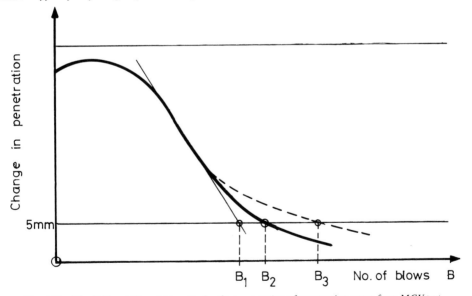

Fig. 6.31 *'English' and 'Scottish' methods of interpretation of penetration curve from MCV test*

Part 4:1990. It is in line with the current TRRL approach, as described in SR522, Section 11 (Appendix). It is sometimes referred to as the 'English' method.

Draw the steepest 'give and take' (average) straight line through the plotted points immediately before the 5 mm change in penetration value. Extend this line, if necessary, to intersect the horizontal line corresponding to 5 mm change. Read off the number of blows (B), to two significant figures, at the intersection point. The procedure is illustrated in Fig. 6.31. The MCV is then defined as $10 \log_{10} B$, in which $B = B_1$.

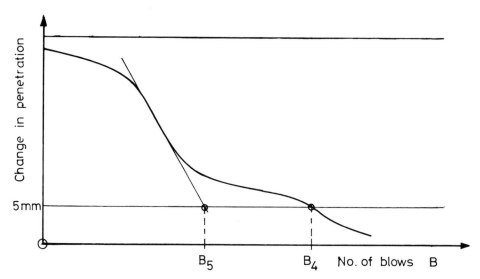

Fig. 6.32   Type of MCV curve sometimes obtained for granular soils

The flattening out of the curve, which for many soils found in the UK occurs after, or only a little before, the 5 mm change in penetration is reached, reflects the increasing difficulty of expelling air from the soil as the state of full compaction is approached. The steepest straight line interpretation helps to minimise this effect.

In Fig. 6.30, an arithmetical scale of 0 to 20, for the range of blows from 0 to 100, has been added along the bottom. This scale enabled the MCV to be read off directly, to the nearest 0.1.

The Note to the above clause in BS1377, draws attention to problems which may arise with interpretation of the graph, some of which are covered in (c) below.

(b)   *Scottish method*

This method is advocated by the TRRL in Scotland because it appears to be more suitable for the granular tills found in Scotland. It was given in the original TRRL document LR750, and also appears in Section 3 of SR522.

Determine the point at which the plotted curve intersects the line representing 5 mm change in penetration. Read off the number of blows ($B = B_2$ in Fig. 6.31) corresponding to this point, and calculate the MCV as described above; or read off the MCV directly from the arithmetical scale.

The 'English' method can give appreciably lower, and therefore more conservative, MCV results than the 'Scottish' method when the flattening out of the graph begins above the 5 mm change line, as illustrated by the dashed curve in Fig. 6.31. This curve gives $B = B_3$, which is significantly greater than B_1.

(c)   *Other interpretations*

When using the 'English' method, the steepest straight line should be the steepest sensible line, obtained by averaging the points if there is some scatter ('give and take' approach). A reasonable interpretation is shown by the full line in Fig. 6.30. The steepest line should not be obtained by joining two of the plotted points, illustrated by the broken line in Fig. 6.30.

With some soil types, notably granular soils, the relationship between change in penetration and number of blows may be of the form shown in Fig. 6.32. Here the slope of the curve decreases, and then increases again, before reaching the 5 mm change line. This is probably due to factors other than reaching full compaction, such as expulsion of water and crushing of soil particles. The latter is very likely with sandy gravel soils. If it is unlikely that crushing of particles will occur during compaction in the field, it would be unreasonable to apply the 'Scottish' method, which gives $B=B_4$, to this type of curve. On the other hand, the 'English' method, giving $B=B_5$, could produce an unreasonably conservative MCV. Interpretation of a curve of this type should be avoided unless adequate field data are also available.

In general, MCV tests on granular soils are difficult to interpret. Fine to medium sands in particular are not amenable to this test. Very low MCV results may be obtained for granular soils, indicating that they are apparently not suitable for earthworks, whereas in fact they may be good free-draining material.

Interpretation of MCV test results is discussed in more detail by Dennehy (1988).

REPORTING RESULTS

The plot of change in penetration against logarithm of the number of blows should normally form part of the test report. The method of interpretation of the graph should be clearly stated.
    The report should also include the following.

The moisture condition value (MCV) of the soil, to the nearest 0.1.
The moisture content at which the soil was tested if required, and whether it represents the natural moisture content.
The proportion by dry mass of particles larger than 20 mm (if any) which were removed from the initial sample.
The method of test (Clause 5.4 of BS1377: Part 4:1990).

6.6.4 MCV/Moisture Content Relationship (BS1377: Part 4:1990:5.5)

This is a calibration procedure for a given soil, and is referred to here as the Moisture Condition Calibration (MCC) test. The MCV test is carried out on the soil over a range of moisture contents.

APPARATUS

As for the MCV test, Section 6.6.3.

PREPARATION OF TEST SAMPLE

If the soil contains particles which are susceptible to crushing during compaction, or includes clay of low permeability requiring at least 24 hours after mixing with water to ensure uniform distribution of water, a separate batch of soil should be prepared for compaction at each moisture content (method (1)). Otherwise a single sample may be prepared, and re-used after mixing with further increments of water (method (2)). The soil should not be dried to a moisture content less than the lowest value required for the test.

(1)   *Separate batches*

Prepare the soil as described in Section 6.6.3 (2), after partially air-drying to the lowest moisture content required for the test. Do not allow the soil to dry completely. At least 10 kg of soil passing the 20 mm sieve is required.
    Sub-divide the soil to give at least 4 test samples each of about 2.5 kg. Mix each sample with a different amount of water to give a suitable range of moisture contents. The range should give MCVs ranging from about 14 to 3. Samples of cohesive soil should be stored in sealed containers for at least 24 hours before testing.

(2) *Single batch*

Prepare the soil as described in Section 6.6.3 (2), to give a sample of about 4 kg. The initial moisture content should produce an MCV of 13 to 15 (full compaction after 20 to 32 blows of the rammer). Reduce the moisture content by partial air drying, or add water and mix well in, if it is already too dry, to achieve this condition.

TEST PROCEDURE

(1) *Separate batches*

Determine the MCV of each sample in turn, using the procedure described in Section 6.6.3.

   After extracting each sample from the mould, take a representative portion for the determination of the moisture content. The remainder of each sample may be discarded.

(2) *Single batch*

Take a representative sample of 1.5 kg ($\pm$ 20 g) of the prepared soil and carry out the MCV test as described in Section 6.6.3.

   After extracting the soil from the mould, take a representative portion for the determination of moisture content. Break up the remainder of the soil and mix with the remains of the original sample. Add a suitable amount of water and mix in thoroughly.

   Repeat the MCV test after each increment of moisture content, to make at least four determinations in all. The moisture contents should give MCVs ranging from about 14 to 3.

PLOTTING

Plot the moisture content of each compacted sample against the MCV for the sample, and draw the best straight line through the points. A typical relationship is of the form shown by the full line in Fig. 6.33, which is plotted above the plots obtained from the individual MCV tests. For clarity only three MCV curves are shown, but in practice 6 or 8 MCV tests would be performed to derive the calibration curve.

   The MCC for a granular soil, or a clayey soil with a high gravel content, may give points marked 'X'. lying on a second calibration line or curve at low moisture content, as indicated by the dashed curve in Fig. 6.33. These points are below an 'optimum MCV' — the whole curve resembles a compaction curve on its side. This branch of the curve represents 'non-effective' calibration, and should not be used.

   The 'effective' calibration for clay soils might be in the form of a curve. This can reflect the extent to which the clay structure is broken down before testing, and often occurs with overconsolidated clays. A linear calibration is more likely if the clay has been worked more before testing, but excessive breaking down and working of clay in the laboratory would not usually represent field working conditions, and a non-linear curve might be more representative.

REPORTING

The graphical calibration plot, which may be added above the plots obtained from the individual MCV tests (see Fig. 6.33), forms part of the test report. The method of test, and other information listed under Reporting for the MCV test (Section 6.6.3), should also be reported.

6.6.5   Assessment of Soil Strength (BS 1377: Part 4:1990:5.6)

This procedure provides a rapid method, which can be carried out on site, for the assessment of the condition of a soil, eg. regarding its suitability for use in earthworks construction. It is based on the results of an MCC test on similar soil, using the procedure given in Section 6.6.4, from which a calibration standard in terms of MCV has already been derived. This

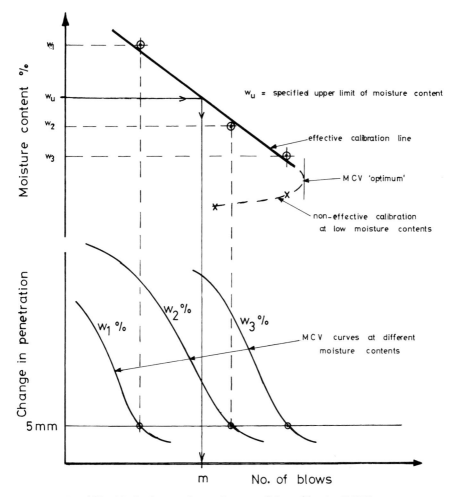

Fig. 6.33   Idealised curves from moisture condition calibration (MCC) tests

rapid test indicates only whether the soil is 'suitable' or 'unsuitable', without indicating the degree to which it exceeds or fails to meet the precalibrated standard. The result is normally insensitive to small variations in soil properties.

The normal procedure, given here, relates to a limit at the wet end of the moisture content range. This method can also be used to relate the moisture condition of the soil to the dry end limit. If penetration measurements are taken after 6, 24 (or 25), 100 blows it can be assessed immediately whether the MCV of the soil lies between 8 and 14.

APPARATUS

As for the MCV test, Section 6.6.3. For a site assessment test, item (2) can be replaced by a balance such as a robust spring balance reading to 20 g, and item (5) is not necessary.

DETERMINATION OF PRECALIBRATED STANDARD

The procedures for deriving a precalibrated standard MCV for a given soil are summarised as follows.

(1)   Carry out the appropriate moisture/density relationship test (2.5 kg or 4.5 kg rammer method) and determine the optimum moisture content (OMC).

(2)   For cohesive soils, determine the plastic limit (w_p).
(3)   Select the upper limit of moisture content (w_u) to be specified for *in-situ* compaction.
(4)   Carry out the MCC test (Section 6.6.4) on a representative sample of the soil.
(5)   From the calibration curve, read off the MCV and initial number of blows corresponding to w_u.

The selected upper limit of moisture content will depend on the type of soil, method of compaction, field conditions and other factors. It can be related to the OMC or plastic limit, and the following relationships have sometimes been used as a general guide.

For clays: $w_u = 1.2 \times w_p \%$
For other soils: $w_u = (OMC + 1.5)\%$

Alternatively, the upper limit may be more easily related to a correlation already established between MCV and CBR value or shear strength. The criterion to be used must be decided by the engineer responsible for the works.

A typical MCC calibration plot is shown in Fig. 6.33. The upper limit of moisture content, w_u, is marked on the moisture content scale, and projected across to the calibration curve. The number of blows corresponding to this value (m) is read off from the horizontal scale. The corresponding MCV can also be derived if required. The number m (blows) corresponds to the precalibrated standard MCV.

PREPARATION OF TEST SAMPLE

Prepare a representative sample of the soil to be tested as for the MCV test, Section 6.6.3.

TEST PROCEDURE

Place the sample in the mould, and apply blows, as in the MCV test, Section 6.6.3, up to the total number of blows (m) equivalent to the MCV of the precalibrated standard, re-setting the height of drop as necessary. Measure the penetration or protrusion of the rammer to 0.1 mm.

Apply additional blows equal to three times the number already applied (i.e. $3\,m$ blows, making a total of $4\,m$ blows), without further adjustment to the height of drop of the rammer. Measure the penetration or protrusion of the rammer as above.

ASSESSMENT

Calculate the difference in penetration or protrusion between the application of m blows and $4\,m$ blows, to the nearest 0.1 mm.

If this difference exceeds 5.0 mm, the soil is stronger than the pre-calibrated standard, and is in a suitable condition for compaction.

If the difference is less than 5.0 mm, the soil is weaker than the standard.

6.7   CHALK CRUSHING VALUE

6.7.1   Scope

This procedure was developed at the TRRL to enable the strength of chalk, in terms of its resistance to crushing, to be measured. In the test, intact lumps of chalk are subjected to crushing by the action of the rammer in the MCV apparatus and the rate at which the chalk lumps are crushed provides the chalk crushing value (CCV). The CCV can be used in conjunction with the saturation moisture content of the intact lumps (Section 2.5.4) to classify the chalk with regard to its behaviour as a freshly placed fill material. (For details see Ingoldby and Parsons, 1977).

6.7.2   Chalk Crushing Value (CCV) Test (BS1377: Part 4:1990:6)

The following procedure is described for a single sample of chalk lumps, but normal practice should be to prepare and test at least 6 representative samples and derive the mean value.

APPARATUS

(1)   Moisture condition apparatus complete with accessories, as described in Section 6.6.3.
(2)   Balance of 2 kg capacity readable and accurste to 1 g.
(3)   Hammer, such as a 2 lb club hammer.
(4)   20 mm and 10 mm sieve and receiver.
(5)   Large metal tray, e.g. $600 \times 600 \times 60$ mm deep.
(6)   Jacking apparatus for extracting compacted soil from the mould (optional).

Preparation test sample

Take a representative sample of the intact chalk lumps and sieve them on the 20 mm and 10 mm sieves. A sample of 1 kg of material passing the 20 mm sieve and retained on the 10 mm sieve is required. Determine the percentage of the whole sample, by mass, of material retained on the 10 mm sieve. If necessary, break down lumps of chalk larger than 20 mm, using the hammer, to provide enough material for the test sample.

   Do not include in the sample any coagulated lumps of chalk fines, fragments of flint, or any other non-chalk material.

   The degree of saturation of the chalk lumps is not significant, but the chalk should not be oven-dried.

Test Procedure

(1)   Place the prepared sample loosely in the clean, dry mould of the MCV apparatus, and place the separating disc on top of the chalk.
(2)   With the rammer held in the raised position by the retaining pin, place the mould in position on the base of the apparatus, and adjust the automatic counter to read zero.
(3)   Hold the rammer steady and remove the retaining pin. Lower the rammer gently onto the separating disc and allow it to penetrate into the mould under its own weight until it comes to rest. Set the height of drop at 250 mm ± 5 mm.
(4)   Apply one blow of the rammer to the sample by raising the rammer until it is released by the automatic catch. Measure the penetration of the rammer into the mould, or the length of rammer protruding from the mould, to 0.1 mm. (See the comment in step (5) of Section 6.6.3). Record the readings on a test form similar to that used for the MCV test (Fig. 6.29) but with the appropriate listing of the number of blows (see below).
(5)   Re-set the height of drop to 250 mm.
(6)   Repeat steps (4) and (5), taking readings of penetration or protrusion after selected accumulated numbers of blows, and resetting the height of drop to 250 mm as necessary.

   The cumulative numbers of blows after which readings are taken should comprise at least the following, which provide a reasonable spacing of points when plotted on a logarithmic scale.

1, 2, 3, 6, 8, 12, 20, 30, 40.

Readings may be taken after intermediate numbers of blows if appropriate.

(7)   Stop the test when water starts to ooze from the base of the mould, or no further penetration occurs, or a maximum of 40 blows is reached. Carefully raise the rammer and insert the retaining pin.
(8)   Remove the mould from the apparatus, take off the base and extract the crushed chalk.

Plotting and Calculations

(1) Plot the penetration or protrusion of the rammer (mm) on a linear scale against the number of blows on a logarithmic scale.
(2) The greater part of the relation should form a straight line, the slope of which represents the rate at which the chalk was crushed. The Chalk Crushing Value (CCV) is taken as one-tenth of the slope of the straight line.

$$CCV = \frac{P_a - P_b}{10(\log a - \log b)}$$

where P_a is the penetration (or protrusion) (mm) after a blows of the rammer as read from the straight line; P_b is the penetration (or protrusion) (mm) after b blows of the rammer as read from the straight line.

For ease of calculation it is convenient to use values of a and b such that $a = 10b$ (e.g. $a = 20$, $b = 2$; or $a = 30$, $b = 3$; etc.). Then $\log a - \log b = 1$ and

$$CCV = \frac{P_a - P_b}{10}$$

The CCV should be expressed as a positive number.

Reporting Results

The test report should include the following.

(1) The chalk crushing value (CCV) to two significant figures.
(2) The plot of penetration against logarithm of the number of blows, if required.
(3) The percentage of material in the original sample retained on a 10 mm BS test sieve.
(4) The saturation moisture content of the chalk, when appropriate.
(5) The method of test (clause 6.4 of BS1377: Part 4:1990).

6.8   COMPACTABILITY TEST FOR GRADED AGGREGATES

This test was developed at the TRRL (Pike, 1972; Pike and Acott, 1975), and is a method for assessing the compactability of graded aggregates, particularly those used in road bases and sub-bases. The standard compaction tests used for soils were found to be unrealiable when applied to some of these materials, and this procedure aims to provide a standardised approach to compactibility testing.

The principle of the test is similar to the vibrating hammer test described in Section 6.5.9. However, a more powerful vibrating hammer is used, in a standardised manner, mounted in a loading frame, and the sample is compacted in a special heavy-duty mould. The test results are presented in the usual form of a moisture content–dry density relationship, but the dry density can also be expressed in items of a volumetric equivalent.

The following special apparatus is required, in addition to standard soil-testing equipment:

(1) Compaction mould, comprising body, base, filter assembly and anvil. The latter covers the whole area of the sample, and may be fitted with an optional vacuum release plug. Any excess water is permitted to drain downwards.
(2) Electric vibrating hammer, power consumption 900 W, frequency about 33 Hz, fitted with a tool to mate with the anvil.
(3) Loading frame to support the hammer and mould, providing a steady downward force of 360 ± 10 kN.

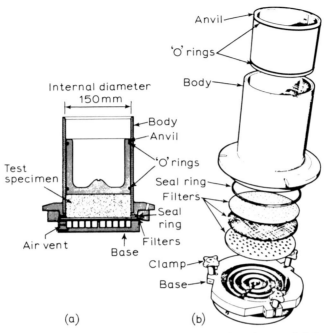

Internal diameter
150 mm

Body
Anvil
'O' rings
Test specimen
Air vent
Seal ring
Filters
Base

Anvil
'O' rings
Body
Seal ring
Filters
Seal ring
Filters
Clamp
Base

(a)   (b)

Fig. 6.34   Mould and anvil for compactability test (*Courtesy Transport and Road Research Laboratory, Crowthorne, Berks.*)

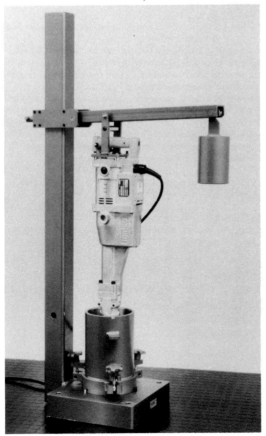

Fig. 6.35   Equipment for compactability test

The mould assembly and its component parts are illustrated in Fig. 6.34. The load frame and mould, which is preferably housed set up for use in a noise-reducing cabinet, are shown in Fig. 6.35.

The procedure is not described here, but it is given in Clause 2.1.5 of BS1924: Part 2:1990.

It is suggested that this apparatus could provide an alternative means for determining the maximum density (minimum porosity) of granular soils, including silty sands, to the procedure described in Sections 3.7.2 and 3.7.3.

BIBLIOGRAPHY

American Association of State Highway Testing Officials (1986). *Standard specifications for highway materials and methods of sampling and testing.* Part II, AASHTO. Designation T99-38, 'Standard laboratory method of test for the compaction and density of soil'. AASHTO, Washington, DC

American Foundrymen's Association (1944). *Foundry Sand Testing Handbook*, Section 4. American Foundrymen's Association, Chicago

BS1924: Part 2:1990. Methods of test for stabilised materials for civil engineering purposes. British Standards Institution, London

BS5835: Part 1:1980. Testing of aggregates — Compactibility tests for graded aggregates. (British Standards Institution, London)

Dennehy, J. P. (1988). 'Interpretation of moisture condition value tests' *Ground Engineering*, January

Little, A. L. (1948). 'Laboratory compaction technique'. *Proc. 2nd Int. Conf. on Soil Mech. and Found. Eng.*, Rotterdam, Paper II g 1

MacLean, D. J. and Williams, F. H. P. (1948). 'Research on soil compaction at the Road Research Laboratory'. *Proc. 2nd Int. Conf. on Soil Mech. and Found. Eng.*, Rotterdam, Paper 1X 5 b

McLeod, N. W. (1970). 'Suggested method for correcting maximum densith and optimum moisture content of compacted soils for oversize particles'. *ASTM STP* 479

Maddison, L. (1944). 'Laboratory tests on the effect of stone content on the compaction of soil mortar'. *Roads and Road Construction*, Vol. 22, Part 254

Markwick, A. H. D. (1944). 'The basic principles of soil compaction and their application'. Inst. Civ. Eng. London. Road Engineering Division, Road Paper No. 16

Odubanjo, T. O. (1968). 'A study of a laboratory compaction test using a Swedish vibratory apparatus'. TRRL Report LR129. Transport and Road Research Laboratory, Crowthorne, Berks.

Parsons, A. W. (1976). 'The rapid measurement of the moisture condition of earthwork material'. TRRL Laboratory Report 750. Transport and Road Research Laboratory, Crowthorne, Berks.

Pike, D. C. (1972). 'Compactibility of graded aggregates: I. Standard Laboratory tests'. TRRL Laboratory Report LR447. Transport and Road Research Laboratory, Crowthorne, Berks.

Pike, D. C. and Acott, S. M. (1975). 'A vibrating hammer test for compactability of aggregates'. TRRL Supplementary Report 140 UC. Transport and Road Research Laboratory, Crowthorne, Berks.

Proctor, R. R. (1933). 'Fundamental principles of soil compaction'. *Engineering News Record*, Vol. 111, No. 9

Taylor, D. W. (1948). *Fundamentals of Soil Mechanics.* Wiley, New York

Transport and Road Research Laboratory (1952). *Soil Mechanics for Road Enginers*, Chapter 9. HMSO, London

Williams, F. H. P. (1949). 'Compaction of soils'. *J. Inst. Civ. Eng.*, Vol. 33, No. 2

Wilson, S. D. (1950). 'Small soil compaction apparatus duplicates field results closely'. *Engineering News-Record*, 2 November

Wilson, S. D. (1970). 'Suggested method of test for moisture-density relations of soils using Harvard compaction apparatus. Special procedures for testing soil and rock for engineering purposes'. *ASTM STP* 479

Chapter 7

Description of soils

7.1 INTRODUCTION

7.1.1 The Nature of Soil

A definition of soil which is appropriate to geotechnical engineering was stated in Section 1.1.5, and is repeated in Section 7.2. Different defintions apply in the fields of geology, pedology, agriculture, and so forth, but in geotechnology the significant features of soils are their particulate nature, and the porous structure which the particles form.

The relatively thin layer of humic topsoil which supports vegetation is usually unimportant unless it extends to exceptionally great depths.

Most natural soils comprise particles known as boulders, cobbles, gravels, sands, silts and clays, as defined in Chapter 4. Extensive deposits of organic matter, notably peats, though usually of fibrous rather than particulate nature, are included within the definition of soils. Natural materials which have been disturbed by man, such as clay or gravel placed in a road embankment, or colliery and quarry waste, are also included. To these may be added man-made materials such as furnace slag; pulverised fuel ash (PFA) from power stations; builders' rubble; domestic refuse.

7.1.2 Origin or Soils

Most natural soils are composed of the breakdown products of rocks which have been attacked by physical, chemical or biological weathering processes. The weathered material may have been transported and deposited elsewhere as sediments, or may remain *in situ* as residual soils.

SEDIMENTS

The principal types of sediments, and their modes of origin, are summarised as follows:

Aeolian deposits are those formed by wind action. They include desert sands and loess silts, and some brick-earths which were probably formed by wind erosion of glacial deposits.

Pyroclastic deposits are granular materials of all sizes blasted from volcanoes and deposited on land or in water. Layers of fine volcanic ash may be found interbedded with loess deposits.

Glacial deposits are materials of all grain sizes, typically unbedded and of various forms and origins deposited by glaciers and their meltwaters in a variety of forms. These included deposits loosely referred to as 'boulder clay' in Britain.

Periglacial deposits are formed in near-glacial conditions by frost heave and solifluxion. An example of these are head deposits (combe rock).

Lacustrine deposits are materials deposited in fresh or salt-water lakes, including glacial lake deposits.

Alluvial deposits are river-laid materials including fine-grained alluvium, levees, river terrace deposits and wadi gravels.

Estuarine deposits are materials deposited in the tidal regions of rivers forming mudflats, and river delta areas.

Marine deposits are varied materials occurring at all levels in the sea, such as deep-sea muds, shell banks, coral reefs and beach deposits.

RESIDUAL SOILS

These soils are formed *in situ* by deep chemical weathering of rocks, usually in a tropical climate. Some components are removed as a result, usually leaving a clay-based deposit. The main soil types are summarised in Table 3.2 of the Geological Society Report on Tropical Residual Soils (1990).

7.1.3   The Need to Identify Soils

Since there is such a wide variety of soil types, it is necessary to be able to describe and classify soils in terms which convey their characteristics clearly and concisely, and which are generally accepted and understood by geotechnical engineers. The engineering properties of a soil are governed to a large extent by its physical properties and behaviour, so a careful visual inspection together with a few simple tests can provide a valuable first appraisal of the soil.

No two sites are alike, and an identification and description system needs to be flexible enough to relate to the relevant critical features and problems which should be clearly identified. However it is essential that, for a given site, the particular details of the description system should be understood and consistently applied.

This chapter covers routine visual description of soils as carried out on samples in a soil laboratory. Some soils of an unfamiliar type may require more detailed study, such as mineralogical analysis to identify clay minerals, but this requires more specialised facilities.

7.1.4   Description in the Laboratory

Before a test on a soil sample is started, the sample should be examined and the observations recorded. This initial description, if carefully prepared, can provide valuable information to which field observations and laboratory test results can be related.

The first requirement is to record the identity of the sample in terms of its location, reference number and other details, as outlined in Section 1.4.2. Every description sheet should bear at least the identification number of the sample, the date, and the name and signature of the person describing. A suitable form for recording sample descriptions is shown in Fig. 7.1.

The sample should be described, briefly and clearly, using the recognised terminology given in the following sections. The description should include not only what is seen, but also what is felt by handling the soil, and, where appropriate, what can be smelled. The approach will depend upon which of the four main categories (Section 7.3) the soil can be assigned to. Detailed procedures for each main type are given in Sections 7.4–7.6. They are based on the principles set out in Clause 41 of BS5930, with some modifications for composite soil types. Soil samples should normally be handled and examined by the person who is responsible for describing the soil. An assistant can do the writing of the descriptions and comments dictated by the one who describes.

7.1.5   Equipment for Soil Description

When a batch of samples are to be examined collectively as a separate operation, special facilities are desirable. Description of soils should be done at a well-lit bench with a generous working area. Natural north daylight is best; direct sunlight should be avoided. Artificial light, if needed, should be from fluorescent tubes of 'colour-matching' standard. Ordinary

Sample Description

Location	BRACKNELL				Location No.	2456

Described by	P D J				Borehole No.	6

Date	3 . 1 . 78				Sheet No.	1 OF 4

Sample No.	Type*	Tube No.	Depth m from	Depth m to	% recovery	VISUAL DESCRIPTION
1	D		G.L.	0.25		Dark brown TOPSOIL
						with plant roots
2	T		0.25	1.40		Yellow brown fine silty
						SAND with a little
						fine to medium gravel
3	D		1.40	1.60		Firm blue-grey CLAY
						occasionally mottled brown
4	U100	18978	1.60	2.05	100	Top: as 3
						Base: firm to stiff blue-grey
						fissured silty CLAY
5	D		2.05	2.1		As base of 4
6	U100	16373	2.1	2.55	60	Top: as 5
						Base: fine to medium grey-green
						SAND with bands of
						brown silty sand
7W	W		2.4			Ground water
8	B		2.7	3.4		Brown SAND and GRAVEL
						with occasional lumps
						of soft grey clay

*U100, U38 etc.: Undisturbed (number indicates dia. mm)	Remarks	
D: Disturbed T : Tin C: Core (state dia. mm)		
B: Bulk W : Water P : Piston	Engineer's signature	P D Jones

Fig. 7.1 *Sample description form*

tungsten electric light bulbs cause distortion of colours, especially of the drab brown, green and blue tints which are frequently encountered in soils. The description bench should be within easy reach of a water-tap and sink, the waste from which should incorporate a silt trap.

Equipment and tools which are useful for soil description are listed below:

Tools for unpacking and unsealing samples.
Extruder for jacking out undisturbed samples.
Sample container racks.
Pocket knife.
Cobbler's knife.
Spatulas — large and small.
Tweezers.
Hand lens (× 10 magnification).
Pocket penetrometer, or hand vane.
Watch-glasses.
Small scoop.
Pestle and mortar.
Sieve, 63 μm, and receiver.
Aluminium or galvanised steel trays.
Small brush.
Beakers — glass and plastics.
Glass stirring rod.

Wash bottle containing distilled water.
Hydrochloric acid (0.2 mol/litre) in bottle with dropper.
Steel rule — 150 mm.
Sample description sheets.
Moisture content sheets.
Clip-board.
Plain white paper.
Waxpot, wax, brushes, muslin.
Aluminium foil.
Cling-film wrap.
Spare polythene bags, glass jars.
Labels, marker pen (waterproof).
Rubber tubing fitted to water-tap.
Wiping cloths.
Fingernail brush.
Waste.bin.

7.2 DEFINITIONS

SOIL   Any naturally occurring deposit forming part of the earth's crust which consists of an assemblage of discrete particles (usually mineral, sometimes with organic matter), that can be separated by gentle mechanical means, together with variable amounts of water and gas (usually air).

IDENTIFICATION   Establishment of the main characteristic features of soil, either visually or as a result of carrying out index tests.

DESCRIPTION   Verbal presentation of soil characteristics based on visual examination, feel, smell and simple hand tests.

CLASSIFICATION   Division of soils into a number of groups, on the basis of standard index tests, each group being defined by fixed limits of certain characteristics.

INDEX TESTS   Relatively simple tests, including those related to density, specific gravity, particle size and plasticity (as distinct from tests to determine mechanical properties such as strength and compressibility).

SEDIMENTS   Soils made up of mterials, derived from pre-existing rocks by weathering, which have been transported by various means and deposited elsewhere.

RESIDUAL SOILS   Soils which have not been transported but are the remains *in situ* from the weathering of rocks.

7.3 IDENTIFICATION OF SOILS

7.3.1 Main Characteristics

The following factors are taken into account when making an engineering description of soils:

(1)   Mass characteristics of the soil formation.
(2)   Material characteristics.
(3)   Geological formation, type and age.
(4)   Classification group.
(5)   Any additional relevant information.

A laboratory description is concerned mainly with item (2), the material characteristics of soils which can be described from undisturbed or disturbed samples. The mass characteristics, item (1), can be described satisfactorily only from exposures on site, although a limited amount of information can be obtained from undisturbed samples. Assignment of soil to a particular geological formation or period, item (3), requires specialist geological knowledge, and any conjecture in this respect should be avoided. Classification groups, item (4), are referred to in Section 2.4.2 for fine soils and in Section 4.4.2 for granular soils.

Under item (5), observations on the general state of the sample, and its packing and preservation, should be recorded at the time of inspection, together with any unusual features or variations in characteristics which may be noticed (see Section 1.4.3).

7.3.2 Soil Groups

Soils can be broadly divided into four main categories:

(1)   **very coarse soils** — boulders and cobbles.
(2)   **coarse soils** — gravels and sands, also called **granular** or **non-cohesive soils**.
(3)   **fine soils** — silts and clays, the latter being **cohesive** soils.
(4)   **organic soils.**

The very coarse materials comprise particles with a mean diameter larger than 60 mm (in practice, particles retained on a 63 mm sieve). Those up to 200 mm across are termed **cobbles**, and particles larger than 200 mm are termed **boulders**.

Before attempting to classify or describe a soil sample, these very coarse materials should first be removed, either physically or visually. The material passing a 63 mm sieve is that which is referred to here as **soil**, and is described according to the principles which follow.

According to the soil classification system summarised in Table 6 of BS5930, coarse soils contain less than 35% (by dry mass) of particles finer than 0.06 mm — that is, less than 35% of fine soil. The main constituents are **gravel** (particles from 60 mm down to 2 mm) and **sand** (2 mm down to 0.06 mm). Fine soils contain more than 35% of particles finer than 0.06 mm. They consist mainly of **silts** (0.06 mm down to 0.002 mm) and **clays** (smaller than 0.002 mm). Organic soils contain organic matter in significant quantity, and include peat soils, which consist mainly of decomposed plant remains.

The arbitrary dividing line of 35% fines can provide general guidance for the description of many soils, but does not take into account other factors such as the type of fines and the overall grading. A description based only on a quantitative criterion will not necessarily reflect the engineering properties and the mass behaviour of the soil *in situ*. An alternative scheme, which provides greater flexibility to allow for differences in soil composition and includes reference to secondary constituents in a consistent manner, was proposed by Norbury, Child and Spink (1986). This scheme is included in the identification and description chart (Table 7.1) in the boxes under the heading 'Composite Soil Types'. Otherwise Table 7.1 is based on Table 6 of BS5930.

Each main soil category is discussed in a separate section below.

7.3.3 Identification Chart

The identification chart given in Table 7.1, provides a key to the visual naming and description of soils. The heavy dividing lines separate soils into the three major categories of coarse, fine and organic soils, as defined in Section 7.3.2. Inorganic soils are further subdivided on the basis of their particle size or their plasticity. These characteristics are described in more detail in Section 7.4 (coarse soils) and 7.5 (fine soils). Organic soils, and other soil types, are covered in Section 7.6.

Table 7.1 may be followed for describing soils on the basis of the following characteristics:

Size of particles.
Plasticity.
Nature of particles.
Compactness, or strength.
Colour.
Structure.
Secondary constituents in mixed basic types.

The terms used in Table 7.1 are those which are generally accepted in the geotechnical engineering sense, and may sometimes differ from geological or colloquial usage.

7.4 DESCRIPTION OF COARSE (GRANULAR) SOILS

7.4.1 Particle Size

The particle size range for **gravel** (from 60 to 2 mm) and for **sand** (from 2 to 0.6 mm), together with their subdivisions into coarse, medium and fine, are indicated in Table 7.1, and are discussed in Section 4.3.2.

Gravel particles can be easily seen and handled, and their shape and surface texture observed. Sand particles are visible to the naked eye. Sand has a gritty feel, and shows very little or no cohesion when dry.

Table 7.1 IDENTIFICATION AND DESCRIPTION OF SOILS

	Basic Soil Type	Particle Size mm	Visual identification	Particle Nature & Plasticity	Composite Soil Types (Mixtures of basic soil types)			
Very Coarse Soils	BOULDERS		Only seen complete in pits or exposures.	Particle Shape:	Scale of secondary constituents with coarse and very coarse soils. Term either before or after principal constituent.			
		—200		Angular Subangular Subrounded Rounded Flat Elongated	Term before	Principal	Term after	Approximate % secondary
	COBBLES		Often difficult to recover from boreholes.					
		— 60						
Coarse Soils	GRAVELS	coarse — 20	Easily visible to naked eye: particle shape can be described: grading can be described.					
		medium			Slightly (sandy*)	With a little (sand*) or occasional (cobbles†)		< 5
		— 6		Texture:	(sandy*)	With some (sand*) or some (cobbles†)		5 to 20‡
		fine — 2						
	SANDS	coarse — .6	Visible to naked eye: very little or no cohesion when dry: grading can be described.	Rough Smooth Polished	Very (sandy*)	With much (sand*) or many (cobbles†)		20 to 40‡
		medium — .2			—	and (sand*) or and (cobbles†)		50‡
		fine — .06						
					* Fine or coarse soil types as appropriate † Very coarse soil type as appropriate ‡ Or described as fine soil depending on mass behaviour			
Fine Soils	SILTS	coarse — .02 medium — .006 fine — .002	Only coarse silt barely visible to naked eye: exhibits little plasticity and marked dilatancy: slightly granular or silky touch. Disintegrates in water: lumps dry quickly, possess cohesion but can be powdered easily between fingers.	Non-plastic or Low Plasticity	Scale of secondary constituents with fine soils. Term either before or after principal constituent.			
					Term before	Principal	Term after	Approximate % secondary
	CLAYS		Dry lumps can be broken but not powdered between the fingers: they also disintegrate under water but more slowly than silt: smooth touch: exhibits plasticity but no dilatancy: sticks to the fingers and dries slowly: shrinks appreciably on drying usually showing cracks. Intermediate and high plasticity clays show these properties to a moderate and high degree, respectively.	Intermediate Plasticity (Lean Clay)	Slightly (sandy*)	With a little (sand*)		< 35
					— (sandy*)	With some (sand*)		35 to 65
				High Plasticity (Fat Clay)	Very (sandy*)	With much (sand*)		65†
					* Coarse soil type as appropriate † Or described as coarse soil depending on mass behaviour			
Organic	ORGANIC CLAY, SILT, OR SAND.	varies	Contains substantial amounts of organic vegetable matter.		Examples of composite types			
	PEATS	varies	Predominantly plant remains usually dark brown or black in colour, often with distinctive smell: low bulk density.		(Indicating preferred order for description) Loose, brown, subangular very sandy, coarse GRAVEL with small pockets of soft grey clay. Firm, brown, thinly laminated SILT and CLAY. Dense, light brown, clayey, fine and medium SAND.			

Reproduced by permission of the Geological Society, London, from 'Soil and Rock Description' by D. R. Norbury, G. H. Child and T. W. Spink, in *Site Investigation Practice*, Engineering Geology Special Publication No. 1, 1986.

Compactness/Strength		Structure			Colour
Term	Field Test	Term	Field Identification	Interval Scales	
Loose	By inspection of voids and particle packing.	Homo-geneous	Deposit consists essentially of one type.	**Scale of Bedding Spacing**	Red Pink
Dense		Inter-stratified	Alternating layers of varying types or with bands or lenses of other materials: Interval Scale for bedding spacing may be used.	Term / Mean Spacing mm	Yellow Brown
				Very thickly bedded — over 2000	
		Hetero-geneous	A mixture of types.	Thickly bedded — 2000–600	Green Blue
Loose	Can be excavated with a spade: 50 mm wooden peg can be easily driven.	Weathered	Particles may be weakened and may show concentric layering.	Medium bedded — 600–200	White
Dense	Requires pick for excavation: 50 mm wooden peg hard to drive.			Thinly bedded — 200–60	Grey Black
				Very thinly bedded — 60–20	etc.
Slightly Cemented	Visual examination: pick removes soil in lumps which can be abraded.			Thickly laminated — 20–6	Supplemented as necessary with:
				Thinly laminated — under 6	Light Dark
Soft or Loose	Easily moulded or crushed in the fingers.	Fissured	Break into polyhedral fragments along fissures (Interval Scale for Spacing of Discontinuities may be used).		Mottled etc. and:
Firm or Dense	Can be moulded or crushed by a strong pressure in the fingers.	Intact	No fissures.	**Scale of Spacing of Other Discontinuities**	
Very soft	Exudes between fingers when squeezed in hand.	Homo-geneous	Deposit consists essentially of one type.	Term / Mean Spacing mm	Pinkish Reddish
Soft	Moulded by light finger pressure.	Inter-stratified	Alternating layers of varying types. Interval Scale for thickness of layers may be used.	Very widely spaced — over 2000	Yellowish
Firm	Can be moulded by strong finger pressure.			Widely spaced — 2000–600	Brownish
Stiff	Cannot be moulded by fingers. Can be indented by thumb.	Weathered	Usually has crumb or columnar structure.	Medium spaced — 600–200	etc.
Very stiff	Can be indented by thumb nail.			Closely spaced — 200–60	
Firm	Fibres already compressed together.	Fibrous	Plant remains recognisable and retain some strength.	Very closely spaced — 60–20	
Spongy	Very compressible and open structure.			Extremely closely spaced — under 20	
Plastic	Can be moulded in hand, and smears fingers.	Amorphous	Recognisable plant remains absent.		

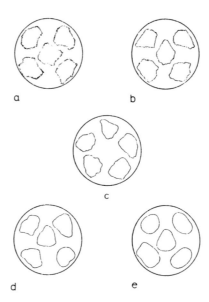

Fig. 7.2   Shape of particles: (a) angular, (b) sub-angular, (c) sub-rounded, (d) rounded, (e) well rounded

The main grading characteristics (that is whether in the engineering sense the soil is uniformly graded, well graded or poorly (gap) graded), as described in section 4.4.2, can usually be estimated by eye, and should be noted as part of the visual description. (These terms relating to grading have different meanings for geologists.)

7.4.2   Nature of Particles

The shape of individual particles should be included in the description of gravel and coarse sand particles. The degree of angularity or roundness is indicated by using the terms *angular, sub-angular, sub-rounded, rounded* or *well rounded*, as depicted in Fig. 7.2. The general form, if other than approximately equidimensional, may be described as *flat, elongated, flat and elongated* or *irregular*.

An indication of the surface texture should be included, the usual terms being *rough, smooth* or *polished*.

The type of rock which forms the soil particles is not normally identified except by a geologist. However, if the particles are of a type which can be recognised easily, such as fragments of chalk or sandstone, this may be included in the description.

7.4.3   Composite Soils

Many soils consist of particle sizes which span two or more basic soil types. The composition of mixtures of particles of sand and gravel size is indicated by using the following terms in the description. The dominant constituent is printed in bold type.

Term	*Approximate constituents by mass*
Slightly sandy **gravel**	up to 5% sand
Sandy **gravel**	5–20% sand
Very sandy **gravel**	20–40% sand
Gravel/sand	about equal proportions of gravel and sand
Very gravelly **sand**	40–20% gravel
Gravelly **sand**	20–5% gravel
Slightly gravelly **sand**	up to 5% gravel

Coarse soils containing fine material (silt or clay, or both) as a secondary constituent are described by the terms *slightly clayey, clayey* or *very clayey* (or similarly with *silty*), as indicated in Table 7.1. If the soil has no cohesive strength when dry, the fines are non-plastic and contain no clay. Occasionally a deposit may have a secondary constituent which is distinctive because of its composition, in which case this can be mentioned. For example, if a sand contained small shiny plates of mica, it would be described as *micaceous*.

Secondary constituents may include organic matter, which is dealt with in Section 7.6.1.

7.4.4 Compactness

The state of compactness (relative density) of granular soils can be assessed only on the basis of *in situ* tests and experience, so this factor is not included in a laboratory description. However, grains may be bonded together, in which case the soil would be described as *cemented*. If lumps can be abraded easily, the soil is *weakly* or *slightly cemented*.

In some sands a coating of fine particles on the grains causes them to stick together and appear to be cemented. On immersion in water the grains rapidly fall apart, and this type of soil should not be described as cemented.

7.4.5 Colour

The colours listed in the right-hand column of Table 7.1, and the qualifying adjectives listed below them, are sufficient for general description purposes. The 'Munsell Soil Colour Chart' may be used for a more detailed classification of colours. Alternatively, reference samples or a made-up colour chart can be used as an aid to consistency in description. The colour of granular soils should be based on overall appearance, not individual particles.

7.4.6 Structure

Typical terms used to indicate the structure of granular soil are given in Table 7.1. These features cannot usually be observed in disturbed samples.

Bedding and laminations are referred to in Section 7.5.8, in connection with fine soils, but similar characteristics may be present in granular soils.

7.4.7 Summary of Terms

A granular soil is usually described in the following sequence:

Compactness, or whether cemented.
Colour.
Structure.
Secondary constituent.
Particle nature.
Grading.
MAIN SOIL TYPE.
Minor constituents.
Other remarks.

The main soil type is written in capital letters, or underlined.

The terms for description of coarse soils are summarised in Table 7.2. This includes in column 8 some typical descriptions relating to minor constituents which might be present.

Three examples of soil descriptions are given at the foot of Table 7.2.

Table 7.2. TERMS AND DESCRIPTION FOR COARSE SOILS

Relative density (1)	Colour (2)	Lithology (3)	Secondary constituents (4)	Particle nature (5)	Grading (6)	Principal soil type(s) (7)	Minor constituents (8)
	light			angular	fine	**boulders**	containing
Very loose	dark	bedded	clayey	subangular		**cobbles**	many crystals
Loose	mottled	laminated	silty	subrounded	medium	**gravel**	
Medium-dense			sandy	rounded		**sand**	with
Dense	pinkish		gravelly	flat	coarse	**(silt)**	scattered
Very dense	reddish			elongated			gravel
	yellowish	very thickly		platey	well-		occasional
	brownish	thickly		spherical	graded		shells
	olive	medium		tubular			pockets of peat
	greenish	thinly			poorly		bands of clay
	bluish	very thinly			graded		lenses of clay
	greyish		peaty				some rootlets
			micaceous	rough	gap-		partings of silt,
	pink			smooth	graded		etc.
	red			polished			
	yellow				uniform		
Weakly	brown						
cemented	olive		very				
	green						
	white		slightly				
	grey						
	black						

Examples:

(a) Loose light grey clayey and peaty fine **sand** with occasional shells
    (1) (2)   (4)   (6) (7)   (8)

(b) Medium-dense dark brownish-yellow subangular well-graded **gravel**
    (1)   (2)   (5)   (6) (7)

(c) Dense mottled yellow and green medium and coarse **sand** and subrounded fine gravel
    (1)   (2)   (6) (7)   (5) (6) (7)

7.5 DESCRIPTION OF FINE SOILS

7.5.1 Size of Particles

Fine soils consist of silts and clays having particles smaller than 0.06 mm. Material known as **silt** consists of particles in the size range 0.06 mm to 0.002 mm, and is subdivided into coarse, medium and fine, as indicated in Table 7.1. Particles of coarse silt can just be discerned by the naked eye, but silt particles generally can be seen under a good hand lens ($\times 10$ magnification).

Particles forming **clay** consist of complex minerals which are mostly flat and plate-like or elongated, and of a size less than 0.002 mm. However, if particles of this size are not true clay minerals, but consist of very fine particles (e.g. rock flour), the material would behave not as a clay but as a silt.

When describing fine soils, it is first necessary to decide whether the soil is effectively a silt or a clay.

7.5.2 Identification of Silt

Silt has little plasticity; dries quite quickly on the hands and can be dusted off. Silt has a smooth or silky touch, but grittiness can be detected between the teeth. Lumps dry quickly, and when dry have a granular appearance and can be powdered easily. A small lump placed in water disintegrates quickly into its individual particles, which will take several minutes to settle. When loose in a sample tube, silt may tend to slump and the water will tend to drain out of the soil.

Silt exhibits *dilatancy* when subject to the following simple hand test. Moisten a pat of soil so that it is soft but not sticky, and place it in the open palm of one hand. Shake the hand, tapping it against the other hand several times. The appearance of a shiny film of water on the surface of the pat indicates dilatancy. Squeeze the pat by pressing with the fingers, and the surface will dull again as the pat stiffens and finally crumbles. These reactions indicate the presence of predominantly silt-sized material or very fine sand, provided that the amount of moisture is not excessive. It is difficult to roll moist silt into threads, which break and crumble easily.

7.5.3 Identification of Clay

The most significant properties of clay are its cohesion and plasticity. If when pressed together in the hands at a suitable moisture content the particles stick together in a relatively firm mass, the soil shows cohesion. If it can be deformed without rupture (i.e. without losing its cohesion), it shows plasticity. Clay dries more slowly than silt and sticks to the fingers; it cannot be brushed off dry. It has a smooth feel, and shows a greasy appearance when cut with a blade. Softer consistencies behave rather like butter, and harder consistencies like cheese. Dry lumps can be broken, sometimes with difficulty, between the fingers, but cannot be powdered. A lump placed in water remains intact.

Clay does not exhibit dilatancy. Lumps shrink appreciably on drying, and show cracks which are the more pronounced the higher the plasticity of the clay. At a moisture content within the plastic range, clay can easily be rolled into threads 3 mm diameter (as in the plastic limit test) which for a time can support their own weight. Threads of high-plasticity clay are quite tough; those of low-plasticity clay are softer and more crumbly.

7.5.4 Plasticity

The plasticity of fine-grained soils, and their subdivision in terms of the liquid limit into non-plastic soil and several ranges of plasticity, are discussed in Section 2.4.2. It may not be possible to assign a clay soil to a particular plasticity range by inspection alone, but it is usually possible to assess whether it is of low plasticity (L) or in the upper plasticity range (U), the latter incorporating groups I, H, V and E of Fig. 2.6.

7.5.5 Composite Soils

Fine soils rarely consist of particles entirely of silt size or clay size, but usually contain a mixture of both. The nature of clay minerals is such that a relatively small proportion of clay in a clay–silt mixture is sufficient to impart to the material the properties of a clay, in which case is is described simply as **clay**. For instance, a medium-plasticity (lean) clay may contain only 10–20% of clay-sized particles, and even a high-plasticity (fat) clay may contain not more than 50–70% clay. Examples are given in Section 4.4.3.

The description of soils consisting of mixtures of fine-grained and coarse-grained materials depends upon which constituent predominates. For example, **sand** may be described as slightly clayey, clayey or very clayey, depending on the approximate percentage of clay present. If the clay predominates, the soil would be described as sandy **clay**, or **clay** with a little (or some) sand. The terms used for various proportions of secondary constituents are summarised in Table 7.1, in both the Coarse Soils and Fine Soils sections.

Secondary constituents may include organic matter, which is dealt with in Section 7.6.1.

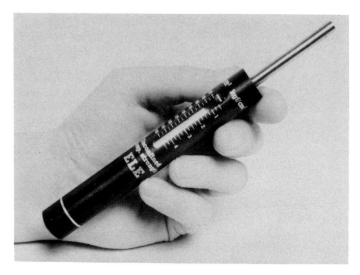

Fig. 7.3   Pocket penetrometer

7.5.6   Strength

The following decriptions provide an indication of the shear strength of undisturbed clay soils.

Very soft: Exudes between fingers when squeezed in the fist. A finger can be pushed 25 mm or so into it quite easily. Can be cut with a knife from a sample tube very easily. Tends to slump on extrusion.
    Soft: Easily moulded in fingers. Finger can be pushed 10 mm or so into it. May tend to slump on extrusion.
    Firm: Can be moulded by strong pressure in fingers. Thumb will make an impression.
    Stiff: Cannot be moulded in fingers and only slight impression possible with thumb. Cutting with a knife requires some effort.
    Very stiff: Brittle or very tough. Difficult to cut with a knife.
    (*Hard*): No impression possible with thumb.

Where a sample appears to be borderline, it can be described as soft to firm, or firm to stiff. Some soft or firm clays when worked in the hand will become significantly softer; the true consistency is the original one, provided the material has not already been disturbed by sampling. Very soft clays are prone to this. For a preliminary assessment the consistency can be assessed by using a pocket penetrometer (Fig. 7.3) or a pocket vane shearmeter (Fig. 7.4), and referring to Table 7.4.
    Silts are described as:

Soft or loose: Easily moulded in fingers, slumps easily and moisture drains out.
Firm or dense: Can be moulded by strong pressure in the fingers. Does not slump.

7.5.7   Colour

See Section 7.4.5.
    In fine soils a mottled pattern probably indicates weathering. A dark-brown, dark-grey or black colour may indicate the presence of organic matter, which would be confirmed by smell. Changes in colour on partial drying can reveal structural (fabric) details which are not obvious at the natural moisture content. Some colours are unstable when exposed to light and air for some time, so colours should be described only from a freshly exposed face, and any rapid change of colour should be recorded.

Fig. 7.4   Pocket vane shear meter

7.5.8 Structure

A deposit consisting essentially of one soil type is described as *homogeneous*. If laminations or fissures are not present and not reported in the description, it is understood that the soil is *intact*.

The presence of layers within a clay deposit is due to *bedding*. If the bedding spacing is more than 20 mm, the clay is described as *bedded* (thinly, medium or thickly). The scale of bedding spacing is shown in Table 7.1. Laminations in clay may be separated by very thin layers of silt or sand of up to 0.5 mm thickness, known as *partings* or *dustings*. If the laminations are all very regular and grade from sand through silt to clay, each representing a seasonal deposition cycle, the clay is described as *varved*, but this term is normally reserved for glacial lake deposits.

If a clay contains discontinuities not necessarily related to bedding surfaces, it is *fissured*. It will tend to break up into irregular blocks along fissure surfaces, which may be described as polished, striated or dull. The nature of any infilling should be reported. Very close fissures may cause the clay to break down to crumbs when worked in the hand, or a high silt content could have the same effect; this type of clay is *friable*. The scale of fissure spacing is given in Table 7.1.

7.5.9 Summary of Terms

A fine-grained soil is usually described in the following sequence:

Strength.
Colour.
Structure.
Discontinuities.
Secondary constituents.
MAIN SOIL TYPE.
Minor constituents.
Other remarks.

The terms used for description of fine-grained soils are summarised in Table 7.3. Three examples of descriptions are given at the foot of the table.

Table 7.3. TERMS AND DESCRIPTIONS FOR FINE SOILS

Strength (1)	Colour (2)	Lithology (3)	Discontinuities (4)	Secondary Constituents (5)	Principal soil type(s) (6)	Minor constituents (7)
Very soft		bedded	fissured			
Soft		laminated		sandy	clay	
Firm	as Table 7.2		very widely	gravelly		as Table 7.2
Stiff		very thickly	widely	silty	(silt)	
Very stiff		thickly	medium	peaty		
Hard		medium	closely	chalky		
		thinly	very closely			
		very thinly	extremely closely			

(a) Very stiff grey thickly bedded silty **clay** with dustings of silt
     /   /   /   / /
   (1)   (2)   (3)   (5) (6)   (7)

(b) Firm olive-brown thinly laminated and closely fissured **clay**
      /   /   /   /
   (1)   (2)   (3)   (4)   (6)

(c) Soft reddish-brown **clay** thickly interbedded with red fine **sand**
    /   / /   /   / /
   (1)   (2)   (6)   (3)   (2) (6)   (7)

Table 7.4. CONSISTENCY OF CLAYS

Consistency of clay	Undrained shear strength (kN/m^2)
Very soft	< 20
Soft	20–40
Soft to firm	40–50
Firm	50–75
Firm to stiff	75–100
Stiff	100–150
Very stiff	> 150

7.6 DESCRIPTION OF OTHER SOIL TYPES

7.6.1 Organic Soils

Organic soils are predominantly plant remains, usually dark brown, dark grey, black or blue-black in colour, often with a distinctive smell and low bulk density. Mineral soils can contain organic matter in finely disseminated form, which produces a similar dark colouring, often oxidising to brown on exposure to air.

Topsoils contain roots or rootlets and living vegetation, but plant rootlets can penetrate much deeper. Wherever rootlets or other organic remains are foumd their attitude and frequency should be reported; for instance, whether as vertical rootlets or as horizontal fibres. These features can have an important effect on drainage characteristics. The diameter of rootlets, and whether they are closed, open or infilled, is also significant.

Organic matter which is partly decomposed may be described as *peat* (see Section 7.6.2). This may be *fibrous*, when the structure of individual leaves, roots, twigs and branches is seen: or *amorphous*, where no structure is visible and the sample looks like a dark-brown or black silt. If it is firm, the material has been compressed and it is not easy to indent it with the thumb. If it is *spongy*, identification is easy and the sample may exude brown water.

Amorphous peat is often plastic. Peat often occurs in conjunction with clay and so may be *clayey*. *Coal* and *lignite* should be described as such and not simply as organic matter, since they are relatively strong and incompressible materials compared with peat.

Organic soils are difficult to form into a thread, which is very weak. Lumps of organic soil crumble easily.

7.6.2 Characteristics of Peat

COMPOSITION

The composition of peats differs significantly from that of inorganic soils, in ways which have an important influence on their engineering properties. Instead of being made up of entirely individual mineral particles, they consist mainly of plant remains in the form of stems, roots, leaf cells and the like which may be in various stages of decomposition by humification. Peats range from coarse fibrous material showing little or no humification, to amorphous highly humified black peat with a granular appearance. The degree of humification according to the 'von Post' system can be assessed by the simple test described below (von Post, 1924; Landva and Pheeney, 1980).

CLASSIFICATION BY THE VON POST TEST

Take a handful of peat and squeeze it in the palm of the hand. Examine the liquid or other material extruded between the fingers and the residue remaining in the hand. By comparing the observations with the description given in Table 7.5 the degree of humification and decomposition on the von Post scale (H_1 to H_{10}) can be assessed.

ENGINEERING PROPERTIES

The particle density of peat ranges from about 1.4 to 2.5, depending upon the amount of mineral matter present. In highly organic peats most of the material is lost when subjected to the loss on ignition test (Section 5.10.3). The mean particle density can be obtained from the loss on ignition by using the following equation:

$$\rho_s = \frac{3.78}{\left(1.3 \times \dfrac{N}{100}\right) + 1.4} \; \text{Mg/m}^3$$

where $N\%$ is the ignition loss. This is based on assumed particle densities of 2.7 for mineral particles and 1.4 for the organic matter (Skempton and Petley, 1970).

Unconsolidated peats typically contain 75% to 95% water by volume. Voids ratios range from about 5–20, and moisture contents of several hundreds percent (sometimes over 1000%) are not uncommon. The voids also include gas which is generated during humification.

7.6.3 Man-made Soils

Where these are basically natural soils which have been reworked by man, the above descriptions apply. Where manufactured material is involved, the terms above should only be used if they will not cause confusion with a natural soil. The fact that the material is unnatural can be emphasised in the description: for example, 'cobble-sized angular **slag**', or alternatively '**slag** consisting of cobble-sized angular pieces'.

Pulverised fuel ash from power stations, known as 'fly ash', is fairly easily recognisable. It is usually light grey in colour, consisting mainly of silt-size spherical particles, and often contains occasional fragments of unburnt coal. The specific gravity of particles is generally low, owing to entrapped air, and some particles ('floaters') may even be less dense than water.

Table 7.5. DEGREE OF HUMIFICATION OF PEAT (von POST SYSTEM)

Degree of humification (von Post scale)	Decomposition	Description	Material extruded between fingers	Residue in hand
H₁	None	Entirely unconverted mud-free peat	Clear, colourless water	
H₂	Insignificant	Almost entirely unconverted mud-free peat	Yellowish water	
H₃	Very slight	Very slightly converted or very slightly muddy peat	Brown, muddy water; no peat	Not pasty
H₄	Slight	Slightly converted or somewhat muddy peat	Dark brown, muddy water; no peat	Somewhat pasty
H₅	Moderate	Fairly converted or rather muddy peat, plant structure still quite evident	Muddy water and some peat	Thick, pasty
H₆	Moderately strong	Fairly converted or rather muddy peat, plant structure indistinct but more obvious after squeezing	About one third of peat squeezed out; water dark brown	Very thick
H₇	Strong	Fairly well converted or markedly muddy peat; plant structure still discernible	About one half of peat squeezed out, consistency like porridge; any water is very dark brown	
H₈	Very strong	Well converted or very muddy peat, very indistinct plant structure	About two thirds of peat squeezed out, also some pasty water	Plant roots and fibres which resist decomposition
H₉	Nearly complete	Almost completely converted or mud-like peat, plant structure almost not recognizable	Nearly all the peat squeezed out as a fairly uniform paste	
H₁₀	Complete	Completely converted or entirely muddy peat, no plant structure visible	All the peat passes between the fingers; no free water visible	

Based on Landva and Pheeney (1980). Table A1.

7.6.4 Composite Soils

Soils which have much secondary material but still retain the basic appearance and behaviour of the principal soil type have been dealt with in the preceding sections.

Some soils, however, particularly *tills* (soils of glacial origin) contain a wide range of particle sizes. The type of glacial till known as *boulder clay* contains particles of all sizes, usually well-graded, ranging from cobbles or gravel down to clay. Meaningful description of these soils requires practice and experience. The principles outlined above still apply but it may be necessary to reappraise the initial description after carrying out field or laboratory tests.

7.6.5 Tropical Soils

The engineering description of soils found in tropical zones follows the same general principles as outlined above, but there are some additional points to be observed, both in description and in the manner in which tests are carried out. Some tropical soils, known as *residual* soils, have been formed by the decomposition of rock *in situ* by chemical decay in humid tropic conditions, and may retain signs of their original structure or fabric. Two of the more important types are laterites and bauxites, but both types are often loosely referred to as *lateritic soils*. *Laterites* are rich in iron oxide, and are characteristically red-brown in colour. *Bauxites* are rich in aluminium, and are usually dirty-white. The clay particles of these soils tend to aggregate into silt-size flocks, and might be difficult to disperse unless a special dispersant is used (see Section 4.5.3).

The *black cotton* soils form another important type which is found extensively in tropical areas. These are usually clays of high plasticity, black or dark brown in colour, and can retain moisture through dry seasons, which is why they are of value for growing crops. The clay fraction contains a large proportion of high-activity minerals of the montmorillonite group, which is responsible for the pronounced shrinkage and swelling capability of these soils. They belong to a category known as expansive clays.

The *sabkha* soils found in the Middle East were formed under arid subtropical conditions. (The Arabic term 'sabkha' refers to the salt-encrusted coastal and inland flats where these soils occur.) Evaporation of groundwater (often sea-water) has left concentrations of salts, mainly chlorides and gypsum, in the pore spaces. The salts contain water of crystallisation which can easily be driven off if dried at too high a temperature, leading to erroneous moisture content and porosity values. (See Section 2.5.2 for special precautions when oven drying.)

Some tropical soils break down further the more they are handled. The results of laboratory tests can vary according to the amount of working, such as by the use of pestle and mortar, sieving or the length of time for which they are worked. Many of these soils are affected by oven drying, or even air drying, and they should never be dried out completely.

If may be necessary, therefore, to carry out a series of comparative tests to assess the efects of sample preparation and methods of testing on the end results, and, if necessary, to adopt modified procedures. Description and identification of tropical residual soils is discussed in detail in the Geological Society Report on Tropical Residual Soils (1990). The procedures laid down in British Standards may not always be suitable, as they stand, for these types of soil.

BIBLIOGRAPHY

BS5930:1981, 'Code of practice for site investigation', British Standards Institution, London

Bridges, E. M. (1970). *World Soils*. Cambridge University Press

Dumbleton, M. J. (1968). 'The classification and description of soils for engineering purposes: a suggested revision of the British system'. TRRL Report LR 182. Transport and Road Research Laboratory, Crowthorne, Berks.

Fookes, P. G. and Higginbottom, I. E. (1975). 'The classification and description of near-shore carbonate sediments for engineering purposes'. *Géotechnique*, Vol. 25, No. 2

Geological Society Engineering Group Working Party Report: Tropical Residual Soils (1990), QJEG, Vol. 23, No. 1

Landva, A. O. and Pheeney, P. E. (1980). 'Peat, Fabric and Structure'. *Canadian Geotech. J.*, Vol. 17, No. 3. pp 416–435

Manual of Applied Geology for Engineers (1976). Institution of Civil Engineers, London

McFarlane, M. J. (1976). *Laterite and Landscape*. Academic Press, London

Munsell Soil Colour Chart. Reference 6-A. Tintometer Limited, Waterloo Road, Salisbury, Wilts.

Norbury, D. R., Child, G. H. and Spink, T. W. (1986), 'A critical review of Section 8 (BS5930) — Soil and Rock Description. Site Investigation Practice: Assessing BS5930', Geological Society Engineering Geology Special Publication No. 2

von Post, L. (1924). 'Das genetische System der organogenen Bildungen Schwedens'. Int. Comm. Soil Sci., IV Commission

Skempton, A. W. and Petley, J. (1970). 'Ignition loss and other properties of peats and clays from Avonmouth, King's Lynn and Cranberry Moss', *Geotechnique*, Vol. 20, No. 4

Symposium on Airfield Construction on Overseas Soils. Proc. Inst. Civ. Eng. Vol. 8, November 1957:

    Clare, K. E. (Paper No. 6243). Part 1, 'The formation, classification, and characteristics of tropical soils'. Part 2, Tropical black clays'.

    Tomlinson, M. J. (Paper No. 6239). Part 3, 'Saline calcareous soils'. Part 4. 'Alluvial sands and silts'.

    Nixon, I. K. and Skipp, B. O. (Paper No. 6258) Part 5, 'Laterite'. Part 6, 'Tropical red clays'.

    Discussion on the above. *Proc. Inst. Civ. Eng.*, Vol. 10, May 1958

APPENDIX: Units and symbols

A.1   METRIC (SI) UNITS

A.1.1   Units in Use

The units and symbols listed in Table A.1 are used or referred to in this volume. This list is taken from the selection of SI Units for Soil Mechanics and Foundation Engineering which is generally accepted within the industry.

SI is the accepted abbreviation for Système International d'Unités (International System of Units), the modern form of the metric system finally agreed at an international conference in 1960.

A.1.2   Multiplying Prefixes

Multiples and submultiples of SI units are formed by placing prefixes in front of the unit symbol. The most commonly used prefixes are given in Table A.2. Recommended prefixes are those representing 10 raised to a power which is a multiple of ± 3. Use of those marked with an asterisk in Table A.2 should be avoided unless the recommended prefixes are inconvenient.

A.1.3   Definitions and Notes

Length

The metre (m) is defined as the length of the path travelled by light in vacuum during a specified time interval (approximately $\frac{1}{3} \times 10^{-9}$ second).

The former prototype metre (a bar of invar metal) is still kept in the custody of the Bureau International des Poids et Mesures (BIPM) at their laboratories at Sèvres, near Paris.

The millimetre (mm) is used in most laboratory measurements:

$$1\,\mathrm{mm} = 10^{-3}\,\mathrm{m}$$

The use of the centimetre (cm) should be avoided.

The micrometre (μm) is often called a micron, but the former is technically correct.

Area

The square millimetre (mm^2) is the unit of area used when areas are calculated from measurements in millimetres.

Volume

The cubic millimetre (mm^3) is the unit of volume derived from volume calculations using linear measurements in millimetres.

Table A.1.

Quantity	Unit	Unit symbol	Application	Conversions
Length	millimetre	mm	Sample measurements, particle size	$1\,\mu m = 10^{-6}\,m$
	micrometre	μm	Sieve aperture and particle size	$= 10^{-3}\,mm$
Area	square millimetre	mm^2	Area of section	
Volume	cubic metre	m^3	Earthworks	
	cubic centimetre	cm^3	Sample volume	$1\,m^3 = 10^6\,cm^3$
	millilitre	ml	Fluid measure	
	cubic millimetre	mm^3	Sample volume as calculated	
Mass	gram	g	Accurate weighings	
	kilogram	kg	Bulk sample and approximate weights	$1\,kg = 1000\,g$
	megagram	Mg	Alternatively known as tonne	$1\,Mg = 1000\,kg$ $= 10^6\,g$
Density	Megagram per cubic metre	Mg/m^3	Sample density and dry density	Density of water $= 1\,Mg/m^3$ $= 1\,g/cm^3$
Temperature	degree Celsius	°C	Laboratory and bath temperatures	Celsius is preferred name for Centigrade**
Time	second	s	Timing of laboratory tests	$1\,minute = 60\,s$
Force	newton	N	Load ring calibrations Small-magnitude forces	$1\,kgf = 9.807\,N$ $1\,N = 101.97$ gramf
	kilonewton	kN	Forces of intermediate magnitude	$1\,kN = 1000\,N$ $= approx.\ 0.1$ tonne f
Pressure and stress	newton per square metre $=$ pascal	N/m^2 Pa	Very low pressures and stresses	$1\,g/cm^2$ $= 98.97\,N/m^2$ $= 98.07\,Pa$
	kilonewton per square metre $=$ kilopascal	kN/m^2 kPa	Pressure gauges Compressive strength and shear strength of soils	$1\,kgf/cm^2$ $= 98.07\,kN/m^2$ $1\,bar = 100\,kN/m^2$
Pressure (vacuum)	torr*	torr	Very low pressure under vacuum	$1\,torr = 133.3\,Pa$ $= 133.3\,N/m^2$ $= 1\,mmHg$
Dynamic viscosity	Millipascal second	mPas	Viscosity of water	$1\,mPas = 1\,cP$ (centipoise)
	$=$ millinewton second per square metre	mNs/m^2		
Amount of substance	mole	mol	Concentration of solutions	(Replaces N or M)

* A non-SI metric unit.
** See Fig. A.1 for temperature conversion chart.
Metric sieve aperture sizes used for soil testing are listed in Table 4.5.

Table A.2

Prefix symbol	Name	Multiplying factor
G	giga	$1\,000\,000\,000 = 10^9$
M	mega	$1\,000\,000 = 10^6$
k	kilo	$1\,000 = 10^3$
h	*hecto	$100 = 10^2$
da	*deca	10
d	*deci	$10^{-1} = 0.1$
c	*centi	$10^{-2} = 0{,}01$
m	milli	$10^{-3} = 0.001$
μ	micro	$10^{-6} = 0.000\,001$
n	nano	$10^{-9} = 0.000\,000\,001$

It usually simplifies calculations if calculated volumes in mm^3 are divided by 1000 at the outset to give cubic centimetres (cm^3):

$$1\,cm^3 = 1000\,mm^3$$

(Although the use of the centimetre is deprecated in SI, the cm^3 is compatible with the recommendation that multiple units should be related by factors of 10^3.)

The litre is recognised as a special name for one cubic decimetre, but should *not* be used to express scientific or high accuracy measurements of volume.

For most practical purposes,

$$1\,litre = 1\,dm^3$$
$$= 1000\,cm^3$$

In precise scientific work

$$1\,litre = 1000.028\,cm^3$$

The millilitre (ml) is one-thousandth of a litre:

$$1\,ml = 1\,cm^3$$

for practical purposes.

Mass

The kilogram (kg) is equal to the mass of the international platinum prototype kept by BIPM at Sèvres. It is the only basic quantity to be a multiple unit:

$$1\,kg = 1000\,g\ (grams)$$

There is no SI unit of 'weight'. When 'weight' is used to mean the force due to gravity acting on a mass, the mass (kg) must be multiplied by $g(9.807\,m/s^2)$ to give the force in newtons (N).

Density

The megagram per cubic metre (Mg/m^3) is the density unit adopted for soil mechanics. It is 1000 times larger than the kilogram per cubic metre, the basic SI unit, and is equal to one gram per cubic centimetre:

$$1 \, Mg/m^3 = 1 \, g/cm^3$$
$$= 1000 \, kg/m^3$$

The density of soil particles (particle density) is expressed in Mg/m^3, which is numerically equal to the specific gravity (now obsolete). Using Mg/m^3, the density of water is unity.

Time

The second (s) is defined in terms of the period of radiation of the caesium-133 atom under specified conditions.

Minutes, hours, days and years are used where appropriate. It is useful to note that

$$1 \, day = 1440 \, minutes$$

$$1 \, year = 31.56 \times 10^6 \, seconds$$

Sedimentation test times are usually measured in minutes.

Force

The newton (N) is that force which, applied to a mass of 1 kilogram, gives it an acceleration of 1 metre per second per second:

$$1 \, N = 1 \, kg \, m/s^2$$

The kilonewton (kN) is the force unit most used in soil mechanics:

$$1 \, kN = 1000 \, N$$

$$= approximately \, 0.1 \, tonne \, f \, or \, 0.1 \, ton \, f$$

Pressure and Stress

The pascal (Pa) is the pressure produced by a force of 1 newton applied, uniformly distributed, over an area of 1 square metre.

The pascal has been introduced as the pressure and stress unit, and is exactly equal to the newton per square metre:

$$1 \, Pa = 1 \, N/m^2$$

In dealing with soils the usual unit of pressure is kilonewton per square metre (kN/m^2), or kilopascal:

$$1 \, kN/m^2 = 1 \, k \, Pa = 1000 \, N/m^2$$

The bar is not an SI unit but is sometimes encountered in fluid pressure:

$$1 \, bar = 100 \, kN/m^2 = 100 \, k \, Pa$$

$$= 1000 \, mb \, (millibars)$$

Standard atmospheric pressure

$$1 \, atm = 101.325 \, kPa$$

$$= 1013.25 \, mb$$

(about atmospheric pressure).

Amount of Substance

The mole (mol) is the amount of substance of a system which contains as many elementary entities as there are atoms in 0.012 kg of carbon 12.

For practical purposes the 'amount of substance' is equal to the molecular mass in grams.

Standard Gravity

The international standard acceleration due to the Earth's gravity is accepted as

$$g = 9.80665 \text{ m/s}^2$$

although it varies slightly from place to place. For practical purposes $g = 9.81 \text{ m/s}^2$, the conventional reference value used as a common basis for measurements made on the Earth.

A.1.4   Conversion Factors, Imperial and SI Units

Table A.3.   CONVERSION FACTORS, IMPERIAL AND SI UNITS

	Imperial to SI				SI to imperial
Length	0.3048	m	:	foot (ft)	3.281
	25.4	mm	:	inch (in)	0.03937
Area	0.09290	m²	:	square foot	10.76
	645.2	mm²	:	square inch	0.001550
Volume	0.02832	m³	:	cubit foot	35.31
	4.546	litre	:	gallon (UK)	0.2200
	3.785	litre	:	gallon (USA)	0.2642
	28.32	litre	:	cubic foot	0.03531
	16.39	ml	:	cubic inch	0.06102
	16387	mm³	:	cubic inch	
Mass	1.016	Mg (tonne)	:	ton	0.9842
	0.4536	kg	:	pound (lb)	2.205
	453.6	g	:	pound	
	28.35	g	:	ounce (oz)	0.03527
Density	0.01602	Mg/m³ (g/cm³)	:	pound per cubic foot	62.43
Force	9.964	kN	:	ton force	0.1004
	4.448	N	:	pound force	0.2248
Pressure	0.04788	kN/m² (kPa)	:	lb f/sq ft	20.89
	6.895	kN/m²	:	lb f/sq in	0.1450
	47.88	N/m² (Pa)	:	lb f/sq ft	0.02089

Examples: Imperial to SI: to convert feet to m, multiply by 0.3048
SI to imperial: to convert m to feet, multiply by 3.281

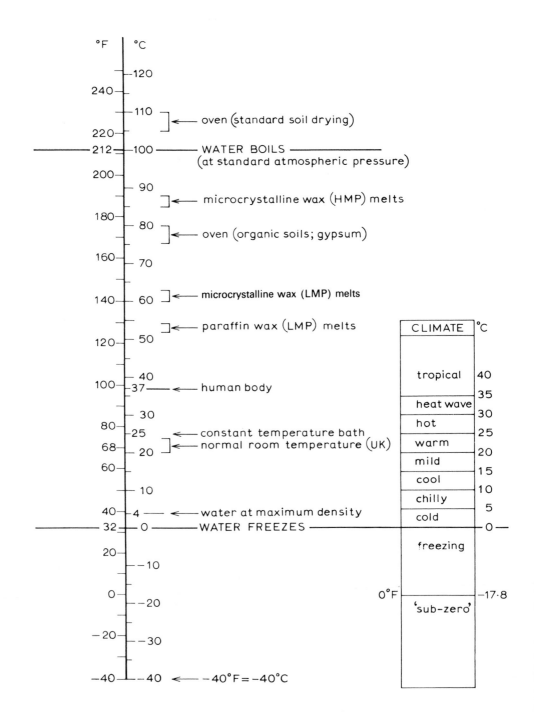

Fig. A.1 Temperature conversion chart

A.2   SYMBOLS

A.2.1   Symbols for Soil and Water Properties

Table A.4.

Measured quantity	Symbol	Unit of measurement
Moisture content	w	%
Liquid limit	w_L	%
Plastic limit	w_P	%
Plastic index	I_P	%
Non-plastic	NP	–
Relative consistency	C_r	–
Liquidity index	I_L	–
Shrinkage limit	w_S	%
Linear shrinkage	L_S	%
Shrinkage ratio	R	–
Unit weight	γ	kN/m^3
Bulk (mass) density	ρ	Mg/m^3
Dry density	ρ_D	Mg/m^3
Saturated density	ρ_s	Mg/m^3
Submerged density	ρ'	Mg/m^3
Minimum dry density	ρ_{Dmin}	Mg/m^3
Maximum dry density	ρ_{Dmax}	Mg/m^3
Density of water	ρ_w	Mg/m^3
Optimum moisture content	OMC	%
Density of soil particles	ρ_S	Mg/m^3
Density of liquid	ρ_L	g/ml or Mg/m^3
Degree of saturation	S	%
Voids ratio	e	–
Porosity	n	–
Percentage air voids	V_a	%
Particle size	D	μm or mm
Percentage smaller than D	P	%
Effective size	D_{10}	mm
'60% finer than' size	D_{60}	mm
Uniformity coefficient	U	–
Dynamic viscosity of water	η	mPas

A.2.2   The Greek Alphabet

Table A.5.

Upper case	Lower case	Name	Upper case	Lower case	Name
A	α	alpha	N	ν	nu
B	β	beta	Ξ	ξ	xi
Γ	γ	gamma	O	o	omicron
Δ	δ	delta	Π	π	pi
E	ε	epsilon	P	ρ	rho
Z	ζ	zeta	Σ	σ	sigma
H	η	eta	T	τ	tau
Θ	θ	theta	Y	υ	upsilon
I	i	iota	Φ	φ	phi
K	κ	kappa	X	χ	chi
Λ	λ	lambda	Ψ	ψ	psi
M	μ	mu	Ω	ω	omega

A.3   USEFUL DATA

Time
1 day = 1440 minutes
1 week = 10 080 minutes
1 year = 0.526×10^6 minutes = 31.56×10^6 seconds

Density

		g/cm^3
Pure water	15°C	0.999 09
(see Table 4.10)	20°C	0.988 20
	25°C	0.997 04
Sea water	20°C	1.04
Waxes	20°C	
Paraffin	(m.p. 52°–54°C	0.912
Microcrystaline	(m.p. 60°–63°C)	0.915
Mercury	20°C	13.546

Viscosity
Dynamic viscosity of water at   1.0019 mPas
20°C
   (see Table 4.13)
   1 mPas = 1 mNs/m^2 = 1 cP (centipoise)

Nominal Container Sizes

	Diameter (mm)	Height (mm)	Volume (cm^3)	Approximate mass of soil contained
Compaction mould	105	115.5	1000	1.8–2.2 kg
CBR mould	152	127	2305	4.0–5.0 kg
U-100 tube (per 100 mm)	100	100	785.4	1.4–1.7 kg
U-100 tube (full)	100	450	3534	6.3–7.7 kg
Sample tube	38	76	86.2	150–190 g

General
Circumference/diameter of circle $\pi = 3.142$
Base of natural logarithms $e = 2.718$
Standard acceleration due to gravity $g = 9.81 \text{ m/s}^2$
Standard atmosphere   1 atm = 101.325 kPa

BIBLIOGRAPHY

Anderton, P. and Bigg, P. H. (1972). *Changing to the Metric System*. National Physical Laboratory, HMSO, London
British Geotechnical Society Sub-committee on the Use of SI units in Geotechnics (1973). 'Report of the sub-committee'. News Item, *Géotechnique*, Vol. 23, No. 4, pp. 607–610
BS3763:1970, 'The International system of units (SI)'. British Standards Institution, London
Metrication Board (1976). *Going Metric — The International Metric System*. Leaflet UM1, 'An outline for technology and engineering'. Metrication Board, London
Metrication Board (1977). *How to Write Metric — A Style Guide for Teaching and Using SI Units*. HMSO, London
Page, C. H. and Vigoureux, P. (1977). *The International System of Units* (approved translation of *Le Systeme International d'Unités*, Paris, 1977). National Physical Laboratory, HMSO, London
Walley, F. (1968). 'Metrication' (Technical Note). *Proc. Inst. Civ. Eng.*, Vol. 40, May 1968. Discussion includes contribution by Head, K. H., Vol. 41, December 1968

A.4 COMPARISON OF BS AND ASTM SIEVE APERTURE SIZES

BS sieve aperture size	Sieves to ASTM D422	
	Nearest designation	Aperture size
75 mm	3 inch	75 mm
63	$2\frac{1}{2}$ inch	63.5
50	2 inch	50.8
37.5	$1\frac{1}{2}$ inch	38.1
28	–	
–	1 inch	25.4
20	$\frac{3}{4}$ inch	19.05
14	–	
10	$\frac{3}{8}$ inch	9.52
6.3	–	
5	No. 4	4.75
3.35	No. 6	3.35
–	No. 8	2.36
2	No. 10	2.00
1.18	No. 16	1.18
–	No. 20	850 μm
600 μm	No. 30	600
425	No. 40	425
300	No. 50	300
–	No. 60	250
212	No. 70	212
150	No. 100	150
–	No. 140	106
75	No. 200	75
63	No. 230	63

Index

A-line, 69
Accident prevention, 48
Accuracy, 40, 53, 73, 235
Acid-soluble sulphates, 261
Acidity, 243, 252
Acids, 51, 243
Activity of clays, 64
Adsorbed water, 61
Aeolian deposits, 357
Aggregations of particles, 46, 178
Air voids lines, 303, 306, 308
Algae, 18
Alkalinity, 243, 252
Alkalis, 51, 243
Alluvial deposits, 357
Andreasen pipette, 212
Angularity of particles, 364
Applications of soil testing, 2, 68, 128, 165, 168, 252, 313
Aqueous extract, 264, 271
Archimedes' principle, 124
Artificial light for soil description, 358
Atomic mass, 242
Atterberg limits, 59, 62, 71, 78

B-line, 69
B.D.H. soil testing outfit, 257
Bacteria, 254
Balances, calibration of, 37, 55
    care and use of, 36
    setting up, 36
    types of, 7, 9, 10, 11
Barometer, 7, 41
Bentomite, 65, 113
Black cotton soils, 373
Blower, warm air, 15
Bottle shaker, 25, 27
Boulder clay, 159, 168, 196, 232, 373
Boulders, 162, 184, 360
British Standard 1377:1990, 4, 58
Bulk density, 117, 128
Bulking of sand, 126
Burette, use of, 6, 239

Calcareous matter, 76, 208
Calculations, 38, 39
Calgon, 172
Calibration, 53–58
Calibration and checking intervals, 57
Calibration of
    balances, 37, 55
    constant temperature bath, 18
    density bottle, 144
    density of liquid, 144

Calibration *continued*
    hydrometer, 219, 225
    oven, 14
    pH meter, 257
    reference standards, 54
    sampling pipette, 217
    sieves, 174–175
    vibrating hammer, 337
    working instruments, 55
Calipers, 6, 34
Capillary water, 61
Carbonate content test:
    Collins calcimeter, 283, 289
    gravimetric, 280
    rapid titration, 278
Carbonates, 248, 253, 278
Casagrande liquid limit apparatus, 78, 89–91
Centrifuge, 26, 206, 209
Ceramic ware, 29, 30, 41
Chalk, 71, 77, 162, 253, 315
    crushing value (CCV) test, 352
    saturation moisture content of, 59, 71, 77
Checks on test equipment, 56–58
Cheesecloth (muslin), 43
Chemical tests, 235–237, 254–300
Chemicals, 32, 51, 220, 238
Chloride content test,
    acid soluble, 295
    Mohr's method, 293
    qualitative, 290
    Volhard's method, 290
Chlorides, 250, 253, 275, 290
Classification of peat, 371
    of soils, 1, 68, 358, 360, 371
Classification according to sulphate content, 253
Classification, by particle size, 162, 165, 166, 167
Clay, 62, 69, 160, 361, 366, 367
Clay fraction, 64, 160, 161
Clay minerals, 60, 366
Cleaning, 33, 41, 48
Climate, 40
Coarse-grained soils, 44, 360, 366
Coarse material, correction for, 76
Cobbles, 162, 171, 184, 191, 360
Cohesionless soils, 44, 360
Cohesive soils, 44, 46, 316, 360
Collins' calcimeter, 283
Colour, 365, 368
Compactability test for aggregates, 354
Compaction, 302, 303, 309, 313
    automatic, 330
    control of, 313
    mould, 58, 309, 319, 321, 332

Compaction *continued*
  of chalk, 315
  rammer, 58, 319, 321, 327
  types of curve, 314
  types of test, 309, 315, 330
Compaction test
  ASTM, 330
  BS 'heavy', 326
  BS 'light', 319
  BS 'vibration', 331
  Harvard, 338
  in CBR mould, 310, 329
  moisture condition, 339
  stony soils, 329
Compactive effort, 305, 308
Compactness, 365
Composite soils, 364, 367, 373
Concentration of solutions, 238
Cone-and-quarter, 47, 182
Cone penetrometer apparatus, 82–85
Consistency limits, 59
Consistency of clays, 62, 368, 370
Constant temperature bath, 18, 141
Container sizes, 382
Containers, drying, 15
Conversion factors, 376, 379
Cooling, 74
Corrosion, 254
Critical density and voids ratio, 117
Crushing, particles susceptible to, 315, 325, 336

Datum reading for hydrometer test, 222
De-airing water, 22
De-ionised water, 20, 82
De-ioniser, 22
Definitions, 3, 60, 117, 160, 241, 303, 360, 375
Degree of saturation, 117, 122
Dense packing, 126
Density, 116–124, 128, 129, 382
Density bottle, 6, 144
Density index, 117, 127–129
Density of pure water, 117, 119, 205
Density test,
  in tube, 133
  in water, 138
  linear measurement, 130
  water displacement, 134
Density values, 129, 382
Depth gauge, 6, 58
Description of soils, 4, 37, 41, 357, 358, 360, 361, 366
Desiccant, desiccator, 16, 17, 74
Desiccation, 60
Dial gauge, 6, 55, 58
Dilatancy, 367
Disaggregation, 46
Dispersing agents, 172, 215
Dispersion, 160, 172, 187, 196, 205, 209–210
Distilled water, 20, 82, 243
Dough-hook paddle, 22
Drains, 50
Dry density/moisture relationship, 302, 304
Dry soil, 60, 119
Drying temperature, 12
Drying time, 15, 74
Drying,
  in air, 46, 60, 79
  in oven, 12, 15, 60, 72, 74

Dust mask, 53

Ear muffs, 52
Effective depth of hydrometer, 222, 227
Effective size of particles, 160, 167
Electrical apparatus, 49
Electrical conductivity, 245
Electrolytic dissociation, 241
Empirical index tests, 59, 109
Equipment, 5, 33, 358
Errors, 38
Estuarine deposits, 358
Eye shield, 53

Filtration, 208, 240
Fine-grained soils, 44, 360, 366, 370
Fineness modulus, 160
Fines in soil, 100, 159
Fire hazard, 49
Flocculation, 160
Flocs, 172
Flow curve, 65, 93
Fluid content, 61, 62
Forms, test, 37, 359
Free swell test, 113
Fuller grading curve, 126, 166, 167
Fume-cabinet, 51, 238

Gap-graded soils, 167
Gas appliances, 50
  jar, 6, 145
Gauge blocks, 54
Glacial deposits, 168, 357, 373
Glass cutting, 52
Glassware, 6, 29, 41, 52, 58
Gloves, 52, 53
Grading curve, 159
Granular soils, description of, 361
Graphs, 38, 39
Gravel, 160, 181, 191, 361
Gravitational water, 61
Gravity, 379, 382
Greek alphabet, 381
Grooving tool, 90
Groundwater sample, 266
Gypsum, 72, 76

Hardware, 30
Harvard miniature compactor, 338
Heavy compaction, 303, 309, 326
Hot-plate, 15
Hydrogen ions, 243
Hydrometer, 6, 218
  calibration of, 219, 225
  corrections, 220
  readings, 219, 221
Hydrometer sedimentation test, 218

Identification of samples, 42
Identification of soils, 356, 360
Ignition, 247, 298
Illite, 65, 163
Imperial units, 5, 379
Implosion, 19
Index tests, 59, 360
Indicators, 32, 241, 244

Indicator papers, 244, 254, 300
Infra-red drying cabinet, 15
Ion, 241
Ion exchange, 20, 268, 271

Kaolinite, 65, 163
Kerosene, 142

Laboratory, 3
Laboratory environment, 7, 40
Lacustrine deposits, 357
Laminated clays, 73, 369
Light compaction, 303, 309, 319
Limiting densities, 117, 126, 129, 149
Linear shrinkage, 59, 60, 68, 109
Linear shrinkage test, 107
Liquefaction, 129
Liquid limit, 59, 60, 78, 95
    Casagrande one-point test, 95
    Casagrande test, 88
    cone penetrometer test, 82
    one point cone test, 88
Liquid limit correlation, 79
Liquid limit values, 70
Liquid state, 62
Liquidity index, 60, 63, 70, 76
Loam, 165
Loose packing, 126
Loss on ignition, 298

Man-made soils, 357, 371
Marine deposits, 358
Mass, measurement of, 7, 54
Mass of soil for test, 45
Maturing of soil, 80
Maximum density, 117, 149
Maximum density test,
    gravelly soils, 154
    sands, 150
    silty sands, 154
Maximum dry density, 303, 326
Maximum mass retained on sieves, 178
Maximum/minimum thermometer, 7, 12, 40
MCV derivation,
    'English' method, 345
    'Scottish' method, 348
Measuring cylinder, 6, 29
Measuring instruments, 5, 6, 7
Mechanical analysis, 159
Medium-grained soils, 44
Meniscus correction, 221
Mercury, 50, 101, 105, 382
Metric (SI) units, 4, 375
Micrometer, 6
Microwave oven, 15
Minimum density, 117, 149
Minimum density test,
    gravelly soils, 156
    sands, 155
Minimum mass for sieving, 170–172
Mixing soil, 22, 82, 84
Moisture condition calibration (MCC) test, 349
Moisture condition value, 339
Moisture condition value (MCV) test, 341, 350
Moisture content, 59–64, 68, 318
Moisture content containers, 15, 72

Moisture content tests, 71–77
Molar solution, 238, 242
Mole, 238, 242
Montmorillonite, 65, 163
Muffle furnace, 26
Muslin, 43

NAMAS requirements, 53
National standards of measurement, 53
Natural moisture content, 60
Noise, 52
Non-cohesive soils, 361
Non-plastic, 60
Normal solution, 238

Observations, 41, 46
One-point liquid limit test factors, 88, 96
Optimum moisture content, 303, 326
Organic content test,
    dichromate oxidation, 272
    peroxide odixation, 277
Organic matter, 72, 76, 205, 207, 253
Organic soils, 46, 70, 125, 207, 247, 360, 370
Oven, 12, 13, 72
Overcompaction, 313

Packing of spheres, 126
Pallet racking, 43
Partially saturated soils, 121
Particle density, 117–118, 124–125, 129
Particle density tests:
    density bottle (small pyknometer), 140
    gas jar, 145
    large pyknometer, 147
Particle shape, 364
Particle size, 159, 160, 222, 361, 366
Particle size distribution curve, 159, 164, 180, 217,
        225, 226
Particle size tests:
    sieving, 175–201, 232
    sedimentation, 205–225
    combined sieving and sedimentation, 228, 232
Particles, nature of, 364
Peat, 72, 125, 129, 162
    classification of, 371
Periglacial deposits, 357
Pestle, 46
pH meter, 255, 257
pH test,
    colorimetric, 257
    electrometric, 255
    indicator papers, 254
    Lovibond, 259
pH value, 241, 243, 252
Pipette, volumetric, 6, 29, 51, 239
Pipette sampling times, 204, 214
Pipette sedimentation test, 211
Plastic limit, 59, 60, 78, 98
Plastic limit test, 96
Plastic state, 62
Plasticity, 69, 70, 367
Plasticity chart, 69, 70
Plasticity index, 60, 62, 68, 99, 109
Plastics ware, 31
Platform scales, 7, 9
Pocket penetrometer, 368

Pocket vane shearmeter, 369
Policeman, 240
Polythene bags, 43
Poorly graded soils, 107
Porosity, 117, 119, 128
Pretreatment, 205, 207–208
Preparation of samples for test, 4, 44, 79
   dry method, 81
   for ASTM compaction tests, 329
   for BS compaction tests, 315, 318
   for chemical tests, 240
   wet method, 80
Presentation of data, 38
Proctor compaction, 302
Prong plate, 104
Protective clothing, 49, 53
Puddle-clay index tests, 110
Pyknometer, 30, 141, 147
Pyroclastic deposits, 357

Quantabs, 300
Quantity of soil for particle size tests, 170–172, 207
Quartering, 47, 160
Quartz, 125

Reagents, 32
Recording data, 37
References, 5
Reference standards of measurement, 53–54
Relative compaction, 303, 314
Relative consistency, 60, 63
Relative density, 127, 365
Relative humidity, 40
Reporting results, 39
Representative sample, 44, 47
Residual soils, 4, 46, 79, 161, 357, 358, 360, 373
Riffle box, 22, 24, 47, 160
Riffling, 47, 160, 182
Roundness, 364
Routine checks on apparatus, 58

Sabkha soils, 373
Safety, 48, 238, 332
Saline water, 61
Sample, 3, 44
Sample divider, 22
Sample splitter, 47
Samples:
   care of, 41–44
   description of, 4, 37, 41, 43, 358–373
   identification of, 37, 42
   opening, re-sealing and protection of, 42
   selection for test, 44, 73
   storage of, 43
   sub-dividing, 47
Sand, 160, 361
Sand-bath test, 76
Saturated soil, 117, 120
Saturation line, 303
Saturation moisture content of chalk, 59, 71, 77
Scales, choice of, 39
Sea-water, 62, 243, 382
Sedimentary soils, 4, 357, 360
Sedimentation, 159, 201
Sedimentation test:
   combined with sieving, 228, 232

Sedimentation test: *continued*
   pretreatment, 205
   quantity of soil for, 207
Sedimentation test hydrometer, 218
Sedimentation test pipette, 211
Sedimentation theory, 201
Semi-logarithmic graph, 38
Semi-solid state, 62
Shaker:
   bottle, 27, 210
   gas jar, 26, 145
   sieve, 24–26, 177, 195
   vibratory, 25
Shape of particles, 364
Sharpness gauge, 82
Shrinkage, 65, 71
Shrinkage limit, 59, 60, 65, 71, 107
Shrinkage limit:
   alternative test, 104
   TRRL test, 99
Shrinkage range, 60, 68, 71
Shrinkage ratio, 60, 66, 71, 107
SI units, 4, 238, 375–379
Sieve aperture sizes:
   ASTM, 163, 164, 383
   BS, 161, 163, 164, 173, 383
   ISO, 161, 164, 173–174
Sieve diameter, 160
Sieve shaker, 24, 25, 177, 195
Sieves, 160, 164, 173–174, 177
Sieving, 159
Sieving tests:
   'boulder clay', 196
   combined with sedimentation, 228, 232
   composite, 191
   simple dry, 175
   very coarse soils, 184
   wet (cohesive soils), 195
   wet (fine non-cohesive soils), 187
   wet (gravelly soils), 191
Sieving time, 178
Silica gel, 16
Silt, 69, 160, 361, 366
Silt trap, 50, 359
Sinks, 50
Sodium hexametaphosphate, 172, 196
Soil, 1, 3, 162, 357, 360
   fabric and structure, 2, 363, 365, 369
Soil mechanics, 1
Soil mixer, 22
Soils:
   categories of, 44
   classification of, 1, 68, 358, 360
   description of, 4, 37, 41, 357–361, 366
   identification of, 37, 356, 360
Solid state, 62
Solubilities, 245
Solutions, 238, 242
Specific gravity, 117
Specific surface, 160
Specimen, 3
   selection of, 73
Spheres, packing of, 126
   surface area of, 163
Spillage of chemicals and mercury, 51
Sticky limit, 59, 60, 71, 114

Still, water, 20, 21
Stirrers, 29, 210
Stokes' law, 184, 201
Stone content, 203, 310
Stony soils, 73, 310, 329
Strength of clay soil, 368, 370
Submerged soil, 123
Sulphate content tests:
  groundwater (ion exchange), 268
  in aqueous extract:
    gravimetric, 266
    ion exchange, 271
Sulphate content tests:
  total, in soils, 261
  water soluble, 264
Sulphates, 245, 253, 260
Sulphides, 254, 275
Surface tension effect, 190
Suspension, 62, 201
Swelling capability, 59, 113
Symbols, 381

Tabulation, 38
Tamper for vibrating hammer, 150, 151, 334
Tape, measuring, 6, 54, 55, 57
Techniques, 5, 33
Temperature conversion chart, 380
Temperature, laboratory, 40
Terminal velocity, 201–203
Test (definition), 3
Testing, advantages and purpose of, 1, 2
Thermometer, 7, 12
Thermostat, oven, 12, 14
Tidiness, 48
Time, measurement of, 7, 378, 382
Titration, 241
Tongs, oven, 14, 49
Tools, 32, 52, 359
Torr, 19
Total dissolved solids test, 297
Traceability of measurements, 53–58
Triangular classification chart, 165
Trimming test specimens, 131
Trolley, 49
Tropical soils, 4, 46, 79, 125, 172, 207, 238, 373

Undisturbed samples, 43
Uniformly graded soils, 166

Uniformity coefficient, 160, 167
Unit mass and weight, 118
Units of measurement, 4, 375
Useful data, 382

Vacuum, 19, 52
Vacuum desiccator, 17, 19
Vacuum filtration, 53, 240
Vacuum gauge, 7, 19
Vacuum pump, 19, 50
Valves, vacuum, 19
Vane shearmeter, 369
Vapours, 49
Ventilation, 50
Vernier calipers, 6, 34, 35
Vernier scale, 34
Very coarse soils, 162, 360
Vibrating hammer, 52, 149, 332
Vibrating hammer compaction, 310, 331
Vibrating hammer verification, 337
Vibro-stirrer, 29, 210, 211
Viscosity of water, 204, 205, 382
Visual observations, 5, 46, 358
Voids ratio, 117, 119, 128
Volume measurement, 118, 130
Volumetric analysis, 239
Von Post classification of peats, 371

Warm air blower, 15
Washing on sieve, 81, 187, 189
Waste, 48, 50, 59
Water content, 60
Water of hydration, 61
Water purification, 20
Water stabiliser, 18
Water trap, 19
Water, density and viscosity of, 119, 205, 382
Water-soluble sulphates, 245, 252, 260, 264
Wax, 21, 382
Weighing, 36, 73
  in water, 139
Weighing bottles, 15, 72
Well-graded soils, 167, 198
Wet sieving, 187, 189, 191, 195, 198, 211
Wet-and-dry bulb thermometer, 7, 12, 41
Winchester bottles, 49

Zero air voids, 122, 303, 308

This book is
bound upside dow[n]